Methods in Molecular Biology

Series Editor
John M. Walker
School of Life and Medical Sciences
University of Hertfordshire
Hatfield, Hertfordshire, UK

For further volumes:
http://www.springer.com/series/7651

For over 35 years, biological scientists have come to rely on the research protocols and methodologies in the critically acclaimed *Methods in Molecular Biology* series. The series was the first to introduce the step-by-step protocols approach that has become the standard in all biomedical protocol publishing. Each protocol is provided in readily-reproducible step-by step fashion, opening with an introductory overview, a list of the materials and reagents needed to complete the experiment, and followed by a detailed procedure that is supported with a helpful notes section offering tips and tricks of the trade as well as troubleshooting advice. These hallmark features were introduced by series editor Dr. John Walker and constitute the key ingredient in each and every volume of the *Methods in Molecular Biology* series. Tested and trusted, comprehensive and reliable, all protocols from the series are indexed in PubMed.

Cryptococcus neoformans

Methods and Protocols

Edited by

Erin E. McClelland

Department of Biomedical Sciences, Marian University College of Osteopathic Medicine, Indianapolis, IN, USA

Humana Press

Editor
Erin E. McClelland
Department of Biomedical Sciences
Marian University College of Osteopathic Medicine
Indianapolis, IN, USA

ISSN 1064-3745 ISSN 1940-6029 (electronic)
Methods in Molecular Biology
ISBN 978-1-0716-3721-0 ISBN 978-1-0716-3722-7 (eBook)
https://doi.org/10.1007/978-1-0716-3722-7

Cover Image: Scanning electron microscopy image of *C. neoformans* cells

This Humana imprint is published by the registered company Springer Science+Business Media, LLC, part of Springer Nature.
The registered company address is: 1 New York Plaza, New York, NY 10004, U.S.A.

Paper in this product is recyclable.

Preface

As the numbers of immunocompromised patients increase, the model yeast, *Cryptococcus neoformans*, is becoming an increasingly important model organism. Due to the expanding number of strains that show resistance to antifungals, scientists are working diligently to understand its pathology and develop new therapeutics. This volume features a large variety of methods that are commonly used by *Cryptococcus* researchers, including a variety of animal models, genetics techniques, virulence factor phenotyping, microscopy, and others. It is hoped that these techniques will aid new discoveries and therapeutics, and further our understanding of this pathogenic yeast.

Indianapolis, IN, USA *Erin E. McClelland*

Contents

Part V Interactions with Macrophages

Part VI Virulence Factors

Part VII Measuring Specific Phenotypes

Contributors

RUCHI AGRAWAL • *Department of Mycology, Translational Mycology Group, Institut Pasteur, Université Paris Cité, National Reference Center for Invasive Mycoses and Antifungals, Paris, France*
ALEXANDRE ALANIO • *Translational Mycology Research Group, Institut Pasteur, Université Paris Cité, National Reference Center for Invasive Mycoses and Antifungals, Mycology Department, Paris, France; Laboratoire de parasitologie-mycologie, AP-HP, Hôpital Saint-Louis, Paris, France*
PATRÍCIA ALBUQUERQUE • *Faculty of Ceilândia, University of Brasília, Brasilia, DF, Brazil*
J. ANDREW ALSPAUGH • *Department of Molecular Genetics and Microbiology, Duke University School of Medicine, Durham, NC, USA; Department of Medicine, Duke University School of Medicine, Durham, NC, USA*
THAÍS SILVA ALVES • *Faculty of Medicine, University of Brasília, Brasilia, DF, Brazil*
GLAUBER R. DE S. ARAÚJO • *Laboratório de Biofísica de Fungos, Instituto de Biofísica Carlos Chagas Filho, Universidade Federal do Rio de Janeiro, Rio de Janeiro, RJ, Brazil*
ANTOINE BODEIN • *Computational Biology Laboratory, CHU de Québec – Université Laval Research Center, QC, Canada*
INDRANI BOSE • *Department of Biology, Western Carolina University, Cullowhee, NC, USA*
VERONICA SOARES BRAUER • *Department of Microbiology and Immunology, Stony Brook University, Stony Brook, NY, USA*
ADRIELLE VELOSO CAIXETA • *Faculty of Medicine, University of Brasília, Brasilia, DF, Brazil*
ALTHEA CAMPUZANO • *Department of Molecular Microbiology and Immunology, University of Texas at San Antonio, San Antonio, TX, USA; South Texas Center for Emerging Infectious Diseases, University of Texas at San Antonio, San Antonio, TX, USA*
ARTURO CASADEVALL • *Department of Molecular Microbiology and Immunology, Johns Hopkins Bloomberg School of Public Health, Baltimore, MD, USA*
RAFAEL F. CASTELLI • *Instituto Carlos Chagas, Fundação Oswaldo Cruz (Fiocruz), Curitiba, Brazil*
NATALIA CASTRO-LOPEZ • *Department of Biology, Texas Christian University, Fort Worth, TX, USA*
NICHOLAS F. CHONG • *School of BioSciences, The University of Melbourne, Parkville, VIC, Australia*
CAROLINA COELHO • *MRC Centre for Medical Mycology, College of Health and Medicine, University of Exeter, Exeter, Devon, UK*
DEVENEY DASILVA • *Department of Microbiology and Immunology, Stony Brook University, Stony Brook, NY, USA*
RAFFAEL J. A. DE CASTRO • *Department of Mycology, Translational Mycology Group, Institut Pasteur, Université Paris Cité, National Reference Center for Invasive Mycoses and Antifungals, Paris, France*
HERDSON RENNEY DE SOUSA • *Faculty of Medicine, University of Brasília, Brasília, DF, Brazil*

MAURIZIO DEL POETA • *Department of Microbiology and Immunology, Stony Brook University, Stony Brook, NY, USA; Institute of Chemical Biology and Drug Discovery (ICB&DD), Stony Brook, NY, USA; Division of Infectious Diseases, School of Medicine, Stony Brook University, Stony Brook, NY, USA; Veterans Administration Medical Center, Northport, NY, USA*

LARISSA VERGINIA DO NASCIMENTO MIRANDA • *Institute of Biology, Department of Cell Biology, University of Brasília, Brasília, DF, Brazil*

ARNAUD DROIT • *Computational Biology Laboratory, CHU de Québec – Université Laval Research Center, QC, Canada; Proteomics platform, CHU de Québec – Université Laval Research Center, QC, Canada; Canadian Proteomics and Artificial Intelligence Consortium, Guelph, ON, Canada*

HELENE C. EISENMAN • *Department of Natural Sciences, Baruch College, The City University of New York, New York, NY, USA*

SUSANA FRASES • *Laboratório de Biofísica de Fungos, Instituto de Biofísica Carlos Chagas Filho, Universidade Federal do Rio de Janeiro, Rio de Janeiro, RJ, Brazil; Rede Micologia RJ, FAPERJ, Rio de Janeiro, Brazil*

BETTINA C. FRIES • *Department of Medicine, Division of Infectious Diseases, Stony Brook University, Stony Brook, NY, USA; Department of Microbiology and Immunology, Renaissance School of Medicine, Stony Brook University, Stony Brook, NY, USA; Veterans Administration Medical Center, Northport, NY, USA*

JENNIFER GEDDES-MCALISTER • *Molecular and Cellular Biology Department, University of Guelph, Guelph, ON, Canada*

CHARLES GIAMBERARDINO • *Duke University School of Medicine, Department of Medicine, Division of Infectious Diseases, Durham, NC, USA*

SETH D. GREENGO • *Department of Molecular Microbiology and Immunology, Johns Hopkins Bloomberg School of Public Health, Baltimore, MD, USA*

TIFFANY E. GUESS • *Middle Tennessee State University, Murfreesboro, TN, USA; Molecular Pathology Laboratory Network, Inc., Maryville, TN, USA*

ALLAN JEFFERSON GUIMARÃES • *Laboratório de Bioquímica e Imunologia das Micoses, Departamento de Microbiologia e Parasitologia, Instituto Biomédico, Fluminense Federal University, Niterói, RJ, Brazil; Pós-Graduação em Imunologia e Inflamação, Instituto de Microbiologia Paulo de Góes, Federal University of Rio de Janeiro, Rio de Janeiro, RJ, Brazil; Programa de Pós-Graduação em Microbiologia e Parasitologia Aplicadas, Instituto Biomédico, Fluminense Federal University, Niterói, RJ, Brazil*

MAUREEN M. HESTER • *Department of Medicine, The University of Massachusetts Chan Medical School, Worcester, MA, USA*

HUNTER J. HOFFMAN • *Department of Biomedical Sciences, Marian University College of Osteopathic Medicine, Indianapolis, IN, USA*

BENJAMIN HOMMEL • *Translational Mycology Research Group, Institut Pasteur, Université Paris Cité, Centre National de Référence Mycoses Invasives et Antifongiques, Département de Mycologie, Paris, France*

CHIUNG-YU HUNG • *Department of Molecular Microbiology and Immunology, University of Texas at San Antonio, San Antonio, TX, USA; South Texas Center for Emerging Infectious Diseases, University of Texas at San Antonio, San Antonio, TX, USA*

ALEXANDER IDNURM • *School of BioSciences, The University of Melbourne, Parkville, VIC, Australia*

LUISA J. JOZEFOWICZ • *Instituto Carlos Chagas, Fundação Oswaldo Cruz (Fiocruz), Curitiba, Brazil*

DIOGO KUCZERA • *Instituto Carlos Chagas, Fundação Oswaldo Cruz (Fiocruz), Curitiba, Brazil*

EMMANUEL LAFONT • *Translational Mycology Research Group, Institut Pasteur, Université Paris Cité, National Reference Center for Invasive Mycoses and Antifungals, Mycology Department, Paris, France*

HOYIN LAI • *SVision LLC, Bellevue, WA, USA*

WOEI C. LAM • *Department of Molecular Genetics and Microbiology, Duke University School of Medicine, Durham, NC, USA; Pfizer STL, W Chesterfield, MO, USA; Department of Molecular Microbiology, Washington University in St. Louis School of Medicine, St. Louis, MO, USA*

FANNY LANTERNIER • *Translational Mycology Research Group, Institut Pasteur, Université Paris Cité, National Reference Center for Invasive Mycoses and Antifungals, Mycology Department, Paris, France; MRC Centre for Medical Mycology, College of Health and Medicine, University of Exeter, Exeter, Devon, UK; Service de maladies infectieuses et tropicales, hôpital Necker-Enfants Malades, Paris, France*

STUART M. LEVITZ • *Department of Medicine, The University of Massachusetts Chan Medical School, Worcester, MA, USA*

JENNIFER K. LODGE • *Department of Molecular Genetics and Microbiology, Duke University School of Medicine, Durham, NC, USA; Department of Molecular Microbiology, Washington University in St. Louis School of Medicine, St. Louis, MO, USA*

LIVIA LIPORAGI LOPES • *Department of Molecular Microbiology and Immunology, Johns Hopkins Bloomberg School of Public Health, Baltimore, MD, USA*

DIANA LOURENCO • *Department of Medicine, The University of Massachusetts Chan Medical School, Worcester, MA, USA*

GABRIEL SOARES MATOS • *Department of Microbiology and Immunology, Stony Brook University, Stony Brook, NY, USA*

BRIAN T. MAYBRUCK • *Department of Molecular Genetics and Microbiology, Duke University School of Medicine, Durham, NC, USA; AstraZeneca, South San Francisco, CA, USA*

ERIN E. MCCLELLAND • *Department of Biomedical Sciences, Marian University College of Osteopathic Medicine, Indianapolis, IN, USA*

SCOTT A. MCCONNELL • *Department of Molecular Microbiology and Immunology, Johns Hopkins Bloomberg School of Public Health, Baltimore, MD, USA*

SUSANA RUIZ MENDOZA • *Laboratório de Bioquímica e Imunologia das Micoses, Departamento de Microbiologia e Parasitologia, Instituto Biomédico, Fluminense Federal University, Niterói, RJ, Brazil; Pós-Graduação em Imunologia e Inflamação, Instituto de Microbiologia Paulo de Góes, Federal University of Rio de Janeiro, Rio de Janeiro, RJ, Brazil*

BEN MUSELIUS • *Molecular and Cellular Biology Department, University of Guelph, Guelph, ON, Canada*

DAVID E. NELSON • *Department of Biology, Middle Tennessee State University, Murfreesboro, TN, USA*

ANDRÉ MORAES NICOLA • *Faculty of Medicine, University of Brasília, Brasilia, DF, Brazil*

BRIDGIT C. NUGENT • *School of BioSciences, The University of Melbourne, Parkville, VIC, Australia*

HAROLDO C. OLIVEIRA • *Instituto Carlos Chagas, Fundação Oswaldo Cruz (Fiocruz), Curitiba, Brazil*

NATALIA K. OLIVEIRA • *Department of Microbiology and Immunology, Renaissance School of Medicine, Stony Brook University, Stony Brook, NY, USA*

STEFÂNIA DE OLIVEIRA FRAZÃO • *Institute of Biology, Department of Cell Biology, University of Brasília, Brasília, DF, Brazil*

HUGO COSTA PAES • *Faculty of Medicine, University of Brasília, Brasília, DF, Brazil*

ILDINETE SILVA PEREIRA • *Institute of Biology, Department of Cell Biology, University of Brasília, Brasília, DF, Brazil*

JOHN R. PERFECT • *Duke University School of Medicine, Department of Medicine, Division of Infectious Diseases, Durham, NC, USA*

BRUNO PONTES • *Laboratório de Pinças Ópticas Moysés Nussenzveig (LPO), Universidade Federal do Rio de Janeiro, Rio de Janeiro, RJ, Brazil; Centro Nacional de Biologia Estrutural e Bioimagem (CENABIO), Universidade Federal do Rio de Janeiro, Rio de Janeiro, RJ, Brazil; Rede Micologia RJ, FAPERJ, Rio de Janeiro, Brazil*

CORINNA PROBST • *Department of Molecular Genetics and Microbiology, Duke University School of Medicine, Durham, NC, USA; Department of Medicine, Duke University School of Medicine, Durham, NC, USA*

SAMYR MACHADO QUEROBINO • *Department of Microbiology and Immunology, Stony Brook University, Stony Brook, NY, USA*

FLAVIA C. G. REIS • *Instituto Carlos Chagas, Fundação Oswaldo Cruz (Fiocruz), Curitiba, Brazil; Centro de Desenvolvimento Tecnológico em Saúde (CDTS), Fiocruz, Rio de Janeiro, Brazil*

AMANDA PEREIRA ROCHA • *Faculty of Medicine, University of Brasília, Brasilia, DF, Brazil*

MARCIO L. RODRIGUES • *Instituto Carlos Chagas, Fundação Oswaldo Cruz (Fiocruz), Curitiba, Brazil; Instituto de Microbiologia Paulo de Góes (IMPG), Universidade Federal do Rio de Janeiro, Rio de Janeiro, Brazil*

FLORENCE ROUX-DALVAI • *Computational Biology Laboratory, CHU de Québec – Université Laval Research Center, QC, Canada; Proteomics platform, CHU de Québec – Université Laval Research Center, QC, Canada; Canadian Proteomics and Artificial Intelligence Consortium, Guelph, ON, Canada*

REBECCA L. SEIPELT-THIEMANN • *Biology Department, Middle Tennessee State University, Murfreesboro, TN, USA*

VANESSA K. A. SILVA • *Department of Medicine, Division of Infectious Diseases, Stony Brook University, Stony Brook, NY, USA*

MARINA DA SILVA FERREIRA • *Laboratório de Bioquímica e Imunologia das Micoses, Departamento de Microbiologia e Parasitologia, Instituto Biomédico, Fluminense Federal University, Niterói, RJ, Brazil; Pós-Graduação em Imunologia e Inflamação, Instituto de Microbiologia Paulo de Góes, Federal University of Rio de Janeiro, Rio de Janeiro, RJ, Brazil*

LINDA M. SIRCY • *Department of Biology, Middle Tennessee State University, Murfreesboro, TN, USA; Vaccine and Infectious Disease Division, Fred Hutchinson Cancer Center, Salt Lake City, UT, USA*

EMILY SMITH-PEAVLER • *Department of Biology, Middle Tennessee State University, Murfreesboro, TN, USA; KIPP Collegiate High School, Nashville, TN, USA*

CHARLES A. SPECHT • *Department of Medicine, The University of Massachusetts Chan Medical School, Worcester, MA, USA*

AUDE STURNY-LECLÈRE • *Department of Mycology, Translational Mycology Group, Institut Pasteur, Université Paris Cité, National Reference Center for Invasive Mycoses and Antifungals, Paris, France*

JENNIFER L. TENOR • *Duke University School of Medicine, Department of Medicine, Division of Infectious Diseases, Durham, NC, USA*

DENA L. TOFFALETTI • *Duke University School of Medicine, Department of Medicine, Division of Infectious Diseases, Durham, NC, USA*

RAJENDRA UPADHYA • *Department of Molecular Genetics and Microbiology, Duke University School of Medicine, Durham, NC, USA; Department of Molecular Microbiology, Washington University in St. Louis School of Medicine, St. Louis, MO, USA*

MICHELE RAMOS VALENTE • *Laboratório de Bioquímica e Imunologia das Micoses, Departamento de Microbiologia e Parasitologia, Instituto Biomédico, Fluminense Federal University, Niterói, RJ, Brazil; Programa de Pós-Graduação em Microbiologia e Parasitologia Aplicadas, Instituto Biomédico, Fluminense Federal University, Niterói, RJ, Brazil*

MAGGIE P. WEAR • *Department of Molecular Microbiology and Immunology, Johns Hopkins Bloomberg School of Public Health, Baltimore, MD, USA*

FLOYD L. WORMLEY JR • *South Texas Center for Emerging Infectious Diseases, University of Texas at San Antonio, San Antonio, TX, USA; Department of Biology, Texas Christian University, Fort Worth, TX, USA*

Part I

Virulence Models

Chapter 1

A *Galleria* Model of *Cryptococcus neoformans* Infection

Helene C. Eisenman

Abstract

Galleria mellonella larvae are a popular and simple model organism for infectious disease research. Last instar larvae can be purchased inexpensively from commercial suppliers and infected with *Cryptococcus*. Injection into the proleg of larvae results in systemic infections. Larvae may then be monitored for survival or homogenized to determine fungal burden. Fixation of infected larvae produces samples suitable for histological staining and analysis.

Key words Insect, Larvae, Infection, Fixation, *Galleria mellonella*, *Cryptococcus*

1 Introduction

The *G. mellonella* infection model has become integral to *Cryptococcus* research. It is one of several "alternate" hosts available to researchers. In the laboratory, a wide range of organisms may be infected with *Cryptococcus*, including insects, nematodes, protozoans, and plants [1–4]. Each system has unique advantages. The reasons for *G. mellonella's* popularity in research include its low cost and ease of use, as well as the recapitulation of many aspects of cryptococcal pathogenesis during larval infection. Multiple *Cryptococcus* species have been evaluated in the *G. mellonella* model, including *neoformans*, *grubii*, and *gattii* [5]. The system is amenable to a wide range of research questions. Recently, it has been used as an infection model for studying antifungal efficacy [6, 7], virulence factor discovery and analysis [8, 9], and strain comparisons [10–12].

In order to be useful, an animal model of infection needs to reproduce the pathogenesis found in human hosts. The importance of the capsule in cryptococcal virulence in mammals is well documented. The capsule undergoes structural changes during larval infection, as it does in mammalian hosts [13]. *C. neoformans* interacts with cells of the larval immune system similarly to how it

Erin E. McClelland (ed.), *Cryptococcus neoformans: Methods and Protocols*, Methods in Molecular Biology, vol. 2775,
https://doi.org/10.1007/978-1-0716-3722-7_1,

interacts with mammalian macrophages. In *G. mellonella*, hemocytes are the functional counterparts to mammalian phagocytes. Capsule inhibition of phagocytosis occurs with hemocytes. Additionally, if phagocytosis does occur, the fungal cells may survive and reproduce within the hemocytes [14].

All experimental systems have limitations that must be taken into consideration when designing experiments and interpreting results. The anatomical differences between insect and mammalian hosts preclude using *G. mellonella* for most research related to dissemination. One must also consider the differences between invertebrate and vertebrate immunity. Unlike vertebrates, *G. mellonella* has no adaptive immunity. The larvae do have innate defenses, such as the aforementioned hemocytes. However, the innate defenses include a phenoloxidase system not found in vertebrates. This system defends against infection using melanin and melanin intermediates to heal wounds and kill pathogens [15]. These differences mean that results obtained with *G. mellonella* need to be validated in a mammalian model. Despite this, the ability to test in an invertebrate first minimizes animal suffering. In addition, use of the larvae provides educational opportunities for infectious disease research in settings where vertebrate models are not feasible. In this chapter detailed methods for infections, survival analysis, and histology are discussed.

2 Materials

2.1 Larva Infections

1. Last-instar *Galleria mellonella* larvae can be purchased from commercial bait and pet suppliers (*see* **Note 1**).
2. A repeating dispenser (Hamilton Company) can be fitted with various syringe types to quickly and easily inject multiple larvae. A single 1 mL syringe is sufficient for injecting 50 larvae (*see* **Note 2**).
3. Disposable syringes (1 mL).
4. 26 ½ gauge needles.
5. Streptomycin sulfate solution: 50 mg/mL in water.
6. Phosphate-buffered saline (PBS) solution: 137 mM sodium chloride, 2.7 mM potassium chloride, 10 mM phosphate, pH 7.4.
7. Ethanol: 70% solution in water.
8. Bleach: 10% solution in water.
9. Screwdriver for assembling the repeating dispenser.
10. Forceps for holding the larvae.
11. Sabouraud Dextrose Agar: 0.5% peptic digest of animal tissue, 0.5% pancreatic digest of casein, 4% dextrose, 1.5% agar.

2.2 Fungal Cultures

1. Sabouraud Dextrose Broth: 0.5% peptic digest of animal tissue, 0.5% pancreatic digest of casein, 2% dextrose.
2. Yeast peptone dextrose (YPD) Broth: 1% yeast extract, 2% peptone, 2% dextrose.
3. Chemically defined minimal medium for *C. neoformans*: 15 mM dextrose, 10 mM magnesium sulfate, 29.4 mM potassium phosphate, 13 mM glycine, 3 μM thiamine, pH 5.5.

2.3 Fungal Burden

1. Homogenizer.
2. Scalpels.
3. Sabouraud Dextrose Agar with antibiotics: 0.5% peptic digest of animal tissue, 0.5% pancreatic digest of casein, 4% dextrose, 1.5% agar, 0.1 mg/mL ampicillin, 0.1 mg/mL streptomycin.

2.4 Larva Fixation

1. High-purity paraformaldehyde: 4% solution diluted in PBS.
2. Sucrose: 30% sucrose dissolved in PBS.
3. Ethanol: 70% solution diluted in water.
4. Solvent-resistant markers.
5. 27 ½ gauge needles.

3 Methods

3.1 Preparation of Larvae

1. Inspect and select the larvae for infection. There is likely to be variation in the appearance of the larvae. Choose larvae of uniform size without any dark pigmentation. Ensure that the larvae are alive and active before proceeding with the infection.
2. Determine the average weight of the larvae. Place ten larvae in a Petri dish and weigh them. Divide by ten to obtain the average weight. If antibiotics are to be administered to prevent bacterial infection, the dosage is based on weight.
3. Divide the larvae into equal-sized groups, and place them in Petri dishes with the lid on to prevent them from escaping.

3.2 Preparation of Inoculum

Cryptococcal strains and culture conditions will depend upon the experimental design (*see* **Note 3**).

1. Culture cryptococcal cells in liquid medium, such as Sabouraud Dextrose Broth or chemically defined minimal medium. Before injection, cells must be washed and counted. To wash, collect cells by centrifugation in disposable centrifuge tubes at 671 ×*g* for 5 min at room temperature. Pour off supernatant with care to avoid dislodging the pellet and losing cells. Add PBS (25 mL) to the pellet, and suspend the cells by gently shaking the tube. Repeat the cell collection, and wash the cell pellet

again in PBS. After the second wash, pour of the supernatant, and suspend the cell pellet in PBS or antibiotic solution (*see* **Note 4**) to the desired cell density.

2. Remove a small aliquot of the cell suspension for determining the cell density by hemocytometer count. Save the aliquot on ice for later plating of the inoculum to determine actual colony-forming units (CFUs) injected.

3.3 Larva Infection

1. Assemble the following items in the biosafety cabinet or laboratory bench where the injections will be performed: dispenser, syringes, needles, cotton swabs, forceps, 70% ethanol, 10% bleach, Petri dishes, laboratory wipes, forceps, and screwdriver.
2. Larva can be injected with a 1 mL disposable syringe fitted in a Hamilton dispenser. The dispenser will deliver 1/50 the volume of the syringe or 20 μL. To set up the dispenser, first insert the barrel of the syringe into the dispenser, followed by the plunger. It may be necessary to trim the barrel handle with a pair of scissors in order to fit it into the dispenser. Tighten all screws, and check that the plunger moves when the button is pressed (Fig. 1).
3. The syringe will be exposed to potential contaminants during assembly. Therefore, it is necessary to disinfect by washing with 10% bleach. To do so, insert the syringe in bleach, pull back on the plunger to fill, and then push down on the plunger to empty the syringe. Repeat the bleach disinfection two more times. Next wash the syringe with sterile water three times to remove all traces of bleach. Once the syringe is disinfected, it can be filled with the desired inoculum, and the needle can be attached. Tap out any large bubbles with your fingers, and press the button a few times until contents are dispensed consistently.
4. Before beginning injections, briefly incubate the larvae in a Petri dish on ice or in the refrigerator to anesthetize them (*see* **Note 5**). To minimize bacterial contamination, it is important to disinfect the cuticle before injecting the fungal inoculum. This can be accomplished by spreading a bit of 70% ethanol on a Petri dish lid with a cotton swab. Gently grasp the larva with a pair of forceps, and dip the ventral side in the ethanol.
5. To inject a larva, hold it with flat forceps so that the prolegs are exposed (Fig. 2). Insert the needle into the last proleg at a 45° angle. Once the needle is inserted, release the forceps, and lift up the larva slightly. The needle will hold the larva (*see* **Note 6**). Press the button on the dispenser to inject. Place the larva in a clean and dry Petri dish with the cover on. Repeat the injections on additional larvae. Many larvae can be injected rapidly with same syringe and needle. However, the needle and syringe can be changed between groups.

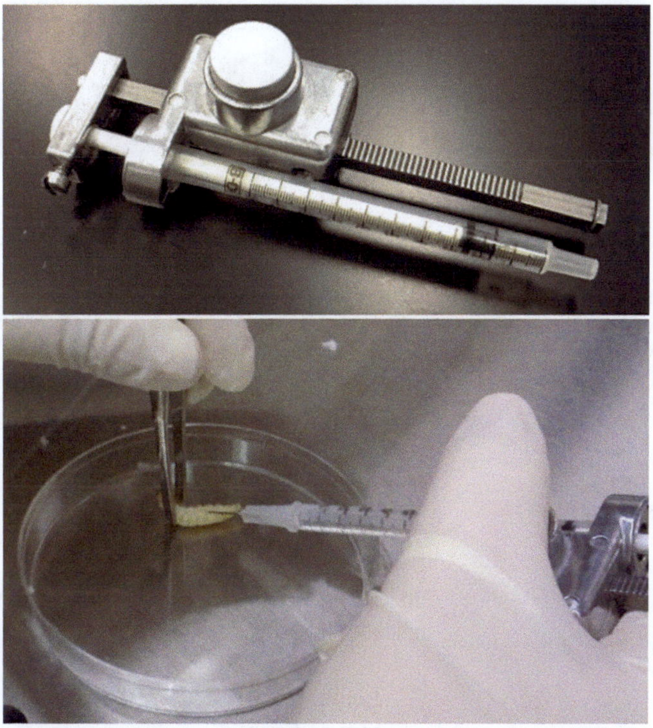

Fig. 1 Injection technique. Repeating dispenser with 1 mL syringe installed correctly (top). Insertion of needle in *G. mellonella* larvae (bottom)

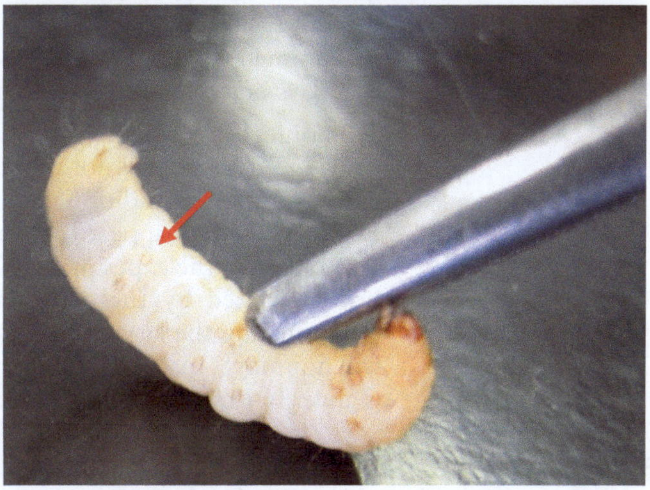

Fig. 2 *G. mellonella* proleg identification. There are four pairs of prolegs on the abdomen of the larva. The red arrow points to the posterior proleg of the larva

6. Once the injections are complete, inspect all the larvae. Be sure that they are moving and alive before incubating. In addition, check that the Petri dish is clean and dry. A lab wipe dipped in a little ethanol can be used to clean the Petri dish if necessary.

Fig. 3 *G. mellonella* pupae appearance. Larvae in various stages of pupation are indicated by asterisks

Leave covered Petri dishes in an incubator at the desired temperature.

7. After injecting the larvae, it is a good idea to plate the fungal inoculum to ensure culture viability and CFU count. Serially dilute the culture to achieve a density of 10^3 CFU/mL. Plate 0.1 mL (100 CFUs) on Sabouraud Dextrose Agar. Plate each inoculum in triplicate for accuracy. Incubate plates at 30 °C for 2–3 days before counting colonies.

8. To monitor survival after infection, larvae should be inspected daily. Dead larvae will not move or respond to touch stimulus with a probe. In addition, they will often turn a dark color. Living larvae will appear off-white in color and move. Larvae may pupate if the experiment is left for too long. Pupation can be recognized by the appearance of a tan-colored cocoon encapsulating the larvae (Fig. 3). Larvae undergoing pupation will not move, making it difficult to determine if the individual is alive. In this case, color may be the best indicator of health.

3.4 Fungal Burden

To determine the number of fungal cells in the infected larvae, a fungal burden assessment may be performed (*see* **Note** 7).

1. Homogenize larva in a small volume of PBS (1 mL). Perform homogenization on ice. Next, prepare tenfold serial dilutions by combining 0.1 mL of homogenate with 0.9 mL of PBS. Plate 0.1 mL of each dilution on Sabouraud agar with antibiotics. Plate each dilution in triplicate. Incubate 2–3 days at 30 °C.

2. After incubating 2–3 days, count the fungal colonies. Determine fungal burden with the formula: CFU/larvae = number of colonies ÷ final dilution factor (FDF), where FDF is equal to the dilution factor of the tube times the volume plated (0.1 mL).

3.5 Larva Fixation

G. mellonella larvae infected with *C. neoformans* may be stained with various stains, including hematoxylin and eosin (H&E) or Gomori methenamine silver after fixation.

1. Label all tubes for fixation with solvent-resistant markers.
2. At the chosen time point, remove larvae from incubator, and place them on ice in a Petri dish or in the refrigerator for a few minutes to anesthetize them.
3. Carefully poke holes in the cuticle of the larvae with a fine-gauge needle, and then transfer to tubes with 1.5 mL of paraformaldehyde solution. Incubate the larva in fixative for 10 days at 4 °C.
4. After 10 days, remove larvae from the paraformaldehyde, and place in a tube with 30% sucrose solution. Incubate 12–24 h at 4 °C.
5. Store larvae in 70% ethanol solution at 4 °C until they can be embedded, sectioned, and stained by the desired method.

4 Notes

1. *G. mellonella* is not maintained for research purposes, and larvae from different suppliers may give variable results. To avoid inconsistencies, purchase larvae from a single supplier. Alternatively, *G. mellonella* has been maintained in the laboratory by some researchers [16].

   Larvae are often sold in small containers with 200–300 organisms. Upon arrival, make sure the larvae appear healthy. They should be moving and off-white in color without visible black spots. Lab temperatures can fluctuate over the course of the year. For consistency, larvae can be stored at 4 °C until ready to use. The larvae will eventually pupate. To avoid this, infections should be carried out within a few days of larval receipt.
2. Many researchers use a Hamilton® syringe (10 μL). However, these will need to be re-filled with inoculum in between each injection.
3. Average survival of infected larvae depends on a number of factors, such as the cryptococcal strain, the incubation temperature, and the number of CFUs injected. Pilot experiments will

help determine the best inoculum size for the experiment. A series of tenfold dilutions of fungal cells can be tested to determine a preliminary survival time. Depending on the experimental design, inoculums that result in survival times of approximately 1–2 weeks enable one to determine differences in survival for different experimental conditions. Larvae incubated for longer times may pupate, making it difficult to interpret the experimental results.

4. The inoculum can be suspended in a solution of streptomycin in PBS to prevent bacterial infection of the larvae. An effective amount is 20 mg/kg of body weight. For example, if the average body weight is 0.2 g, then 0.004 mg per larvae is needed. If larvae receive an inoculum of 20 μL, then the streptomycin stock solution (50 mg/mL) can be diluted 250-fold (4 μL per 1 mL of inoculum suspension).
5. Do not overdo the cold incubation. A few minutes is sufficient. The larvae should be moving slowly. Larvae that are very limp will be difficult to inject.
6. Holding the larva with the needle instead of forceps reduces the chances of crushing the larva while injecting it. No hemolymph should leak out during or after the injection. If it does, discard that larva.
7. The insect cuticle can make it difficult to homogenize the larvae. The homogenizer may become clogged and require cleaning during the homogenizing process. It is helpful to cut the larvae in small pieces with a scalpel or razor blade before homogenization.

References

1. Xue C, Tada Y, Dong X et al (2007) The human fungal pathogen *Cryptococcus* can complete its sexual cycle during a pathogenic association with plants. Cell Host Microbe 1:263–273
2. Steenbergen JN, Shuman HA, Casadevall A (2001) *Cryptococcus neoformans* interactions with amoebae suggest an explanation for its virulence and intracellular pathogenic strategy in macrophages. Proc Natl Acad Sci U S A 98: 15245–15250
3. de Souza PC, Morey AT, Castanheira GM et al (2015) *Tenebrio molitor* (Coleoptera: Tenebrionidae) as an alternative host to study fungal infections. J Microbiol Methods 118:182–186
4. Apidianakis Y, Rahme LG, Heitman J et al (2004) Challenge of *Drosophila melanogaster* with *Cryptococcus neoformans* and role of the innate immune response. Eukaryot Cell 3:413–419
5. Mylonakis E, Moreno R, El Khoury JB et al (2005) *Galleria mellonella* as a model system to study *Cryptococcus neoformans* pathogenesis. Infect Immun 73:3842–3850
6. Sangalli-Leite F, Scorzoni L, Alves de Paula E, Silva AC et al (2016) Synergistic effect of pedalitin and amphotericin B against *Cryptococcus neoformans* by in vitro and in vivo evaluation. Int J Antimicrob Agents 48:504–511
7. Palanco AC, Lacorte Singulani J, Costa-Orlandi CB et al (2017) Activity of 3′-hydroxychalcone against *Cryptococcus gattii* and toxicity, and efficacy in alternative animal models. Future Microbiol 12:1123–1134
8. McClelland EE, Ramagopal UA, Rivera J et al (2016) A small protein associated with fungal energy metabolism affects the virulence of *Cryptococcus neoformans* in mammals. PLoS Pathog 12:e1005849

9. Desalermos A, Tan X, Rajamuthiah R et al (2015) A multi-host approach for the systematic analysis of virulence factors in *Cryptococcus neoformans*. J Infect Dis 211:298–305
10. Aminnejad M, Cogliati M, Duan S et al (2016) Identification and characterization of VNI/VNII and novel VNII/VNIV hybrids and impact of hybridization on virulence and antifungal susceptibility within the *C. neoformans/C. gattii* species complex. PLoS One 11:e0163955
11. Gago S, Serrano C, Alastruey-Izquierdo A et al (2017) Molecular identification, antifungal resistance and virulence of *Cryptococcus neoformans* and *Cryptococcus deneoformans* isolated in Seville, Spain. Mycoses 60:40–50
12. Firacative C, Duan S, Meyer W (2014) *Galleria mellonella* model identifies highly virulent strains among all major molecular types of *Cryptococcus gattii*. PLoS One 9:e105076
13. Garcia-Rodas R, Casadevall A, Rodriguez-Tudela JL et al (2011) *Cryptococcus neoformans* capsular enlargement and cellular gigantism during *Galleria mellonella* infection. PLoS One 6:e24485
14. Trevijano-Contador N, Herrero-Fernandez I, Garcia-Barbazan I et al (2015) *Cryptococcus neoformans* induces antimicrobial responses and behaves as a facultative intracellular pathogen in the non mammalian model *Galleria mellonella*. Virulence 6:66–74
15. Cerenius L, Soderhall K (2004) The prophenoloxidase-activating system in invertebrates. Immunol Rev 198:116–126
16. Firacative C, Khan A, Duan S et al (2020) Rearing and maintenance of *Galleria mellonella* and its application to study fungal virulence. J Fungi (Basel) 6:130

Chapter 2

The Rabbit Model of Cryptococcal Meningitis

Charles Giamberardino and John R. Perfect

Abstract

Cryptococcal meningitis (CM) is a fungal disease caused by the invasion of *Cryptococcus* yeast cells into the central nervous system. The organism is thought to enter the body through the lungs and then escape due to dysregulation of the immune response. Multiple animal species have been used to model the infection and characterize CM including mice, rats, dogs, guinea pigs, and rabbits. The rabbit model has over 40 years of data and has been used to study host–pathogen interactions and the efficacy of antifungal therapeutics. The model begins with immune suppression to eliminate the lymphocytic cell population followed by direct infection of the central nervous system via an injection of a suspension of yeast cells into the cisterna magna. The organism remains in the CNS during the course of infection, and cerebrospinal fluid can be repeatedly sampled to quantify the burden of organism, measure drug levels in the CSF, profile the immune response in the CSF, and/or characterize the yeast cells. The rabbit model of infection is a robust experimental model for better understanding CM and *Cryptococcus* cellular behavior.

Key words *Cryptococcus*, Rabbit model, Meningitis, Drug testing, Virulence

1 Introduction

The protocol below describes the steps for the general structure of the rabbit cryptococcal meningitis (CM) model. The protocol is only appropriate for researchers who have training and experience working with rabbits. Five to six team members are recommended for inoculations, cerebrospinal fluid (CSF) collections, and euthanasia and two team members for daily observations and drug dosing (*see* Table 1). The rabbit model, when executed correctly, is highly reproducible, and we have used it to study antifungal regimens [1, 2], the efficacy of new antifungal agents [3–7], virulence, and the molecular biology of *Cryptococcus* cells [8–15]. For studies comparing different strains of *C. neoformans*, to statistically show a difference in fungal burden of 1 $\log_{10}$ colony-forming units (CFU)/mL of CSF, a minimum of three rabbits per group should be used. Using four rabbits is preferred to ensure adequate sample numbers at all experimental time points. Methods for drug dosing

Erin E. McClelland (ed.), *Cryptococcus neoformans: Methods and Protocols*, Methods in Molecular Biology, vol. 2775,
https://doi.org/10.1007/978-1-0716-3722-7_2,

Table 1
Personnel requirements for inoculations and CSF collections

Task	Personnel
Sedations[a]	1 team member to draw up reagents 1 team member to calculate doses and record the agent volumes 1 team member to inject
IC injection or CSF collection	1 team member to secure the rabbit 1 team member to perform the injection or collection
Recovery	1 team member to administer yohimbine and recovery fluids
Documentation/scribe	1 team member to record sedation times, inoculation times, CSF collection times, collection volumes, CSF quality
Vitals checks after sedations	1 team member to record sedation times, check vitals, monitor recovery status
Daily health checks, weighing, and HC injections	2 team members, one to weigh and inject and one to scribe

[a]The team members performing sedations can later assist with recovery or documentation once all rabbits are sedated; therefore, five to six total team members are required

or blood collection are not included, as these are standard techniques when working with rabbits. The protocol begins with inoculum preparation and rabbit immune suppression using hydrocortisone acetate (HC). The inoculum can range from ~1×10^6 cells for drug efficacy, lasting up to 14 days postinfection, to 1×10^9 cells for short-duration molecular biology studies, lasting a maximum of 4 days postinfection. The rabbits are then sedated and inoculated with the organism. The immune suppression continues, daily, through the end of the study. The protocol includes techniques for collecting CSF, quantifying the fungal burden, executing necropsies, and processing brain tissue for fungal burden.

This experimental cryptococcal model allows a dynamic removal of yeasts from the subarachnoid space in an immunocompromised host. It has been and continues to be used effectively to understand serial yeast RNA expression at a very relevant site of infection. It allows the evaluation of mutants and other yeast strains in their survival at this relevant host site. Furthermore, it has elegantly and critically predicted yeast outcomes during treatment with a variety of drugs for cryptococcal meningitis that were validated in humans. The rabbit model is an ideal rigorous screen for new treatment modalities and drugs prior to human studies for cryptococcal meningitis.

2 Materials

2.1 Materials

1. Yeast peptone dextrose (YPD) broth.
2. YPD agar.
3. YPD + C: yeast peptone dextrose + chloramphenicol 50–100 μg/mL.
4. YPD + C agar petri dishes, 4/CSF collection/rabbit, 4/brain region/rabbit.
5. Glass beads or L-shaped spreaders.
6. 5 mL round-bottom polystyrene tubes with snap caps.
7. Chloramphenicol.
8. 4% chlorhexidine gluconate.
9. 70% ethanol or 70% isopropyl alcohol.
10. Clippers with No. 10 blade.
11. 2"× 2" or 4" × 4" non-woven gauze.
12. 1 mL syringes with 25G 5/8" needle.
13. 1 mL syringes with 27G 1/2" needle.
14. 3 mL syringes with 25G 5/8" needle.
15. Ketamine 100 mg/mL.
16. Xylazine 100 mg/mL.
17. Euthanasia agent: pentobarbital 390 mg/mL + phenytoin 50 mg/mL.
18. Nutritional supplement (e.g., Oxbow, Critical Care Herbivore Apple-Banana).
19. Scale (0–20 kg).
20. Supplemental heat lamp.
21. Veterinary blankets.
22. Ophthalmic ointment.
23. 0.9% saline, 5% dextrose, or lactated Ringer's solution for intravenous (IV) infusion.
24. Petri dishes, 1/rabbit.
25. Isoflurane.
26. Isoflurane vaporizer with non-rebreather mask.
27. 100% oxygen.
28. Yohimbine HCl.
29. 25G butterfly needle venous blood collection sets.
30. Rongeurs.
31. Toothed forceps.

32. Bone saw.
33. 0.9% saline or phosphate-buffered saline (PBS) in a squirt bottle.
34. Wide-bore filter pipet tips.
35. Dry ice in a styrofoam container (~ 2–5 kg).
36. Absorbent blue bench pads.
37. Preparation of hydrocortisone acetate suspension: Prepare the buffer. Prepare all buffers in a BSC to ensure sterility. For each milliliter of buffer, combine 9 mg NaCl, 5 mg sodium carboxymethylcellulose, and 4 mg Tween 20, and adjust the volume to 1.0 mL with sterile water. Add benzyl alcohol to a final concentration of 0.9% as a preservative, autoclave the buffer, and store at room temperature until use. Use hydrocortisone 21-acetate >/= 98% (HPLC), and store stock at room temperature. Use a sterile 100 mL griffin beaker with autoclaved magnetic stirrer. Add 2.5 grams of hydrocortisone 21-acetate powder (non-sterile) to 40 mL of sterile buffer. Mix on a magnetic stirrer for 3–4 h to generate a fine suspension. Bring the volume to 50 mL using the sterile buffer. The hydrocortisone suspension is heat sensitive and cannot be autoclaved and, because it is a fine suspension, cannot be filtered. For quality control, inoculate 10 μL of the suspension onto a Luria–Bertani (LB) agar plate, and incubate at 37 °C for 2–3 days to check for microbial contamination.
38. Preparation of sedatives for inoculations, CSF collections, and euthanasia: It is recommended to weigh the rabbits and draw up all sedatives 1 h prior to use. Label each syringe with the rabbit identification (ID) number and agent. Labels ~ 1” × 0.5” are very useful for labeling. A tray with dividers is ideal for this purpose. Draw up the reversal agent as well or euthanasia reagents if at the study endpoint.

2.2 Documentation

Documentation for rabbit experiments is critical for reproducibility and accurate reporting. Each rabbit should have a separate study form. The rabbit ID should be indicated on each page of the form. All events, along with the date, time, and person who executed the action, should be recorded. For dosing, the drug, route, time, stock concentration, volume administered, and individual performing the dosing should be recorded. During sedations, it is important to check and record vital signs every 15 min, including mucous membrane color, response to toe pinch, and evidence of respiration noted by the rise and fall of the chest. For CSF collections, the time of collection, volume of CSF, clarity of the CSF, and presence of blood in the CSF should be recorded. Additionally, daily health checks and supplemental care, such as nutritional supplementation, should be recorded. *See* Tables 2, 3, 4, 5 and 6 for examples of study records.

Table 2
Example inoculation record table

Intracisternal inoculation events (Day 0)	
Date	
Organism	
Strain	
Inoculum amount (number of cells)	
Inoculum volume, mL	
Eye ointment application time	
Inoculation time	
Inoculated by	
Notes/comments	

Table 3
Example inoculation sedation and recovery table

Inoculation sedation & recovery events (Day 0)	
Date	
Rabbit weight, kg	
Ketamine dose, mg/kg	
Xylazine dose, mg/kg	
Ketamine volume, mL	
Xylazine volume, mL	
[ketamine]	100 mg/mL
[Xylazine]	100 mg/mL
Sedation time	
Sedated by	
Route	IM
Yohimbine volume, mL	
Yohimbine [stock]	
Yohimbine administration time	
Yohimbine administered by	
Yohimbine route	IV
Saline administration time	
Saline volume, mL	
Saline administered by	
Route	SC

Table 4
Example vitals check record

Inoculation Vitals and Recovery Monitoring						
Rabbit ID				Date		
Time	Breathing (Y/N)	Mucous Membrane Pink (Y/N)	Toe Pinch Reflex (Y/N)	Sternal Recumbent (Y/N)	Notes/Comments	Scribe Initials

Table 5
Example CSF collection record

CSF collection
Date
Eye ointment application time
CSF collect time
CSF volume collection, mL
CSF appearance
Collected by

Table 6
Example health check log and daily dose record

	Hydrocortisone Stock 50 mg/mL		Amphotericin B Stock 2.5 mg/mL
Date	Dose, mg/kg	5	Dose, mg/kg
Wt, kg	Route	IM	Route
Health	Time		Time
Time	Volume, mL		Volume, mL
Initials	Initials		Initials

Abbreviations: *H* healthy, *L* lethargic, *S* severe symptoms, *FD* found dead, *NS* neurological symptoms, *RS* respiratory symptoms

3 Methods

3.1 Inoculum Preparation

1. Five to seven days before the inoculation, streak out the organism onto a YPD agar petri dish from a glycerol freezer stock.
2. Incubate the petri dish for 3–4 days at 30 °C.
3. After culturing the organism for 3–4 days, in a biosafety cabinet (BSC), prepare 10–25 mL of YPD broth in a container that is 5× the target broth volume. For example, for 10 mL of broth, use a 50 mL container. For inocula of 1×10^8 to 1×10^9 cells, use a 125 mL Erlenmeyer flask. For inocula of 1×10^6 to 1×10^7 cells, a 50 mL conical is sufficient.
4. In a BSC, select 1–10 individual colonies from the petri dish to inoculate the culture.
5. Incubate the culture for 1–2 days at 30 °C in a shaking incubator at 225 rpm. If an Erlenmeyer flask is used, cover the top with aluminum foil. If a 50 mL conical tube is used, ensure it is loosely capped to allow for adequate oxygenation.
6. Remove the culture from the incubator, and working in a BSC, transfer the culture to a 50 mL conical tube, and adjust the volume to 50 mL using PBS.
7. Centrifuge the culture at 500–2000× *g* for 3–5 min to pellet the cells, and then decant the supernatant.
8. Resuspend the pelleted cells in 50 mL of PBS and repeat the centrifugation.
9. Decant the supernatant and repeat the resuspension and centrifugation.
10. Resuspend the pellet in 10 mL of PBS. Serially dilute 100 μL of the stock suspension, count the serial dilutions, and then calculate the cell concentration.
11. Calculate the volume of suspension needed for the experiment as follows: total volume, mL, = (number of rabbits +3) × 0.3 mL/rabbit. More may be prepared to ensure an adequate amount.
12. Calculate the volume of the stock cell suspension and volume of PBS diluent to prepare the injection suspension as follows (using a 1×10^6 inoculum as an example): volume of stock suspension, mL, = (total volume for inoculations, mL) × ($3.3–3.5 \times 10^6$ cells/mL) /(cell concentration, cells/mL). Volume of PBS, mL, = (total volume for inoculations, mL) – (volume of stock suspension, mL).
13. Add the volume of PBS to a 15 mL round-bottom tube, add the volume of stock suspension, cap tightly, and vortex the tube. Reserve at least 0.2 mL of the inoculum, serially dilute 1:10, and then plate the suspension onto YPD agar. Incubate

for 3–5 days at 30 °C, to quantify the CFU/mL and to check for any signs of contamination.

14. Load 0.3 mL of the inoculum into 3 mL syringes with 25 G 5/8” needles, and carefully clear any air from the syringes. Note that the safest method to execute the filling involves using one needle to draw up the solution, discarding the needle, and then placing a new 25 G needle on the syringe to avoid recapping.
15. Pack the syringes into a hard-sided container with adequate absorbent material, and transport the syringes to the location where the rabbits are housed.

3.2 Immune Suppression

1. One day before the infection, weigh each rabbit, record the weight in kg to the nearest 0.1 or 0.01 kg, and then round the weight to the nearest 0.1 kg.
2. Calculate the dose as follows: dose, mg, = (weight, kg) × (5 mg/kg). Volume, mL, = (dose, mg) / (50 mg/mL).
3. Using a 1 mL syringe with 25 G 5/8” needle, draw the appropriate volume from the HC stock, and clear any air bubbles from the syringe.
4. Secure the rabbit on a solid surface, and inject the HC suspension intramuscularly (IM) by inserting the needle into the upper quadriceps or lower lumbar region, gently drawing back on the plunger to ensure no blood is present. Then, depress the plunger and quickly remove the syringe.
5. Note the time and location of the dose. Alternate the injection site each day, ensuring sequential elections are performed contralaterally.

3.3 Sedations for Inoculations and CSF Collections

1. Weigh the rabbit and administer the daily dose of HC.
2. Calculate the ketamine dose (30–40 mg/kg) and the xylazine dose (3–5 mg/kg).
3. If using the lower range (30 mg/kg of ketamine and 3 mg/kg of xylazine), then inject, wait 10 min, check that the rabbit is fully sedated, and then proceed to the infection (*see* **Note 1** on ketamine/xylazine dosing).
4. The higher dose will result in a deeper and longer sedation, but can also lead to complications, particularly respiratory arrest (*see* **Note 1**).
5. Draw the ketamine dose into a 3 mL syringe with a 25 G 5/8” needle, and draw the xylazine dose into a 1 mL syringe with a 25 G needle.
6. Immediately before dosing, carefully remove the 25 G needle from the 3 mL syringe that contains the ketamine, and draw back on the plunger. Then, inject the xylazine into the 3 mL syringe containing the ketamine.

7. Place a new 25 G 5/8" needle on the 3 mL syringe. Invert three times to mix and then clear any air bubbles.
8. Place the rabbits on a solid surface, inject the ketamine/xylazine mixture IM, and return the rabbit to its home cage. Note the time and location of the ketamine/xylazine injection.
9. Once the rabbit appears fully sedated, about 10 min, check for a response to deep toe pinch.
10. If the rabbit is refractory to sedation, either administer isoflurane, 2 L/min, 1–2% for 2–3 min or give an additional 50% dose of ketamine (15–20 mg/kg) (*see* **Note 2**).

3.4 Inoculation

1. Sedate the rabbits in groups of three to six animals per group.
2. Place the rabbit on a flat solid surface, preferably a stainless steel table with an absorbent pad, with the rabbit positioned perpendicularly to the long axis of the table.
3. Using clippers, remove the fur over the back of the head to the C1 vertebra.
4. Prep the area over the cisterna magna (the space between the base of the skull and the C1 vertebra that consists of a relatively large reservoir of CSF) with three alternating washes of 4% chlorhexidine and 70% ethanol.
5. Place the rabbit in the prone position, with its hind limbs extended out from its body and its head slightly over the edge of the table, ensuring the spine is straight.
6. A holder secures the rabbit's head by placing the fore and middle fingers over the snout and the thumbs behind the ears and then flexing the head to approximately 90°. The rabbit should be held in this position for a maximum of 20 s.
7. The researcher performing the injection secures a 3 mL syringe with 25 G 5/8" needle that contains 0.3 mL of the *Cryptococcus* cell suspension and removes the cap.
8. The researcher performing the injection locates the cisterna magna by feeling for a depression behind the ears and above C1, which is in line with the spine, and then carefully inserts the needle, keeping the syringe oriented vertically, while drawing back very slowly on the plunger.
9. Once CSF is observed entering the syringe, the researcher performing the injection quickly depresses the plunger and removes the needle. Note if any blood is observed in the syringe during the inoculation.
10. The rabbit is immediately placed in lateral recumbency and transitioned to a recovery area. Note the time and quality of the inoculation.

11. Administer yohimbine, 0.2–1 mg/kg, IV, and then observe the color of the mucous membranes and the rise and fall of the chest every 15 min until the rabbit is fully recovered (approximately 1–2 h if yohimbine is used).
12. OPTIONAL: Administer supplemental fluids, 50–100 mL, subcutaneous (SC), of 0.9% saline, 5% dextrose solution (D5W), or lactated Ringer's solution (LRS).

3.5 Daily Observation

1. Weigh the animal and record the weight; use this weight to calculate the dose of HC.
2. Note if the animal is lethargic or has any respiratory or neurological symptoms, including head tilt (*see* **Notes 3** and **5**).
3. Consult a veterinarian if the animal has lost 10% of its maximum body weight or if fecal output appears reduced (*see* **Note 4**).

3.6 CSF Collection

1. Sedate the rabbits in groups of three to six animals per group (see Subheading 3.3).
2. Place the rabbit in the prone position on a solid steel table with an absorbent pad below it.
3. Prep the area over the cisterna magna with three alternating washes of chlorhexidine and 70% ethanol.
4. Flex rabbit's head to 90° and hold it in position for no more than 20 s.
5. Gently insert the 3 mL syringe with 25 G 5/8" needle into the cisterna manga while gently withdrawing the plunger.
6. Once CSF is observed entering the syringe, hold it in place until approximately 0.3–0.6 mL of CSF is collected, and then quickly remove the syringe.
7. Immediately position the rabbit in lateral recumbency, and transition it to a recovery area. Note the time, color, and volume of the CSF sample. Using hemostats, carefully remove the needle, discard it into a sharps container, and cap with a sterile Luer–Lok syringe cap.
8. Administer yohimbine, 0.2–1 mg/kg, IV, and then observe the color of the mucous membranes and the rise and fall of the chest every 15 min until the rabbit is fully recovered (approximately 1–2 h if yohimbine is used).
9. OPTIONAL: Administer supplemental fluids, 50–100 mL, SC, of 0.9% saline, D5W, or LRS.

3.7 Euthanasia

1. Sedate the rabbits with 40 mg/kg of ketamine and 5 mg/kg of xylazine, working in groups of three to six animals.
2. Draw 1.4 mL of 390 mg/mL of pentobarbital (a minimum of 390 mg/4.5 kg of pentobarbital is required) into a 3 mL Luer–

Lok syringe with attached 25 G butterfly needle, and then fill the dead space in the tubing with the euthanasia agent. The typical dead volume for a butterfly catheter is 0.4 mL.

3. Place the rabbit in the prone position on a solid steel table with an absorbent pad below it.
4. Prep the area over the cisterna magna with three alternating washes of chlorhexidine and 70% ethanol.
5. Flex rabbit's head to 90° and hold it in position for no more than 20 s.
6. Gently insert the 3 mL syringe with 25 G 5/8" needle into the cisterna manga while gently withdrawing the plunger.
7. Once CSF is observed entering the syringe, hold it in place until approximately 0.3–1.0 mL of CSF is collected, and then quickly remove the syringe. Note the time, color, and volume of the CSF sample. Using hemostats, carefully remove the needle, discard it into a sharps container, and cap with a sterile Luer–Lok syringe cap.
8. Place the rabbit in lateral recumbency, and dilate the ear vein by holding pressure on the proximal portion of the vein for 10–20 s.
9. Carefully insert the 25 G butterfly needle into the vein, and gently pull back on the plunger to ensure proper placement, evidenced by blood entering the tubing.
10. Slowly depress the plunger, administer the entire volume, and then check that the heart has stopped beating and the pupils are fixed and dilated.
11. If needed, use a 2 1/2" needle (18–21 G) connected to a 3–5 mL Luer–Lok syringe to collect a final blood sample via cardiac puncture.
12. Perform a second method of euthanasia, i.e., bilateral thoracotomy, and proceed to necropsy or disposal.

3.8 Necropsy Preparation

1. Fill tubes for homogenization with 1–3 mL of PBS, weigh the tubes, and record the weight of the tubes without tissue and with the volume of PBS added. Tubes can be filled and weighed up to 1 week before use.
2. Place absorbent bench pads on the necropsy surface.
3. Secure 1 disposable scalpel per rabbit and 1 petri dish per rabbit.
4. Fill a 500 mL squirt bottle with 1× PBS.
5. Ensure there is an adequate amount of 70% ethanol and non-woven gauze, either 2" × 2" or 4" × 4" (~ 5–10 per rabbit).

6. Place necropsy tools, bone saw, rongeurs, and toothed forceps in the necropsy area.
7. Place ~5 pounds (2.2 kg) of dry ice in a Styrofoam container if collecting brain tissue for drug levels, tissue protein analysis, or tissue metabolite analysis.

3.9 Necropsy

1. Place the carcass in the prone position and wet the skull with 70% ethanol.
2. Use scissors and toothed forceps to remove the skin over the head and neck, and then remove muscle over the cervical vertebrae.
3. Use nonwoven gauze to dry the skull.
4. Use a bone saw to cut through the skull, taking care not to damage the meninges.
5. Use rongeurs to carefully remove pieces of bone and expose the meninges, brain, and cervical vertebrae.
6. If collecting the meninges, secure it with toothed forceps, and then cut it away with scissors or a scalpel.
7. Use a spatula to carefully separate the brain from the skull, and then cut the spinal cord, preserving approximately 1 cm of the spinal cord.
8. Remove the brain and spinal cord, place it in a clean petri dish, then rinse the tissue with 1× PBS to remove blood, and carefully blot with nonwoven gauze.
9. Dissect the brain per the needs of the experiment. Place the pieces either in tubes containing 1–3 mL of 1× PBS for fungal burden or in empty tubes for drug levels, tissue proteomics, or metabolomics.
10. Weigh the tubes and record the weights.
11. Freeze any of the tissue in tubes containing only tissue for drug levels, proteomics, or metabolomics by placing them on dry ice.

3.10 CSF Processing

1. For fungal burden, fill four to five round-bottom polystyrene tubes with 0.9 mL of 1× PBS for serial dilutions. The dilution tubes can be made up to 1 week in advance.
2. Create a "neat" sample by placing 0.1 mL of the undiluted CSF on a YPD + C petri dish, and spread the CSF with glass beads or an L-shaped spreader.
3. Use an additional 0.1 mL to make serial dilutions by adding it to 0.9 mL of 1× PBS in 5 mL round-bottom tubes.
4. Plate the dilutions on YPD + C, and then incubate all YPD + C petri dishes at 30 °C for 3–5 days.

5. Count the colonies, and calculate the CFU/mL by multiplying the CFU counted by the dilution and then dividing by 0.1 mL. Ideally, the dilution with 30–300 colonies should be used for counting CFU.
6. Centrifuge the remaining CSF at 500–2000× *g* to pellet the cells, and then collect the supernatant.
7. Store the cell pellet at −80 °C for later analysis, or lyse the cells for molecular analyses.

3.11 Brain Tissue Processing

1. Store the tissue for protein or metabolite analysis at −80 °C until use.
2. Homogenize the brain tissue using a probe or tube homogenizer.
3. Create a "neat" sample by placing 0.1 mL of the undiluted homogenate onto a YPD + C petri dish using a wide-bore pipet tip, and then spread the tissue with glass beads or an L-shaped spreader.
4. Use an additional 0.1 mL to make serial dilutions by adding it to 0.9 mL of 1× PBS in 5 mL round-bottom tubes, using wide-bore pipet tips for the first two dilutions.
5. Incubate the petri dishes for 3–5 days at 30 °C.
6. Count the colonies, and calculate the CFU/g as follows: Calculate the CFU/mL by multiplying the CFU counted by the dilution and then dividing by 0.1 mL (or the volume added to the petri dish). Calculate the net tissue weight, g, by subtracting the tube weight without tissue from the tube weight with tissue. Calculate the total volume, mL, by adding the volume of PBS in the tube to the net weight of the tissue. Assume a tissue density of 1 g/mL. Calculate the total CFU by multiplying the total volume, mL, × the CFU/mL. Calculate the CFU/g by dividing the total CFU by the net tissue weight, g. Ideally, the dilution with 30–300 colonies should be used for counting CFU.

4 Notes

1. Ketamine/xylazine dosing and negative responses to ketamine/xylazine. For terminal collections prior to euthanasia, a higher dose of ketamine (40–50 mg/kg) and xylazine (5–6 mg/kg) is preferred. Utilizing the lower dose for inoculations and CSF collections limits complications. Rabbits can occasionally go into respiratory arrest almost immediately after ketamine/xylazine administration, and this occurs in approximately 0.5–1% of rabbits. It is important to closely observe the rabbits after administering the ketamine/xylazine. In the event

of shallow breathing or a lack of breathing, quickly administer the yohimbine reversal agent, and apply supplemental heat and supplemental 100% oxygen at 2 L per minute.

2. Rabbits refractory to ketamine/xylazine. Use inhaled isoflurane administered via face mask to ensure full sedation.
3. Neurological symptoms. Head tilt is very common in the model. Note the date and time of observation. If the head tilt appears to interfere with the health and well-being of the animal, consult a veterinarian.
4. Weight loss. If weight loss exceeds 85% of an animal's maximum observed body weight, provide supplemental nutrition. Example: 40 mL of dry Oxbow Critical Care mixed with 40 mL of water. It can be helpful to provide additional daily fresh vegetables to encourage eating and limit the amount of weight loss.
5. Severe neurological symptoms, respiratory symptoms, or lethargy. If an animal develops respiratory symptoms, lethargy (evidenced by a reluctance to move), or severe neurological symptoms, such as difficulty righting, terminate the study, and adjust future experiments to end the study before these outcomes are observed.

References

1. Lestner J, McEntee L, Johnson A et al (2017) Experimental models of short courses of liposomal amphotericin B for induction therapy for cryptococcal meningitis. Antimicrob Agents Chemother 61(6). https://doi.org/10.1128/AAC.00090-17
2. Perfect JR, Wright KA (1994) Amphotericin B lipid complex in the treatment of experimental cryptococcal meningitis and disseminated candidosis. J Antimicrob Chemother 33(1): 73–81. https://doi.org/10.1093/jac/33.1.73
3. Giamberardino CD, Schell WA, Tenor JL et al (2022) Efficacy of APX2039 in a rabbit model of cryptococcal meningitis. MBio 13(6): e0234722. https://doi.org/10.1128/mbio.02347-22
4. Perfect JR, Cox GM, Dodge RK et al (1996) *In vitro* and *in vivo* efficacies of the azole SCH56592 against *Cryptococcus neoformans*. Antimicrob Agents Chemother 40(8): 1910–1913. https://doi.org/10.1128/AAC.40.8.1910
5. Schell WA, De Almeida GM, Dodge RK et al (1998) *In vitro* and *in vivo* efficacy of the triazole TAK-187 against *Cryptococcus neoformans*. Antimicrob Agents Chemother 42(10): 2630–2632. https://doi.org/10.1128/AAC.42.10.2630
6. Wright KA, Perfect JR, Ritter W (1990) The pharmacokinetics of BAY R3783 and its efficacy in the treatment of experimental cryptococcal meningitis. J Antimicrob Chemother 26(3):387–397. https://doi.org/10.1093/jac/26.3.387
7. Perfect JR, Wright KA, Hobbs MM et al (1989) Treatment of experimental cryptococcal meningitis and disseminated candidiasis with SCH39304. Antimicrob Agents Chemother 33(10):1735–1740. https://doi.org/10.1128/AAC.33.10.1735
8. Cox GM, McDade HC, Chen SC et al (2001) Extracellular phospholipase activity is a virulence factor for *Cryptococcus neoformans*. Mol Microbiol 39(1):166–175
9. Cox GM, Mukherjee J, Cole GT et al (2000) Urease as a virulence factor in experimental cryptococcosis. Infect Immun 68(2):443–448
10. Lee A, Toffaletti DL, Tenor J et al (2010) Survival defects of *Cryptococcus neoformans* mutants exposed to human cerebrospinal fluid result in attenuated virulence in an experimental model of meningitis. Infect Immun 78(10):

4213–4225. https://doi.org/10.1128/IAI.00551-10

11. Sephton-Clark P, Tenor JL, Toffaletti DL et al (2022) Genomic variation across a clinical *Cryptococcus* population linked to disease outcome. MBio 13(6):e0262622. https://doi.org/10.1128/mbio.02626-22
12. Steen BR, Zuyderduyn S, Toffaletti DL et al (2003) *Cryptococcus neoformans* gene expression during experimental cryptococcal meningitis. Eukaryot Cell 2(6):1336–1349. https://doi.org/10.1128/EC.2.6.1336-1349.2003
13. Toffaletti DL, Nielsen K, Dietrich F et al (2004) *Cryptococcus neoformans* mitochondrial genomes from serotype A and D strains do not influence virulence. Curr Genet 46(4):193–204. https://doi.org/10.1007/s00294-004-0521-9
14. Wang P, Nichols CB, Lengeler KB et al (2002) Mating-type-specific and nonspecific PAK kinases play shared and divergent roles in *Cryptococcus neoformans*. Eukaryot Cell 1(2):257–272
15. Yu CH, Sephton-Clark P, Tenor JL et al (2021) Gene expression of diverse *Cryptococcus* isolates during infection of the human central nervous system. mBio 12(6):e0231321. https://doi.org/10.1128/mBio.02313-21

Chapter 3

Models for Inducing Experimental Cryptococcosis in Mice

Natalia Castro-Lopez and Floyd L. Wormley Jr

Abstract

Cryptococcus neoformans and *Cryptococcus gattii* are the predominant etiological agents of cryptococcosis, a particularly problematic disease in immunocompromised individuals. The increased clinical use of immunosuppressive drugs, the inherent ability of *Cryptococcus* species to suppress and evade host immune responses, and the emergence of drug-resistant yeast support the need for model systems that facilitate the design of novel immunotherapies and antifungals to combat disease progression. The mouse model of cryptococcosis is a widely used system to study *Cryptococcus* pathogenesis and the efficacy of antifungal drugs in vivo. In this chapter, we describe three commonly used strategies to establish cryptococcosis in mice: intranasal, intratracheal, and intravenous inoculations. Also, we discuss the methodology for delivering drugs to mice via intraperitoneal injection.

Key words Cryptococcosis, Fungal infection, Intranasal inoculation, Intravenous inoculation, Intratracheal inoculation, Intraperitoneal injection, Animal model, *Cryptococcus neoformans*, *Cryptococcus gattii*

1 Introduction

Cryptococcosis is primarily caused by fungal pathogens of the *Cryptococcus neoformans* and *C. gattii* species complex [1, 2]. Cryptococcal meningoencephalitis is the most common disseminated fungal disease in AIDS patients [3]. Global estimates suggest that approximately 152,000 cases of cryptococcal meningitis occur annually, with 112,000 cryptococcal-related deaths comprising 19% of AIDS-related deaths [4, 5]. *C. neoformans* is the most clinically prevalent *Cryptococcus* species [6] and causes severe pneumonia and disease of the central nervous system (CNS), particularly in immunocompromised patient populations. *C. gattii* occurs in immunocompromised and healthy individuals with no apparent underlying condition [7–9], but typically presents as a fungal pneumonia [8]. *Cryptococcus* infection typically starts via the inhalation of desiccated spores or basidiospores that are ubiquitous throughout the environment into the lung alveoli [10]. While the

Erin E. McClelland (ed.), *Cryptococcus neoformans: Methods and Protocols*, Methods in Molecular Biology, vol. 2775, https://doi.org/10.1007/978-1-0716-3722-7_3,

overwhelming majority of *Cryptococcus* exposures do not result in disease, dissemination to the CNS and other tissues (bone, skin) may occur in those at a high risk for developing cryptococcosis (i.e., solid organ transplant recipients, otherwise healthy HIV-positive persons, and immune-competent persons in endemic areas observed to contain *C. gattii*) [6, 11]. Access to the three mainstay antifungal agents used to treat cryptococcosis (i.e., amphotericin B, flucytosine, and fluconazole) is hampered in resource-limited settings due to high drug costs, inadequate drug stocks or supply chains, and difficulty managing adverse side effects of conventional non-lipid formulations of amphotericin B in patients [12]. Also, no vaccine is clinically available to prevent or treat cryptococcosis in humans. The ready exposure of high-risk populations, increased use of essential immune suppressive therapies that enhance susceptibility of patients to fungal disease, lack of a well-resourced pipeline for the discovery of novel antifungal drugs, and lack of an effective vaccine underline the urgency to develop effective model systems that facilitate the development of novel measures to treat and/or prevent cryptococcosis [2].

Various model systems (nematodes [13–17], mice [18–28], rats [29–32], primates [33–35], zebrafish [36–40], and wax moths [14, 41–43]) have been developed over the years to study cryptococcosis [44] to determine the role of immune cells and the efficacy of new antifungal drugs to mitigate disease and the role of different cryptococcal virulence factors that facilitate pathogenesis. Mouse models have been among the most widely used animal model systems due to the ample availability of murine strains, gene-deficient strains, humanized mouse models, reproducibility of the system, low costs of maintenance compared to that of other models (i.e., primates, rabbits), and ample selection of reagents and techniques available to examine host immune responses [24, 45–49].

Various routes of inoculation are used to induce cryptococcosis, and care should be taken in considering which route of inoculation is most appropriate for the question under investigation. Duration of disease and symptomology depends on several factors including, but not limited to, route of infection, infectious dose, *Cryptococcus* species, and mouse strain employed. The most probable route of infection in humans is via nasal inhalation of the infectious propagule into the lung alveoli [50]. Consequently, the most common infection route in mice is via intranasal inoculation (IN) of cryptococcal yeast into the nares. However, inhalation of yeast into the nasal cavity may result in some cells also directly reaching the brain. Another route of initiating cryptococcal disease is via intratracheal inoculation (IT). Initiating cryptococcosis via IT inoculation results in yeast reaching the lung alveoli via the trachea, avoiding the nasal cavity and possibly seeding the brain via the nasal cavity. IT infection is typically conducted via a simple surgery, but a nonsurgical technique has been developed and is described here. Finally, the use of intravenous infections allows the study of the bloodstream and CNS infections. Each method of inoculation has its advantages and disadvantages. Here we describe the different techniques to experimentally infect mice.

2 Materials

2.1 General Equipment and Material

1. 70% ethanol.
2. Class II biosafety cabinet (BSC) for handling BSL2-level pathogens.
3. Phosphate buffer saline (PBS): To make 1 L of 10× PBS, dissolve 17.8 g of Na_2HPO_4, 2.4 g of KH_2PO_4, 80 g of NaCl, and 2 g of KCl in 700 mL of water, and adjust final volume to 1 L. Autoclave or filter sterilize.
4. Refrigerated centrifuge.
5. 1.5–2 mL microcentrifuge tubes.
6. 50 mL tubes.
7. 15 mL tubes.
8. Disposable pipettes.
9. Vortex.
10. Pipettes.
11. Barrier tips (100 μL).
12. Absorbing pads.
13. Weight balance.

2.2 Preparation of Inoculum

1. Hemocytometer.
2. Light microscope.
3. Trypan blue.
4. Yeast–peptone–dextrose (YPD) media: To make 1 L of YPD media, dissolve 20 g of peptone, 10 g of yeast extract, and 20 g of dextrose in 1 L of water, and autoclave or filter sterilize.
5. YPD agar: To make 1 L of YPD, dissolve 20 g of peptone, 10 g of yeast extract, 20 g of dextrose, and 15 g of agar in 1 L of water, and autoclave or filter sterilize.
6. Chloramphenicol: Prepare a stock solution of 20 mg/mL, and use at a final concentration of 20 μg/mL in YPD media. To make the stock solution, weigh 1 g of chloramphenicol, and dissolve in 50 mL of ethanol, filter sterilize, and keep refrigerated until use (*see* **Note 1**).

7. Inoculation loops (sterilized wire or sterile plastic).
8. Shaking incubator.
9. Incubator.

2.3 Intranasal Infection

1. Micropipette 10–100 μL (alternative 20–200 μL).
2. Barrier tips (10–100 μL).
3. Access to rodent anesthesia device.
4. Isoflurane.
5. Oxygen.

2.4 Intratracheal Infection

1. Blunt-end forceps (*see* **Note 2**).
2. Access to rodent anesthesia device.
3. 45°-incline intubating platform (plexiglass blackboard 45° with wire strings) (*see* **Note 3**).
4. Warming pad.
5. Headlamp with magnifier.
6. Barrier tips (10–100 μL).
7. PE-10 tubes.
8. 0.5 mL insulin syringes (28G).

2.5 Intravenous Injection

1. Mouse restrainer, commercially available (*see* **Note 4**).
2. A heat lamp or heat source.
3. 0.5 mL insulin syringes (28G).
4. Sterile alcohol prep pads.
5. Sharps container.

2.6 Inoculum Determination

1. YPD agar plus chloramphenicol plates.
2. 30 °C incubator.
3. 100 μL tips.
4. Pipettes.
5. Tubes containing 900 μL of PBS.
6. Spreaders (sterile glass or plastic).

2.7 Intraperitoneal Injection

1. Syringes (1 mL).
2. Needles (23–27G).
3. Drugs to be administered.
4. Sharp disposal container.

3 Methods

3.1 General Recommendations Before Start

1. Ensure the animal vivarium is properly equipped and lab personnel are appropriately trained to operate at biosafety level 2 and animal safety level 2 levels. All pertinent permits and animal and biosafety protocols should be approved prior to beginning work, noting that each institution and country has different regulations.
2. Mice of various strains or with various gene deficiencies are available from multiple commercial sources. C57BL/6 and BALB/c mice are common mouse strains used to study cryptococcosis. However, a literature review is recommended since different mouse strains used for this work have various levels of susceptibility to the different *Cryptococcus* strains [18, 20, 23, 25, 51]. Using both male and female mice is also recommended, unless the research question dictates otherwise, to address any biological variables that my impact research results.
3. Once the mice arrive at the vivarium, allow 5–7 days for the mice to acclimate to the new environment to help mitigate stress-induced responses. One or two days before inoculation, arrange the mice according to the experimental groups specified in your research plan (*see* **Note 5**).
4. Mice are typically housed in groups of five maximum per cage, in 12 h light/dark cycles, and receive food and water ad libitum, unless the experimental plan necessitates otherwise, and nesting material (shredded paper or similar material) is provided for enrichment [52–54]. One should avoid housing mice individually, they should be separated only in cases of misbehavior.

3.2 Preparation of the Inoculum

1. From frozen stock inside the biosafety hood, take a loop of the *Cryptococcus* strain to be used for infection, grow it using the four-quadrant streak method on YPD plus chloramphenicol (*see* **Note 6**), and incubate at 30 °C for 48 h.
2. Isolated colonies should be visible at the end of the four-quadrant streak following 48 h of growth.
3. Within the biosafety cabinet, add 10 mL of YPD media to a 50 mL conical tube (optional: a sterilized 100 mL flask may be used instead). A sterile inoculation loop should be used to transfer a colony from the culture plate grown in Subheading 3.2, **step 1** to the conical tube. The cap of the conical tube should be loosely attached to allow air into the tube (a piece of tape can be applied over the cap to help secure the lid to the tube). Grow the culture at 30 °C in a shaker incubator at 180–200 rpm for 14–16 h (*see* **Note 7**).

4. Following 16 h, wash the culture 3× using 10 mL of sterile PBS.
5. To wash the culture, centrifuge the 50 mL tube (*see* **Note 8**), for 5–7 min at 2465× *g* to pellet the yeast.
6. Discard the supernatant into a waste container (*see* **Note 9**), add 10 mL of 1× PBS, and resuspend the cells by vortexing or pipetting up and down.
7. Centrifuge the conical tube again for 5–7 min at 2465× *g* to pellet the yeast. Then, repeat this process twice, for a total of three washes.
8. After the last wash, add 5–10 mL of 1× PBS (*see* **Note 10**), and prepare the counting solution.
9. For the counting solution, in a 1.7 mL microcentrifuge tube, add 940 μL of 1× PBS and 50 μL of trypan blue. Resuspend cells by vortexing the tube for 10–20 s, and add 10 μL of the cell suspension to the counting solution (*see* **Note 11**).
10. An automated counter or manual counting may be used to enumerate the yeast. For manual counting using a glass hemocytometer, very gently add 10 μL of the counting solution to both chambers underneath the glass coverslip, allowing the cell suspension to be drawn out by capillary action. Count the number of live, unstained cells (live cells do not take up the trypan blue) in the top and bottom chambers using a light microscope, and calculate the cell number (*see* **Note 12**).
11. Once finished counting the cells, determine the concentration by calculating the average of cells counted × the dilution factor (100 if you prepare it as step 7) × 10^4 = total number of cells per mL in your solution.
12. Once the total cell number is determined, calculate the amount needed for the inoculum. Depending on the experiment and route of infection, different volumes may be needed. For intranasal and intratracheal infections, 30 μL/per mouse is typically used (*see* **Note 13**); for intravenous infections, 100 μL/per mouse is generally used. Calculate the total volume needed depending on the number of mice being infected (*see* **Note 14**).

3.3 Anesthesia Procedure

1. The anesthesia procedure described here may be used for both IN and IT methods described below. An inhalation anesthetic (i.e., isoflurane plus oxygen anesthetic gas mixture dispensed using a rodent anesthesia system equipped with a calibrated vaporizer) or an injectable anesthetic (such as a ketamine/xylazine cocktail) [55] is widely used (*see* **Note 15**).
2. We recommend that mice undergoing IN or IT inoculations be anesthetized with 3–5% isoflurane plus 0.5 L/min O_2 using a

rodent anesthesia system, since the time the mice are under anesthesia and the recovery time are minimal, but both methods are described here.

3. Mice may also be anesthetized by intraperitoneal (IP) injections (described below in Subheading 3.10).
4. All preparations should be mixed and stored in sterile vials, and care should be taken not to exceed the expiration dates of all reagents. All controlled substances should be appropriately secured, dispensed, and usage logged. Mice anesthetized using a ketamine/xylazine cocktail are typically anesthetized for 20–40 min, depending on the ketamine/xylazine formulation used, and require longer monitoring to ensure a safe recovery.
5. Anesthesia formulations using ketamine vary, and consultation with appropriately trained and experienced veterinary staff is recommended prior to initiating any work involving anesthesia. Formulations include ketamine hydrochloride (20.0 mg/kg) and xylazine hydrochloride (2 mg/kg) mixed in sterile PBS [56–58] or a mixture of ketamine (8 mg/mL)/Dexdomitor (0.05 mg/mL) in 200 μL of sterile PBS followed by 200 μL of atipamezole (0.25 mg/mL) to reverse the anesthesia effects [56].
6. Mice may also be anesthetized by inhalation using a vaporizer system in which liquid isoflurane is converted to gas and mixed with oxygen to a preferred percentage (we use 3–5% isoflurane plus 0.5 L/min O_2) that is delivered to mice within an induction chamber or a nose cone. Waste anesthetic gas scavenging may be managed actively via the use of a vacuum exhaust or passively through the use of activated charcoal canisters (*see* **Note 16**). Consult with vivarium staff on proper use and maintenance of isoflurane anesthesia gas and scavenging systems. Weigh the activated charcoal canister prior to beginning the study.
7. Add mice to the induction chamber, turn on the anesthesia machine by turning on the oxygen (flow rate 300–600 mL/min), and adjust the isoflurane to 3–5% for induction. It takes approximately 2–3 min for the mice to fall asleep. Once all the mice are under anesthesia, the isoflurane may be dropped to 1–2% for maintenance.

3.4 Intranasal Infection

1. While the mice are under anesthesia, vortex the tube containing the inoculum to allow for a homogenous distribution of the sample. Using a pipette fitted with a 100 μL barrier pipette tip, withdraw 30 μL of the inoculum into the pipette tip.
2. Take one mouse from the induction chamber, and hold the mouse in a secure position in one hand. Bring the pipette close

to the left or right nostril, and, without touching the pipette tip to the nostril, slowly apply the liquid drop by drop into the nostril allowing the mouse to inhale the inoculum slowly (*see* **Note 17**).

3. Hold the mouse upright for 10–15 s until all the inoculum has been completely inhaled. Place the mouse back into the cage while it recovers, and continue to the next mouse.
4. Once finished, turn off the isoflurane and oxygen while keeping the chamber closed until all the isoflurane has been scavenged through the activated charcoal canister. Clean the chamber, and weigh and record the weight of the activated charcoal canister after use (*see* **Note 18**).
5. Check that all the mice are awake and breathing well.

3.5 IT Infection

1. Once the mouse is under anesthesia, suspend the mouse in the supine position by the incisors on the platform (Fig. 1).
2. Ensure that the inoculum is ready for IT inoculation. A volume of 30–50 μL per mouse may be utilized for IT inoculation (do not exceed this volume).
3. If a headlamp with a magnifier is available, turn on the headlamp to aid in the visibility of the trachea (*see* **Note 19**).

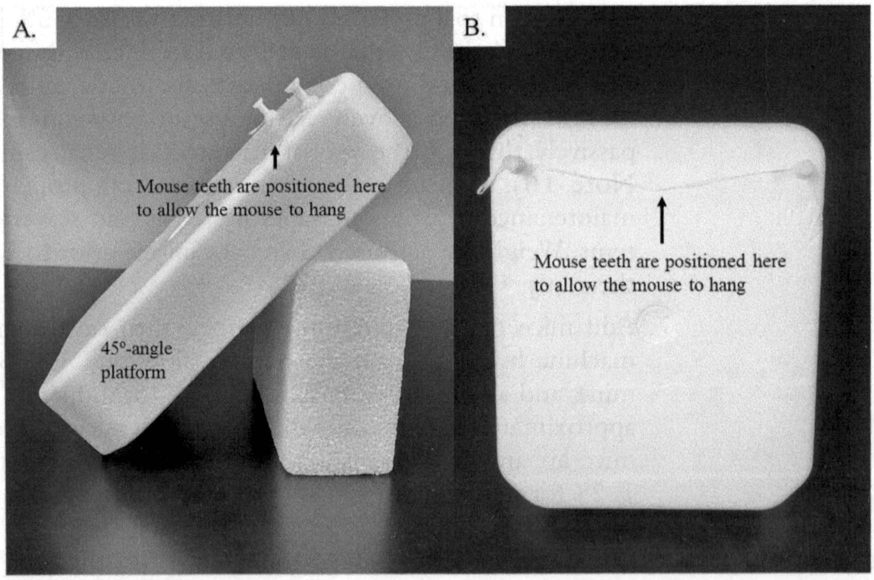

Fig. 1 Homemade platform for intratracheal inoculation. (**a**) Side view of the intratracheal platform and (**b**) front view of the intratracheal platform. This platform is made with Styrofoam; other materials (i.e., wood, plexiglass) may be used. The pins of which the string is hung from should be securely attached to the platform using superglue or similar material. Nylon or dental floss may be used for the string. The string should not be attached too tightly but should allow for enough space to suspend the mice by their incisors

4. Once the mouse is suspended, open the mouth, grab the tongue with care using blunt-end forceps, and move towards one side, as this will allow more visibility of the larynx.
5. Different options are available to deliver the inoculum. First, using a pipette fitted with a 100 μL barrier pipette tip, carefully insert the end of the pipette tip on top of the larynx to gently drop the inoculum. For the second delivery method, insert a PE-10 tube 0.5–1 cm long, attached to an insulin syringe, into the trachea, and deliver the solution. Leave the tube inside for 5 s to avoid escape of the inoculum.
6. After the inoculum has been delivered, maintain the mouse on the platform for 20–30 s to ensure complete delivery of the inoculum.
7. Carefully remove the mice from the platform, and allow them to recover on a heated pad before returning to the cage.

3.6 Intravenous Injection

1. Place the mice in a cage without the cover, and turn on the heat lamp (Fig. 2a) to increase the temperature of the mice (*see* **Note 20**). Additionally, a method to warm and restrain the mice is described by Yano et al. [59].
2. Vortex the tube containing the inoculum to allow for a homogenous distribution of the suspension.
3. Carefully remove the cap of the syringe and draw the inoculum into the syringe. Carefully remove any bubbles within the syringe (*see* **Note 21**).
4. Place the mouse within the restraint device (Fig. 2b–g). Hold the tail with your nondominant hand, and locate the lateral vein starting closer to the tip of the tail (*see* **Note 22**).
5. Once the vein is located, disinfect the injection site with an alcohol prep pad. Insert the needle parallel to the tail (*see* **Note 23**). When inserting the needle or while the plunger is pressed to inject the inoculum (100 μL volume of inoculum works best), no resistance should be encountered, and the vein should appear clear—two signs of a good injection.
6. Encountering resistance while injecting the inoculum and observing swelling of the tail are likely signs that the needle is not properly situated within the tail vein. Remove the needle and proceed further up the tail towards the body and try again.
7. After inoculation, remove the needle and press the injection site with gauze (*see* **Note 24**).
8. Discard the syringe and needle after use in a sharps container. Thoroughly clean the restraining device after use to avoid cross-infection or pheromonal-induced stress [60].

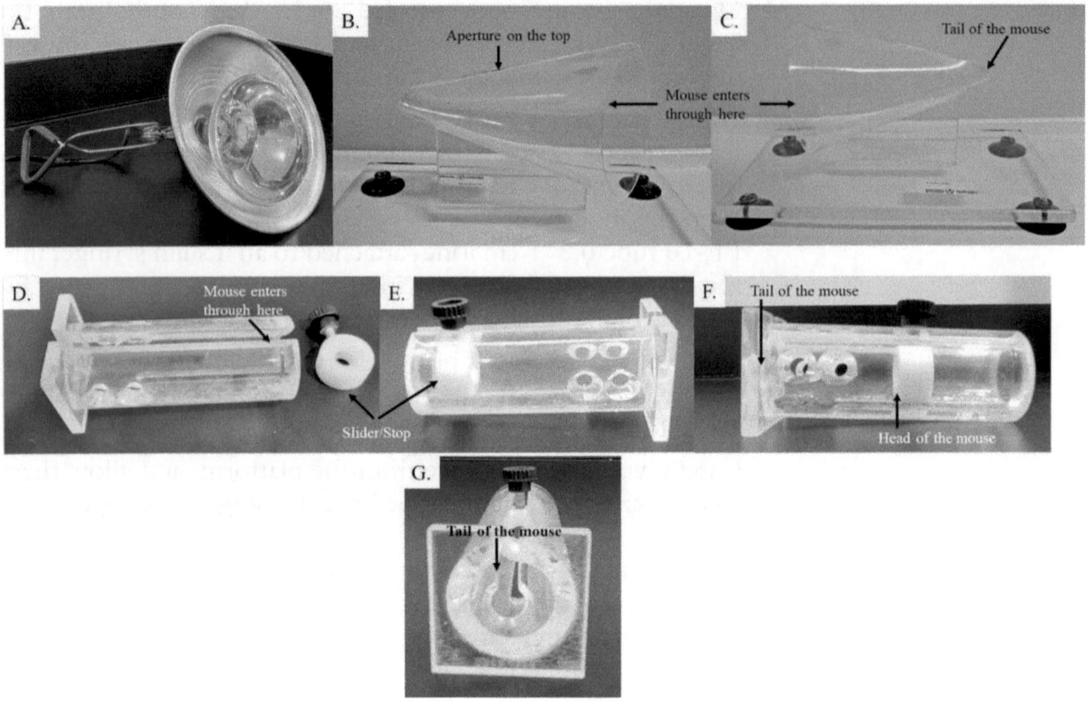

Fig. 2 Examples of heat lamps and types of restrainers that may be used for intravenous injections. (**a**) Clamp heat lamp; standing options are also available. (**b**, **c**) Injection cone views showing where mouse enters and tail protrudes out. (**d**–**g**) Mouse restrainer. Tubular acrylic mouse holder with slider. (**d**) Slider is disengaged from the holder to allow the mouse to enter. (**e**, **f**) Side view of the mouse restrainer. (**e**) The slider is on the furthest part; the slider is moved towards the head of the mouse to hold the mouse in place. (**f**) Slider on the stop position with arrows indicating the position of the head and the position of the tail. (**g**) Front view of the mouse restrainer. In this view, the injection is performed with the tail facing the individual performing the injection

3.7 Inoculum Determination

1. Immediately after infection, take the inoculum back to the lab to confirm the concentration of viable yeasts (*see* **Note 25**).
2. Take 100 μL of the inoculum, and make serial dilutions (1:10, 1:100, and 1:1000) in sterile 1.5 mL tubes containing 900 μL of sterile PBS. Be sure to change tips after each dilution and vortex between dilutions.
3. Plate 100 μL of the dilutions on YPD plus chloramphenicol agar plates, one plate for each dilution, and use a sterile or sterilized spreader to spread the liquid evenly over the surface of the agar plate.
4. Incubate plates at 30 °C for 48 h, and count the colonies to determine colony-forming units (CFUs) per mL of the inoculum.

3.8 Mouse Monitoring

1. Mice should be monitored twice a day (morning and evening during the light cycle) for any signs of disease.
2. When monitoring the mice, check how alert the mice are, their mobility, posture, and breathing.
3. You may monitor their weight, since 20% or more weight loss is indicative of illness; mice should be humanely sacrificed at this point to avoid suffering.

3.9 Fungal Burden

1. Once a mouse has reached the moribund stage or the experiment has reached a pre-scheduled time or end point, mice should be sacrificed according to a method that is consistent with the recommendations of the Panel of Euthanasia of the American Veterinary Medical Association [60] or a similar governing body.
2. Immediately after the mouse is euthanized, clean the appropriate area with 70% ethanol, and proceed with dissection to retrieve tissues of interest (i.e., spleen, lung, and brain tissues).
3. Place each organ in a 14 mL tube with 1 mL of sterile PBS and keep it on ice.
4. Clean the tissue homogenizer with 70% ethanol and then 1× PBS before use and between samples. Dissociate the tissue using the tissue homogenizer.
5. Take 100 μL of the sample, and make serial dilutions (*see* **Note 26**) in 1.5 mL tubes containing 900 μL of sterile PBS. Change tips after each dilution, and vortex tubes between dilutions.
6. Plate 100 μL of the dilutions on YPD plus chloramphenicol agar plates, one plate for each dilution, and evenly spread the liquid over the surface of the agar plate using a sterile or sterilized spreader. Be sure to plate each dilution. Also, dilutions may be kept in the refrigerator for further dilution, if necessary.
7. Incubate the plates at 30 °C for 48 h. Subsequently, count the colonies to calculate the CFUs per mL of the inoculum.

3.10 Intraperitoneal (IP) Injection

IP injection is frequently used to administer drugs or recombinant proteins, such as cytokines, antibodies, or other therapies. In some instances, infectious organisms can be delivered via IP injection [61–63]. As with any procedure involving needles, careful handling and a steady hand are required.

1. Prepare the material to be injected before heading to the vivarium. Most drugs are administered based on body weight, with most mice weighing between 16 g and 20 g. Male and female mice of the same age typically differ in weight. Therefore, we recommend weighing each mouse to customize the

amount of treatment that will be given to each mouse. Be sure to prepare extra doses and take extra needles (*see* **Note 27**).

2. Hold the mouse using the thumb and index finger of your nondominant hand to grab the loose skin found below the head of the mouse. Hold securely to restrict movement of the head. With your free hand, secure the tail, and transfer the tail to be held with the little finger or between the little finger and the ring finger, whichever gives you a better grip on the mouse. One should be able to securely hold the mouse with four fingers (*see* **Note 28**).
3. Locate the peritoneal region of the mouse, located in the lower quadrant of the abdomen (*see* **Note 29**); IP injection is done on the right or left side of this region. To avoid compromising any organs, tilt the mouse backwards with the head downwards.
4. Insert the needle into the abdomen at a 45° angle or lower. The needle should be angled towards the peritoneum to minimize penetration and avoid any organs. Lift the needle, and deliver the reagent or treatment; by raising the needle, you help to avoid puncturing any organs.
5. Place the mouse back in the cage, and dispose of the needle appropriately in a sharps container.
6. The formation of a bubble within the skin at the injection site suggests that a subcutaneous, not IP, injection was performed.
7. If multiple or daily injections are being performed, it is acceptable to vary the site of injection between the right and left sides of the abdomen, assuming the approval of relevant animal care committees.
8. If compromise of the peritoneum occurs, check the mice for sepsis, bleeding, or any signs of illness, and contact veterinary staff.

4 Notes

1. If the chloramphenicol solution will be made at a concentration of 20 mg/mL, use 1 mL per liter to obtain a 20 mg/mL concentration or 500 μL in 500 mL.
2. Preferably use blunt forceps without ridges, or use blunt-end forceps with silicone on the tips to avoid any damage to the tongue.
3. The intubation platform may be at 60° or 45° angles. There are commercially available platforms made with plexiglass, or you may prepare a homemade platform using Styrofoam.
4. There are different types of restrainers, but for an experienced researcher and when a larger number of mice are being used, a

rodent injection cone will be more helpful and easier for handling the mice. A rodent restrainer is recommended for a researcher new to working with mice, since the animal will have less movement. When using a rodent restrainer, identify the appropriate size to avoid any unnecessary distress to the animals.

5. Great care should be taken during the planning stage of each experiment to calculate the number of mice needed for the experiment. There are several resources available online that aid in calculating the correct sample size. As a guide, eight to ten mice per group for survival studies and three to five mice per group with an appropriate number of experimental replicates are recommended for pre-scheduled time points.
6. Chloramphenicol prevents bacteria contamination from forming on plates, especially when tissue homogenates are being plated. When working with a mutant strain with a selection marker, also incorporate the agent required for selection into the agar to grow the strain.
7. Do not close the tubes tightly. Instead, loosely attach the cap to the tube to keep it covered but allow air interchange.
8. If the culture was grown in a 50–100 mL flask, transfer the culture to a 50 mL conical tube.
9. Washes are necessary to remove any extracellular proteins, extracellular vesicles, or any other elements found in the culture media that may interfere with the results after infection [64, 65].
10. The volume required to resuspend yeast pellets will vary depending on the size of the pellet. Adjust the amount of PBS needed for resuspension accordingly.
11. Live cells will exclude the trypan blue; cells stained in blue should be excluded from your count of viable cells. When counting budding cells, count the daughter cells (buds) as separate cells if they are at least 75% the size of the mother cells. If you want to determine the population of viable vs. dead cells, count both live and dead cells.
12. The dilution factor is 100 if the counting solution is prepared as described here. If the suspension of cryptococcal cells appears to be very concentrated, the counting solution may be modified (945 μL of PBS, 50 μL of trypan blue, and 5 μL of cells will result in a dilution factor of 200). Keep in mind that changing the volumes in the counting solution may impact the dilution factor needed to calculate the approximate number of yeast in suspension.
13. The concentration required for the infection depends on different factors, the strain of the mice, strain of *Cryptococcus*,

route of infection, the purpose of the experiment (immunization study or morbidity analysis), and the question being answered. Most IN/IT studies aim for 10^4–10^6 cells/mouse in 30 μL of inoculum. A typical survival study using *C. neoformans* strain H99 in BALB/c or C57BL/6 mice results in mice succumbing to disease around day 22–28 post-inoculation.

14. Calculate for an extra amount of inoculum. For example, an infection study with 20 mice receiving a volume of 30 μL per mouse may require 1.8 mL as your final volume, but preparing 3–5 mL is recommended.
15. Regardless of the anesthesia used, mice should be monitored during the induction of anesthesia through recovery for signs resembling depression of cardiac and respiratory functions [60].
16. Use a fume hood or a system with a gas scavenging system to minimize exposure to the gas when using inhalant anesthesia.
17. Avoid the formation of bubbles in the nostril while the mouse inhales the inoculum because it may result in loss of inoculum and inaccurate infection.
18. Replace the charcoal canister once the canister increases by 50 g in weight from the first usage.
19. While the mouse is suspended, the nose cone may be used to sustain the mice under anesthesia.
20. To warm the mouse to stimulate dilation of the tail vein, place them under a heat lamp while in the cage, or place the tail in warm water or in a warm cloth (30–35 °C). If using the heat lamp, be sure to monitor the temperature next to the cage to avoid overheating the mice.
21. Since this method requires the use of a syringe/needle with an infectious agent, care is required when handling the mice and the syringe. Anesthesia is not required for IV or tail injection. However, if the researcher doing the injection is not proficient, anesthesia may be used to minimize the movement of the mouse.
22. Make sure the restrainer device is held in place. Most commercially available injection cones have suction cups that will hold it in place on the workbench. For restrainer devices that do not, it might be easier to tape the devise to the workbench.
23. Starting closer to the tip of the tail will allow you to have more chances of a successful injection if you fail on the first try. Do not soak the tail in alcohol because it may cause vasoconstriction. If alcohol prep pads are not available, cotton swabs with ethanol may be used to disinfect.

24. We recommend practicing tail vein injection using sterile PBS or a mix of sterile PBS with blue Evans. Perform the tail vein injection as described above; if the injection is successful, the paws and tail of the mouse will turn blue if using blue Evans in the solution.
25. While viable cell counts within the cell suspension were calculated using a light microscope and trypan blue, it's prudent to plate the inoculum to corroborate the viable inoculum and check for signs of any possible contamination.
26. The serial dilutions required to quantify the fungal burden accurately will vary depending on the initial inoculum, the route of infection, the time point for the fungal burden, and the organ site.
27. It is recommended to use a new needle/syringe for each mouse to avoid cross-contamination.
28. Do not exceed the recommended volume for IP injections, 20 mL/kg body weight (400 μL for a 20 g mouse), maximum: 40 mL/kg (800 μL for a 20 g mouse). Most IP injections are conducted using 100–200 μL per mouse.
29. You may guide yourself using the legs as a marker, making the injection parallel to the hips of the mouse.

References

1. Casadevall A, Perfect JR (1998) Cryptococcus neoformans. ASM Press, Washington, DC
2. May RC, Stone NRH, Wiesner DL et al (2016) *Cryptococcus*: from environmental saprophyte to global pathogen. Nat Rev Microbiol 14(2): 106–117. https://doi.org/10.1038/nrmicro.2015.6
3. Mirza SA, Phelan M, Rimland D et al (2003) The changing epidemiology of cryptococcosis: an update from population-based active surveillance in 2 large metropolitan areas, 1992–2000. Clin Infect Dis 36(6):789–794. https://doi.org/10.1086/368091
4. Rajasingham R, Govender NP, Jordan A et al (2022) The global burden of HIV-associated cryptococcal infection in adults in 2020: a modelling analysis. Lancet Infect Dis 22(12): 1748–1755. https://doi.org/10.1016/S1473-3099(22)00499-6
5. Rajasingham R, Smith RM, Park BJ et al (2017) Global burden of disease of HIV-associated cryptococcal meningitis: an updated analysis. Lancet Infect Dis 17(8): 873–881. https://doi.org/10.1016/S1473-3099(17)30243-8
6. Maziarz EK, Perfect JR (2016) Cryptococcosis. Infect Dis Clin N Am 30(1):179–206. https://doi.org/10.1016/j.idc.2015.10.006
7. Byrnes EJ 3rd, Bartlett KH, Perfect JR et al (2011) *Cryptococcus gattii*: an emerging fungal pathogen infecting humans and animals. Microbes Infect 13(11):895–907. https://doi.org/10.1016/j.micinf.2011.05.009
8. Saidykhan L, Onyishi CU, May RC (2022) The *Cryptococcus gattii* species complex: unique pathogenic yeasts with understudied virulence mechanisms. PLoS Negl Trop Dis 16(12): e0010916. https://doi.org/10.1371/journal.pntd.0010916
9. Byrnes EJ 3rd, Li W, Lewit Y et al (2010) Emergence and pathogenicity of highly virulent *Cryptococcus gattii* genotypes in the Northwest United States. PLoS Pathog 6(4): e1000850. https://doi.org/10.1371/journal.ppat.1000850
10. Kwon-Chung KJ, Fraser JA, Doering TL et al (2014) *Cryptococcus neoformans* and *Cryptococcus gattii*, the etiologic agents of cryptococcosis. Cold Spring Harb Perspect Med 4(7): a019760. https://doi.org/10.1101/cshperspect.a019760

11. Negroni R (2012) Cryptococcosis. Clin Dermatol 30(6):599–609. https://doi.org/10.1016/j.clindermatol.2012.01.005
12. Loyse A, Thangaraj H, Easterbrook P et al (2013) Cryptococcal meningitis: improving access to essential antifungal medicines in resource-poor countries. Lancet Infect Dis 13(7):629–637. https://doi.org/10.1016/S1473-3099(13)70078-1
13. Mylonakis E, Ausubel FM, Perfect JR et al (2002) Killing of *Caenorhabditis elegans* by *Cryptococcus neoformans* as a model of yeast pathogenesis. Proc Natl Acad Sci U S A 99(24):15675–15680. https://doi.org/10.1073/pnas.232568599
14. Merryman M, Crigler J, Seipelt-Thiemann RL et al (2020) A mutation in *C. Neoformans* mitochondrial NADH dehydrogenase results in increased virulence in mice. Virulence 11(1):1366–1378. https://doi.org/10.1080/21505594.2020.1831332
15. Tang RJ, Breger J, Idnurm A et al (2005) *Cryptococcus neoformans* gene involved in mammalian pathogenesis identified by a *Caenorhabditis elegans* progeny-based approach. Infect Immun 73(12):8219–8225. https://doi.org/10.1128/IAI.73.12.8219-8225.2005
16. Lee A, Toffaletti DL, Tenor J et al (2010) Survival defects of *Cryptococcus neoformans* mutants exposed to human cerebrospinal fluid result in attenuated virulence in an experimental model of meningitis. Infect Immun 78(10):4213–4225. https://doi.org/10.1128/IAI.00551-10
17. Kitisin T, Muangkaew W, Sukphopetch P (2022) *Caenorhabditis elegans* DAF-16 regulates lifespan and immune responses to *Cryptococcus neoformans* and *Cryptococcus gattii* infections. BMC Microbiol 22(1):162. https://doi.org/10.1186/s12866-022-02579-x
18. Irokanulo EA, Akueshi CO (1995) Virulence of *Cryptococcus neoformans* serotypes A, B, C and D for four mouse strains. J Med Microbiol 43(4):289–293. https://doi.org/10.1099/00222615-43-4-289
19. Kawakami K, Koguchi Y, Qureshi MH et al (2001) Anti-CD11 b monoclonal antibody suppresses brain dissemination of *Cryptococcus neoformans* in mice. Microbiol Immunol 46(3):181–186. https://doi.org/10.1111/j.1348-0421.2002.tb02684.x
20. Rivera J, Casadevall A (2005) Mouse genetic background is a major determinant of isotype-related differences for antibody-mediated protective efficacy against *Cryptococcus neoformans*. J Immunol 174(12):8017–8026. https://doi.org/10.4049/jimmunol.174.12.8017
21. García-Barbazán I, Trevijano-Contador N, Rueda C et al (2016) The formation of titan cells in *Cryptococcus neoformans* depends on the mouse strain and correlates with induction of Th2-type responses. Cell Microbiol 18(1):111–124. https://doi.org/10.1111/cmi.12488
22. Lim TS, Murphy JW (1980) Transfer of immunity to cryptococcosis by T-enriched splenic lymphocytes from *Cryptococcus neoformans*-sensitized mice. Infect Immun 30(1):5–11. https://doi.org/10.1128/iai.30.1.5-11.1980
23. Kwon-Chung KJ, Edman JC, Wickes BL (1992) Genetic association of mating types and virulence in *Cryptococcus neoformans*. Infect Immun 60(2):602–605. https://doi.org/10.1128/iai.60.2.602-605.1992
24. Buchanan KL, Fidel PL Jr, Murphy JW (1991) Effects of *Cryptococcus neoformans*-specific suppressor T cells on the amplified anticryptococcal delayed-type hypersensitivity response. Infect Immun 59(1):29–35. https://doi.org/10.1128/iai.59.1.29-35.1991
25. Rhodes JC, Wicker LS, Urba WJ (1980) Genetic control of susceptibility to *Cryptococcus neoformans* in mice. Infect Immun 29(2):494–499. https://doi.org/10.1128/iai.29.2.494-499.1980
26. Morgan MA, Blackstock RA, Bulmer GS et al (1983) Modification of macrophage phagocytosis in murine cryptococcosis. Infect Immun 40(2):493–500. https://doi.org/10.1128/iai.40.2.493-500.1983
27. Leopold Wager CM, Hole CR, Wozniak KL et al (2014) STAT1 signaling is essential for protection against *Cryptococcus neoformans* infection in mice. J Immunol 193(8):4060–4071. https://doi.org/10.4049/jimmunol
28. Hamed MF, de Araújo GRS, Munzen ME et al (2023) Phospholipase B is critical for *Cryptococcus neoformans* survival in the central nervous system. MBio 14(2):e0264022. https://doi.org/10.1128/mbio.02640-22
29. Feldmesser M, Casadevall A, Kress Y et al (1997) Eosinophil-*Cryptococcus neoformans* interactions *in vivo* and *in vitro*. Infect Immun 65(5):1899–1907. https://doi.org/10.1128/iai.65.5.1899-1907.1997
30. Krockenberger MB, Malik R, Ngamskulrungroj P et al (2010) Pathogenesis of pulmonary *Cryptococcus gattii* infection: a rat model. Mycopathologia 170(5):315–330. https://doi.org/10.1007/s11046-010-9328-z

31. Goulart L, Rosa e Silva LK, Chiapello L et al (2010) *Cryptococcus neoformans* and *Cryptococcus gattii* genes preferentially expressed during rat macrophage infection. Med Mycol 48(7): 932–941. https://doi.org/10.3109/13693781003677494
32. Shibuya K, Hirata A, Omuta J et al (2005) Granuloma and cryptococcosis. J Infect Chemother 11(3):115–122. https://doi.org/10.1007/s10156-005-0387-x
33. Li H, Li Y, Sun T et al (2019) Unveil the transcriptional landscape at the *Cryptococcus*-host axis in mice and nonhuman primates. PLoS Negl Trop Dis 13(7):e0007566. https://doi.org/10.1371/journal.pntd.0007566
34. Pal M, Dube GD, Mehrotra BS (1984) Pulmonary cryptococcosis in a rhesus monkey (*Macaca mulatta*). Mykosen 27(6):309–312. https://doi.org/10.1111/j.1439-0507.1984.tb02035.x
35. Linares G, Daker RD (1972) Cryptococcal dermotropism in the rhesus monkey. Mycopathol Mycol Appl 46(1):17–32. https://doi.org/10.1007/BF02051893
36. Tenor JL, Oehlers SH, Yang JL et al (2015) Live imaging of host-parasite interactions in a zebrafish infection model reveals cryptococcal determinants of virulence and central nervous system invasion. mBio 6(5):e01425-15. https://doi.org/10.1128/mBio.01425-15
37. Bojarczuk A, Miller KA, Hotham R et al (2016) *Cryptococcus neoformans* intracellular proliferation and capsule size determines early macrophage control of infection. Sci Rep 6: 21489. https://doi.org/10.1038/srep21489
38. Davis JM, Huang M, Botts MR et al (2016) A zebrafish model of cryptococcal infection reveals roles for macrophages, endothelial cells, and neutrophils in the establishment and control of sustained fungemia. Infect Immun 84(10):3047–3062. https://doi.org/10.1128/IAI.00506-16
39. Evans RJ, Pline K, Loynes CA et al (2019) 15-keto-prostaglandin E2 activates host peroxisome proliferator-activated receptor gamma (PPAR-γ) to promote *Cryptococcus neoformans* growth during infection. PLoS Pathog 15(3): e1007597. https://doi.org/10.1371/journal.ppat.1007597
40. Nielson JA, Davis JM (2023) Roles for microglia in cryptococcal brain dissemination in the zebrafish larva. Microbiol Spectr 11(2): e0431522. https://doi.org/10.1128/spectrum.04315-22
41. Mylonakis E, Moreno R, El Khoury JB et al (2005) *Galleria mellonella* as a model system to study *Cryptococcus neoformans* pathogenesis. Infect Immun 73(7):3842–3850. https://doi.org/10.1128/IAI.73.7.3842-3850.2005
42. Stączek S, Zdybicka-Barabas A, Wiater A et al (2020) Activation of cellular immune response in insect model host *Galleria mellonella* by fungal α-1,3-glucan. Pathog Dis 78(9): ftaa062. https://doi.org/10.1093/femspd/ftaa062
43. Stempinski PR, Smith DFQ, Casadevall A (2022) *Cryptococcus neoformans* virulence assay using a *Galleria mellonella* larvae model system. Bio Protoc 12(15):e4480. https://doi.org/10.21769/BioProtoc.4480
44. Hohl TM (2014) Overview of vertebrate animal models of fungal infection. J Immunol Methods 410:100–112. https://doi.org/10.1016/j.jim.2014.03.022
45. Chen GH, Teitz-Tennenbaum S, Nealet LM et al (2016) Local GM-CSF-dependent differentiation and activation of pulmonary dendritic cells and macrophages protect against progressive cryptococcal lung infection in mice. J Immunol 196(4):1810–1821. https://doi.org/10.4049/jimmunol.1501512
46. Capilla J, Clemons KV, Stevens DA (2007) Animal models: an important tool in mycology. Med Mycol 45(8):657–684. https://doi.org/10.1080/13693780701644140
47. Hole CR, Leopold Wager CM, Castro-Lopez N et al (2019) Induction of memory-like dendritic cell responses *in vivo*. Nat Commun 10(1):2955. https://doi.org/10.1038/s41467-019-10486-5
48. Nelson BN, Daugherty CS, Sharp RR et al (2022) Protective interaction of human phagocytic APC subsets with *Cryptococcus neoformans* induces genes associated with metabolism and antigen presentation. Front Immunol 13:1054477. https://doi.org/10.3389/fimmu.2022.1054477
49. Bürgel PH, Marina CL, Saavedra PHV et al (2020) *Cryptococcus neoformans* secretes small molecules that inhibit IL-1β inflammasome-dependent secretion. Mediat Inflamm 2020: 3412763. https://doi.org/10.1155/2020/3412763
50. Botts MR, Hull CM (2010) Dueling in the lung: how *Cryptococcus* spores race the host for survival. Curr Opin Microbiol 13(4): 437–442. https://doi.org/10.1016/j.mib.2010.05.003
51. Zaragoza O, Alvarez M, Telzak A et al (2007) The relative susceptibility of mouse strains to pulmonary *Cryptococcus neoformans* infection is associated with pleiotropic differences in the immune response. Infect Immun 75(6):

2729–2739. https://doi.org/10.1128/IAI.00094-07

52. Windsor Z, Bate ST (2019) Assessing the safety and suitability of nesting material for singly housed mice with surgically fitted head plates. Heliyon 5(7):e02097. https://doi.org/10.1016/j.heliyon.2019.e02097
53. Jennings M, Batchelor GR, Brain PF et al (1998) Refining rodent husbandry: the mouse. Report of the rodent refinement working party. Lab Anim 32(3):233–259. https://doi.org/10.1258/002367798780559301
54. Staff of the Jackson Laboratory (1966) In: Green EL (ed) Biology of the laboratory mouse. Dover Publications Inc, New York
55. Ahmadi-Noorbakhsh S, Farajli Abbasi M, Ghasemi M et al (2022) Anesthesia and analgesia for common research models of adult mice. Lab Anim Res 38(1):40. https://doi.org/10.1186/s42826-022-00150-3
56. Upadhya R, Lam WC, Hole CR et al (2021) *Cryptococcus neoformans* Cda1 and Cda2 coordinate deacetylation of chitin during infection to control fungal virulence. Cell Surf 7: 100066. https://doi.org/10.1016/j.tcsw.2021.100066
57. Lee HH, Carmichael DJ, Ríbeiro V et al (2023) Glucuronoxylomannan intranasal challenge prior to *Cryptococcus neoformans* pulmonary infection enhances cerebral cryptococcosis in rodents. PLoS Pathog 19(4):e1010941. https://doi.org/10.1371/journal.ppat.1010941
58. de Campos GY, Oliveira-Brito PKM, Guimarães JG et al (2023) Adjuvant Pam3CSk4 does not improve the immunization against *Cryptococcus gattii* infection in C57BL/6 mice. PeerJ 11:e14778. https://doi.org/10.7717/peerj.14778
59. Yano J, Lilly EA, Noverr MC et al (2020) A contemporary warming/restraining device for efficient tail vein injections in a murine fungal sepsis model. J Vis Exp 2020(165). https://doi.org/10.3791/61961
60. National Research Council Committee for the Update of the Guide for the Care and Use of Laboratory Animals (2011) National Academies Press (US), Washington, DC
61. Zhang L, Yan K, Zhang Y et al (2007) High-throughput synergy screening identifies microbial metabolites as combination agents for the treatment of fungal infections. Proc Natl Acad Sci U S A 104(11):4606–4611. https://doi.org/10.1073/pnas.0609370104
62. Casadevall A, Scharff MD (1994) Serum therapy revisited: animal models of infection and development of passive antibody therapy. Antimicrob Agents Chemother 38(8):1695–1702. https://doi.org/10.1128/AAC.38.8.1695
63. Doyle TC, Nawotka KA, Kawahara CB et al (2006) Visualizing fungal infections in living mice using bioluminescent pathogenic *Candida albicans* strains transformed with the firefly luciferase gene. Microb Pathog 40(2): 82–90. https://doi.org/10.1016/j.micpath.2005.11.003
64. Chen LC, Pirofski LA, Casadevall A (1997) Extracellular proteins of *Cryptococcus neoformans* and host antibody response. Infect Immun 65(7):2599–2605. https://doi.org/10.1128/iai.65.7.2599-2605.1997
65. Bielska E, Sisquella MA, Aldeieg M et al (2018) Pathogen-derived extracellular vesicles mediate virulence in the fatal human pathogen *Cryptococcus gattii*. Nat Commun 9(1):1556. https://doi.org/10.1038/s41467-018-03991-6

Chapter 4

In Vivo Modeling of *Cryptococcus neoformans* Infection and Collection of Murine Samples

Ben Muselius and Jennifer Geddes-McAlister

Abstract

In vivo models provide advantages to study the progression of disease and to identify potential biomarkers to detect and monitor infections. For the human fungal pathogen *Cryptococcus neoformans*, murine intranasal models aim to recapitulate natural infection from inhalation of desiccated fungal cells from the environment and permit monitoring of disease over time. In this chapter, we describe the establishment of a murine model for cryptococcosis and the subsequent collection of organs, tissues, and fluids for sampling. These samples may support novel diagnostic strategies and opportunities to monitor dissemination of the fungal cells throughout the host and propose new treatment options to combat disease.

Key words *Cryptococcus neoformans*, Murine model, Intranasal model, Cryptococcosis, Sample collection

1 Introduction

In vivo modeling provides an opportunity to examine the course of infection across multiple axes and gain information that is not readily available from other experimental designs, such as in vitro assays [1]. For instance, the coupling of in vivo models with tissue collection and analysis provides an ideal strategy to study the progression of disease as a pathogen infects the host, on both the temporal and spatial axes [2]. Defining the relationship between a host and pathogen may identify new druggable targets from both perspectives, reveal novel virulence factors used by the pathogen, and detect biomarkers to track and monitor disease over time and throughout the host system [3].

For the human fungal pathogen *Cryptococcus neoformans*, which is found ubiquitously within the environment, infection occurs upon inhalation of desiccated fungal cells followed by initial colonization of the alveolar spaces of the lung [4]. Within an immunocompetent individual, the innate immune system will

Erin E. McClelland (ed.), *Cryptococcus neoformans: Methods and Protocols*, Methods in Molecular Biology, vol. 2775,
https://doi.org/10.1007/978-1-0716-3722-7_4,

respond effectively and efficiently to remove fungal cells from the host and prevent further infection [5]. However, within immunocompromised individuals, fungal cells will withstand destruction by macrophages to disseminate throughout the body, cross the blood–brain barrier, and cause swelling of the brain and its lining. Such circumstances lead to meningitis and meningoencephalitis, fatal conditions if not treated in a timely manner with the proper antifungal drugs [5]. Importantly, rising rates of antifungal resistance make treatment of the disease (cryptococcosis) difficult, leading to elevated mortality [6]. Gaining a better understanding of the strategies used by *C. neoformans* to promote infection and the response of the host may lead to the discovery of novel antifungal agents to combat disease. Moreover, the identification of host biomarkers indicating infection progression supports the development of personalized and effective treatment options [3].

In this chapter, we present an in vivo model of *C. neoformans* infection via intranasal inoculation within a murine host, followed by the collection of infected organs, tissues, and fluids (Fig. 1). The

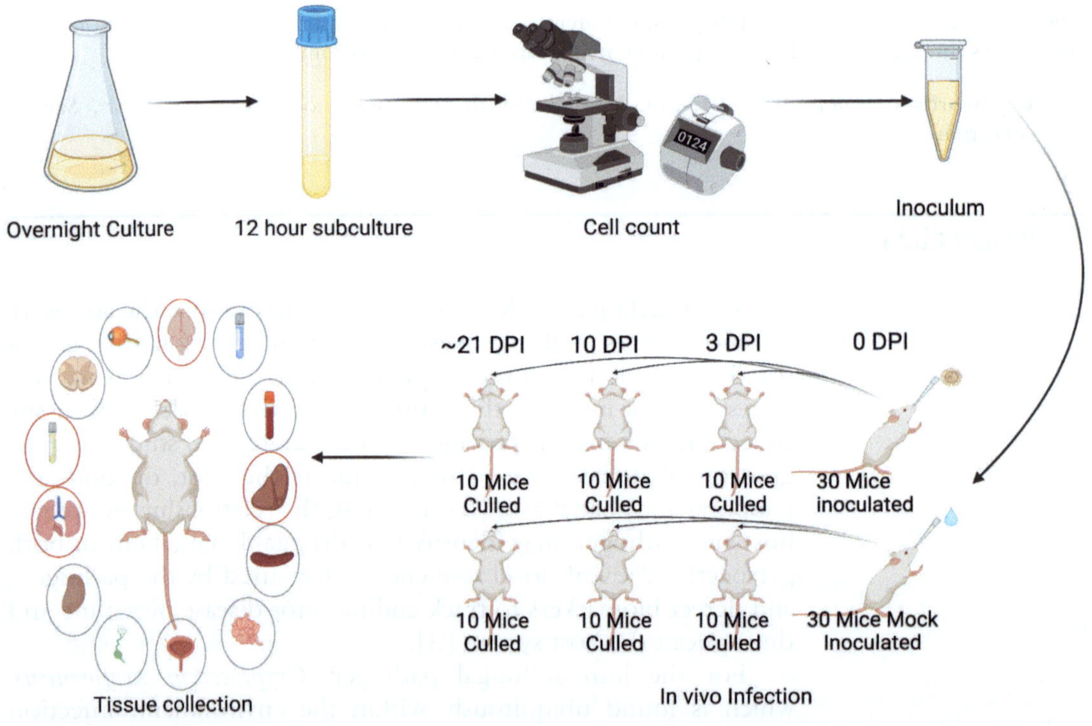

Fig. 1 Overview of murine infection model of cryptococcosis and sample collection. Briefly, inoculum is prepared from a subculture, and mice are infected via intranasal inoculation alongside control uninfected partners receiving only phosphate-buffered saline (PBS). Mice are monitored for the duration of the trial with groups of ten infected and ten uninfected mice being culled at designated time points. Thirteen clinically significant tissues (i.e., brain, lung, spleen, spinal cord, cerebral spinal fluid, liver, eye, kidney, bladder, lung lavage, blood, small intestine, and mesenteric lymph nodes) are collected immediately postmortem to be analyzed further. *DPI* days post-inoculation

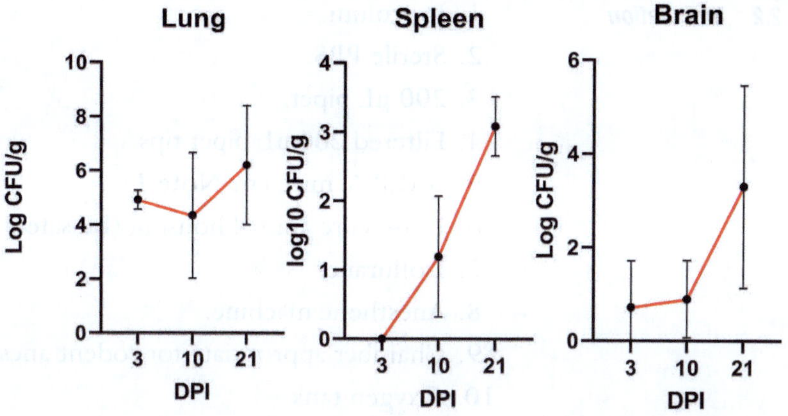

Fig. 2 Temporal and spatial fungal burden over time for the lung, spleen, and brain tissues. To quantify fungal load, a portion of the collected tissue was homogenized in phosphate-buffered saline (PBS) and plated on yeast–peptone–dextrose (YPD) for 48 h at 30 °C, followed by enumeration of colony-forming units (CFU). Counted CFU were standardized against the weight of tissue plated. *DPI* days post-inoculation

samples are collected across a temporal axis, representing early, mid, and late stages of infection to provide a guide for tracking disease progression throughout the host. Cryptococcosis impacts the host on a global scale, and therefore, we describe the collection of 13 sample types relevant for *C. neoformans* infection (i.e., brain, lung, spleen, spinal cord, cerebral spinal fluid, liver, eye, kidney, bladder, lung lavage, blood, small intestine, and mesenteric lymph nodes); however, experiment-specific selections may be made. Together, exploration across temporal and spatial axes of infection provides novel information toward the clinical significance of cryptococcosis (Fig. 2).

2 Materials

2.1 Preparation of the Inoculum

1. *Cryptococcus neoformans* H99 wild-type strain (ATCC 208821).
2. Yeast–peptone–dextrose (YPD) broth.
3. 10 mL test tubes.
4. 15 mL conical tubes.
5. Sterile phosphate-buffered saline (PBS).
6. 30 °C shaking incubator.
7. 1.5 mL microcentrifuge tubes.
8. Hemocytometer.
9. Microscope.

2.2 Inoculation

1. Inoculum.
2. Sterile PBS.
3. 200 μL pipet.
4. Filtered 200 μL pipet tips.
5. BALB/c mice (*see* **Note 1**).
6. Bio-secure animal housing (biosafety level 2).
7. Isoflurane.
8. Anesthetic machine.
9. Chamber appropriate for rodent anesthesia.
10. Oxygen tank.
11. Recovery cage.

2.3 Tissue Collection

1. 1.5 mL microcentrifuge tubes.
2. 300 μL ethylenediaminetetraacetic acid (EDTA)-treated microcuvette blood collection tubes.
3. Liquid nitrogen.
4. Surgical instruments.
5. Carbon dioxide tank.
6. Isoflurane.
7. Anesthesia machine.
8. Chamber for anesthesia.
9. Chamber for euthanasia.
10. Dissecting pad (8" × 4").
11. Six 1-inch pins per mouse.
12. 22-gauge 1" hypodermic needles.
13. 3 mL Luer lock syringes.
14. Sterile PBS.
15. 200 μL pipet.
16. 200 μL gel-loading pipet tips.
17. No. 10 scalpel blade.
18. Scalpel handle.
19. Blunt surgical probe.
20. Adson forceps.
21. Surgical scissors.
22. 70% ethanol.

3 Methods

3.1 Preparation of Inoculum

C. neoformans is a biosafety level 2 organism, and all recommended protocols for working with a pathogen of this classification and/or infected animal specimens should be observed. Any work with infected animals is conducted within a biosafety cabinet. *C. neoformans* H99 cultures are stored on a YPD plate at 4 °C between use.

1. Use a single colony from a YPD agar plate to inoculate 5 mL of YPD in a 10 mL test tube (in duplicate).
2. Grow overnight at 30 °C with 200 rpm shaking.
3. Subculture 1:100 (overnight/fresh culture) in fresh YPD.
4. Grow for 12 h at 30 °C with 200 rpm shaking.
5. Transfer culture to a 15 mL tube and centrifuge at 3500 × *g* for 10 min.
6. Discard supernatant.
7. Wash with 2 mL of PBS.
8. Centrifuge at 3500× *g* for 10 min.
9. Discard supernatant.
10. Repeat **steps 7–9** above.
11. Calculate cell density using a hemocytometer by pipetting 10 μL of cell suspension into the hemocytometer, and add a cover slip. Count cells within one set of 16 corner squares, and then proceed to count all 4 sets of 16 corner squares. Using the average cell count from each of the sets of 16 corner squares, multiple by 10^4 to calculate the number of cells/mL.
12. Collect 2×10^5 cells per 50 μL in PBS to create the inoculum.
13. Prepare an aliquot of sterile PBS as a mock inoculum (*see* **Note 2**).
14. Samples may be stored at room temperature (RT) if inoculation will occur within the next hour. Otherwise store at 4 °C. If samples are to be stored at 4 °C, ensure they return to room temperature before administration to an animal.

3.2 Inoculation

1. Mice are removed from the home cage and placed in an anesthesia tank. Oxygen flow rate is set appropriate to the size of the tank with an isoflurane concentration of 5%. Once mice reach surgical plane, the isoflurane flow rate is adjusted to 0.8% to maintain the anesthetized state.
2. For inoculation, remove mouse from the anesthesia tank; hold mouse along the back vertically with the nose pointing straight up. Draw 50 μL of the inoculum in a 200 μL pipet tip, and place directly on the nose of the mouse. One drop is released

slowly to allow the mouse to breathe in the inoculum before the next drop is released. Release 4 × 50 μL drops (200 μL total volume). After the entire inoculum has been administered, the mouse is held in the vertical position for one additional minute to prevent any loss of inoculum during recovery.

3. Place mouse in a recovery cage until mobile and steady; return to the home cage.
4. Repeat **steps 1–3** in Subheading 3.2 for each mouse in the infected group. Mice in the uninfected control group undergo the same procedure but only receive PBS (4 × 50 μL aliquots) as prepared in Subheading 3.1, **step 13** (*see* **Note 3**).
5. For temporal experiments, infected mice at the early- and mid-endpoints are culled along with an uninfected control group partner at 3 and 10 days post-inoculation (d.p.i.), respectively. Mice in the late temporal group are monitored for symptoms, including weight loss, respiratory distress, posture, behavior, and coat condition. Mice with >20% loss of body mass from time of inoculation or that demonstrated symptoms beyond set humane parameters are culled immediately. All infected mice are culled alongside an uninfected control group partner.

3.3 Tissue Collection (See Note 4)

1. Mice are anesthetized using 5% isoflurane with tank-specific oxygen flow rates. Upon reaching the surgical plane, the mouse is transferred to a tank containing tank-specific CO_2 flow rates and remains within this tank until no vital signs are present (approximately 7 min).
2. Place mouse on the dissecting surface in dorsal recumbency.
3. Insert a 22-gauge needle caudal to the xiphoid sternum pointing the needle cranially toward the heart. Insert needle ~1.5 cm to puncture the heart, and then draw back to collect blood sample (*see* **Note 5**).
4. Rotate mouse into sternal recumbency.
5. Make a dorsal midline incision through the skin and subcutaneous layer from occiput to the second cervical vertebra (C2), exposing the foramen magnum.
6. Insert a 22-gauge needle approximately 2 mm (or until clear fluid is visible) into the foramen magnum, and withdraw the needle.
7. Insert a gel-loading pipet tip attached to a 200 μL pipet into the puncture made by the needle, and draw back to collect the cerebral spinal fluid (*see* **Note 6**).
8. Rotate mouse into dorsal recumbency pinning the pectoral and pelvic limbs in extension.

9. Using a No. 10 scalpel blade, make a midline incision from the ventral aspect of the lower lip to the xiphoid sternum exposing the trachea and ribcage.
10. Remove the rib cage, exposing the pluck.
11. Cut along the dorsal aspect of the pluck to free it from the body wall.
12. Insert a gel-loading pipet tip with 200 μL of PBS into the trachea via the larynx. Use the 200 μL of PBS to lavage the lungs by pipetting up and down, and then recover as much liquid as possible for the lavage sample.
13. Remove the lungs from the pluck and deposit with forceps into a sterile tube.
14. Extend midline incision from the xiphoid to the pubis.
15. Gently separate the liver from the diaphragm, and sever any remaining connections to the body wall.
16. Collect the liver into a sterile tube.
17. Expose the caudal aspect of the stomach, and locate the spleen as it is attached to the greater curvature of the body of the stomach.
18. Excise the spleen and collect into a sterile tube.
19. Excise the small intestine and mesentery severing the omental attachment to the dorsal body wall and connection at the stomach and large intestine.
20. Examine the excised small intestine and mesentery. Using a probe or other surgical tool, examine the mesentery locating the mesenteric lymph nodes.
21. Remove and collect the mesenteric lymph nodes into a sterile tube.
22. Collect the small intestine into a sterile tube.
23. Locate the kidneys on the dorsal body wall in the cranial sublumbar region.
24. Excise the kidneys preserving the ureters and collect into a sterile tube.
25. Follow the ureter caudally to the bladder.
26. Excise the bladder at the neck and collect into a sterile tube.
27. Rotate mouse into ventral recumbency.
28. Extend the dorsal midline incision rostrally to the tip of the nose and caudally to the base of the tail.
29. Sever the intervertebral space above the first thoracic and below the last thoracic vertebrae.
30. To remove, cut a paravertebral incision on either side of the vertebral column.

31. Use a gel-loading tip to push the spinal cord out of the excised section of vertebral canal of the excised section, and collect the spinal cord into a sterile tube (*see* **Note 7**).
32. From the incision that extends to the rostrum, remove the facial skin exposing the orbit of both eyes.
33. Insert Adson forceps behind the globe of the eye, and push the globe out of the orbit, removing both the eye and optic nerve.
34. Collect both eye and optic nerve into a sterile tube.
35. Remove any remaining skin covering the cranium.
36. Insert scissors at the foramen magnum, and cut along the border of the cranial vault, removing the skull and exposing the brain.
37. Insert a probe under the most rostral aspect of the brain at the midline, and gently lift up to remove the brain from the skull, exposing the connection to the spinal cord at the foramen magnum.
38. Sever connection to the spinal cord and collect the brain into a sterile tube.
39. Dispose of any waste or remains by incineration per biosafety and animal handling protocols.

4 Notes

1. Female BALB/c mice, 5–7 weeks of age (upon arrival to the facility), are used. Mice are acclimatized for 1 week prior to experiment commencement. Mice are housed in groups of three or four with all cohabiting mice belonging to the same treatment group. No dietary enrichments are provided.
2. A total volume of 200 μL PBS is required per uninfected mouse. We recommend preparing 1.5 times the required volume to avoid any sample loss.
3. Mice may receive supportive care (e.g., high-nutrient gel diet, mash diet, or heating pucks) depending on the experimental parameters.
4. Between each mouse, change and thoroughly disinfect the scalpel blades and tools with 70% ethanol. Use a new set of tools between infected and uninfected specimens.
5. Collect blood into 300 μL Microvette EDTA-treated tubes, and invert tubes 2–3 times to ensure mixing of anticoagulant and blood.
6. Place tissue samples in individual 1.5 mL microcentrifuge tubes upon collection. Liquid samples, excluding the blood, are collected in 1.5 mL microcentrifuge tubes containing 1 mL of

PBS. Samples are immediately flash frozen in liquid nitrogen and stored at −80 °C until needed.

7. Any sterile object that will fit down the vertebral canal may be used in lieu of a gel-loading tip.

Acknowledgments

The authors thank the members of the Geddes-McAlister lab for their informative and constructive feedback on project design and manuscript preparation, as well as Samiha Kazi and Kiyan Kheradvar for technical assistance.

Author Contributions B.M. and J.G.-M. conceptualized and designed the study. B.M. performed the experiments and data analysis and designed and developed the figures. B.M. wrote the first draft of the manuscript. B.M. and J.G.-M. edited the manuscript and wrote the final version. All authors have read and approved the submitted manuscript.

Funding

This project is supported, in part, by the University of Guelph, the Canadian Foundation for Innovation (JELF 38798), and the Canadian Institutes of Health Research (Project Grant) for J.G.-M.

References

1. Vandamme T (2014) Use of rodents as models of human diseases. J Pharm Bioallied Sci 6(1): 2–9. https://doi.org/10.4103/0975-7406.124301
2. Muselius B, Roux-Dalvai F, Droit A et al (2023) Resolving the temporal and spatial splenic proteome during fungal infection for putative dual perspective biomarker discovery. J Am Soc Mass Spectrom 34(9):1928–1940. https://doi.org/10.1021/jasms.3c00114
3. Muselius B, Durand SL, Geddes-McAlister J (2021) Proteomics of *Cryptococcus neoformans*: from the lab to the clinic. Int J Mol Sci 22(22): 12390. https://doi.org/10.3390/IJMS222212390
4. Setianingrum F, Rautemaa-Richardson R, Denning DW (2019) Pulmonary cryptococcosis: a review of pathobiology and clinical aspects. Med Mycol 57(2):133–150. https://doi.org/10.1093/MMY/MYY086
5. Perfect JR, Bicanic T (2015) Cryptococcosis diagnosis and treatment: what do we know now. Fungal Genet Biol 78:49–54. https://doi.org/10.1016/J.FGB.2014.10.003
6. Bermas A, Geddes-McAlister J (2020) Combatting the evolution of anti-fungal resistance in *Cryptococcus neoformans*. Mol Microbiol 114(5):721–734. https://onlinelibrary.wiley.com/doi/full/10.1111/mmi.14565

Part II

Transformation

Chapter 5

Biolistic Transformation of *Cryptococcus neoformans*

Dena L. Toffaletti, Jennifer L. Tenor, and John R. Perfect

Abstract

Biolistic transformation of *Cryptococcus neoformans* is used as a molecular tool to genetically alter or delete targeted genes. The DNA is introduced into the yeast on DNA-coated gold beads by a helium shock wave produced using a biolistic particle system. The procedure often involves insertion of a dominant selectable marker into the desired site by homologous recombination. To increase the likelihood of homologous recombination, large fragments of overlapping DNA are used. The two most used dominant selectable markers are nourseothricin and Geneticin. With the need to generate multiple gene deletions in the same strain, there are recyclable marker systems, such as the bacteriophage P1 Cre-*loxP* system or CRISPR that provide additional useful molecular tools. While newer strategies exist to generate deletions and introduce markers and other gene modifications, biolistic transformation has remained a viable tool to facilitate the construction of genetically modified yeast strains. This chapter provides a working protocol on how to delete and restore a gene in *C. neoformans.*

Key words *Cryptococcus neoformans*, Biolistic transformation, Homologous recombination, Gene deletion, Cre-recombinase, *loxP*, Gene reconstitution, Drug selection cassette, Geneticin, Nourseothricin

1 Introduction

Cryptococcus neoformans is a pathogenic yeast that causes serious systemic disease in immunocompromised individuals. The yeast has a unique ability to cross the blood–brain barrier causing life-threatening meningoencephalitis. Amphotericin B, flucytosine, and fluconazole are current treatments, but additional antifungal drugs are needed. Molecular studies are important in the understanding of virulence mechanisms of this yeast and could lead to the development of novel drug targets. There are many different approaches and strategies to delete, mutate, or tag genes of interest. These DNAs can be delivered into *C. neoformans* by biolistic transformation or electroporation using strategies including CRISPR [1–5], recyclable marker systems [6–8], and traditional methods [9–11]. Transformants carrying altered or deleted genes generated

Erin E. McClelland (ed.), *Cryptococcus neoformans: Methods and Protocols*, Methods in Molecular Biology, vol. 2775,
https://doi.org/10.1007/978-1-0716-3722-7_5,

by these methods have been examined in the murine and/or rabbit models for their ability to cause disease. In *Cryptococcus*, gene replacement transformation efficiency varies from 0.1% to 10% depending upon the recipient strain, gene locus, transforming DNA, and DNA delivery system [12–16]. Serotype A strains, which cause the majority of human infections, are more receptive to efficient transformation using biolistics than by electroporation, with either circular or linear DNAs [9]. Hence, we routinely use biolistic transformation as a molecular tool for genetic manipulation of this yeast. Gene disruption involves inserting a dominant selectable marker into the coding region of the gene of interest. To promote homologous recombination at the target locus by a double cross-over event, a large segment of cryptococcal DNA (800–1500 bp) upstream and downstream of the target gene should flank each side of the drug selection cassette. Selectable markers that we use commonly for gene disruption/deletion are the drug selection cassettes carrying genes that confer resistance to nourseothricin (NAT) [17] and Geneticin (G418, also known as NEO) [18]. We have also used the bacteriophage P1 Cre-*loxP* system that utilizes a NEO drug selection cassette for gene replacement from plasmid, pJL517, and a Cre-recombinase driven by the galactose-inducible promoter from the plasmid, pJL519, for the excision of the NEO drug selection cassette to generate multiple sequential gene deletions in the same strain [6, 7].

Gene deletion constructs are generated by polymerase chain reaction (PCR) amplification of the 5′ flank of the target gene, the drug selection cassette (NAT or NEO), and the 3′ flank of the target gene with primer overhangs for the adjacent DNA. To join the PCR fragments together, the PCR fragments are cloned into a linearized pUC19 plasmid. Plasmids from the *Escherichia coli* (*E. coli*) transformants are screened for the presence of all three fragments. For biolistic transformation, yeast cells are grown overnight in yeast peptone dextrose (YPD) broth, harvested, resuspended in spent medium, and spread onto nonselective media containing sorbitol and mannitol. The DNA-coated gold bead mixture is then prepared and delivered into *C. neoformans* using biolistic transformation. The cells are allowed to recover, collected with phosphate-buffered saline (PBS), and spread onto drug selection plates. Transformants appear 3–5 days after incubation. After patching individual transformants onto drug selection plates, they are screened by colony PCR for loss of the gene. Transformants that show loss of the gene are processed for genomic DNA isolation to use as a template in PCRs to assess for homologous recombination at both 5′ and 3′ flanks, across the gene locus, and loss of the entire gene. The recyclable NEO drug selection cassette can be flipped out of the transformants through the activity of the Cre-recombinase present on pJL519 that is controlled by the galactose promoter and later passaged in nonselective media to eliminate the nonintegrating pJL519 plasmid. Once flipped out, these transformants can be transformed again with a different gene deletion construct.

Unfortunately, transformants that show integration at the target gene site by the gene deletion construct sometimes have additional ectopic integrative events that may affect the virulence of the transformant. Southern blot analysis can be used to determine if ectopic integrations have occurred. Ectopic integration can also be ruled out by transformation of the wild-type gene back into the native locus marked with a different drug selection cassette and then using PCR to determine the complete loss of the original transformation cassette. In addition, we routinely generate two independent transformants that carry the same gene deletion by performing two biolistic transformations (A and B). These transformants are then assessed for similar phenotypes in our studies. In cases where the phenotypes do not match, additional independent transformants are generated, and/or restoring a wild-type copy of the gene into the deletion strains is completed.

In summary, biolistic transformation is a vital molecular tool in understanding the pathogenicity of this fungus. Targeted genes can be mutated/deleted by gene replacement events, and these events can be assessed in transformants by phenotypic analyses and virulence studies.

2 Materials

2.1 C. neoformans Strains and Positive Selectable Markers

1. Serotype A strains: Wild-type *C. neoformans* var. *grubii* (H99) or mutant background.
2. Nourseothricin (NAT).
3. Plasmid pAI3 [17].
4. Geneticin (G418, also known as NEO).
5. Plasmid pJAF1 [18].
6. Recyclable marker Cre-lox system derived from pJL517 (NEO) and pJL519 (NAT) [7] cloned into pUC19 (New England Biolabs).

2.2 Media

Media is prepared in deionized water. For solid media, add 2% (w/v) bacteriological agar prior to autoclaving, and store at 4 °C. For liquid media, sterilize by filtration (0.22 micron filter), and store at room temperature (RT). Alternatively, broth can be autoclaved.

1. YPD broth: 1% (w/v) yeast extract, 2% (w/v) peptone, 2% (w/v) glucose.
2. YPD agar: 1% (w/v) yeast extract, 2% (w/v) peptone, 2% (w/v) glucose, 2% (w/v) agar.

3. YPD agar +200 μg/mL Geneticin (G418).
4. YPD agar +100 μg/mL nourseothricin (NAT).
5. YPD agar +0.75 M sorbitol and 0.75 M mannitol: 1% (w/v) yeast extract, 2% (w/v) peptone, 2% (w/v) glucose, 0.75 M sorbitol, 0.75 M mannitol.
6. YPG broth (used with recyclable marker): 1% (w/v) yeast extract, 2% (w/v) peptone, 2% (w/v) galactose.
7. YPG agar +100 μg/mL nourseothricin (used with recyclable marker protocol): 1% (w/v) yeast extract, 2% (w/v) peptone, 2% (w/v) galactose, 2% (w/v) agar.
8. YPG agar +0.75 M sorbitol and 0.75 M mannitol (used with the recyclable marker protocol): 1% (w/v) yeast extract, 2% (w/v) peptone, 2% (w/v) galactose, 0.75 M sorbitol, 0.75 M mannitol.
9. Luria-Bertani (LB) broth +100 μg/mL carbenicillin (CRB) disodium salt and store at 4 °C.
10. LB agar +100 μg/mL CRB.

2.3 Cloning

1. HiFi DNA Assembly Cloning Kit.
2. QIAprep Spin Miniprep Kit.
3. iProof High-Fidelity DNA Polymerase can be used for all PCRs. The cost is offset by the advantage of its rapid amplification rate (15 s/1 kilobase [kb]), which significantly reduces the cycling times. Other polymerases can be used such as Hot Start *Taq* DNA Polymerase.

2.4 DNA Primers for PCR and Sequencing

PCR is used to generate the DNA products for cloning into pUC19 and for the identification of positive transformants. Primers are shown in Figs. 1 and 2.

PCRs to construct a plasmid carrying the deletion construct.

1. PCR1: 5′ flank, primers 1 and 2.
2. PCR2: Drug selection cassette, primers 3 and 4.
3. PCR3: 3′ flank, primers 5 and 6.

PCRs to identify and confirm yeast deletion strain.

4. PCR4: Screening for loss of gene, primers 7 and 8 (colony PCR, *see* **Note 1**).
5. PCR5: Assessment of 5′ integration at the target gene allele, primers 9 and 10.
6. PCR6: Assessment of 3′ integration at the target gene allele, primers 11 and 12.
7. PCR7: Assessment of target gene locus, primers 9 and 12.

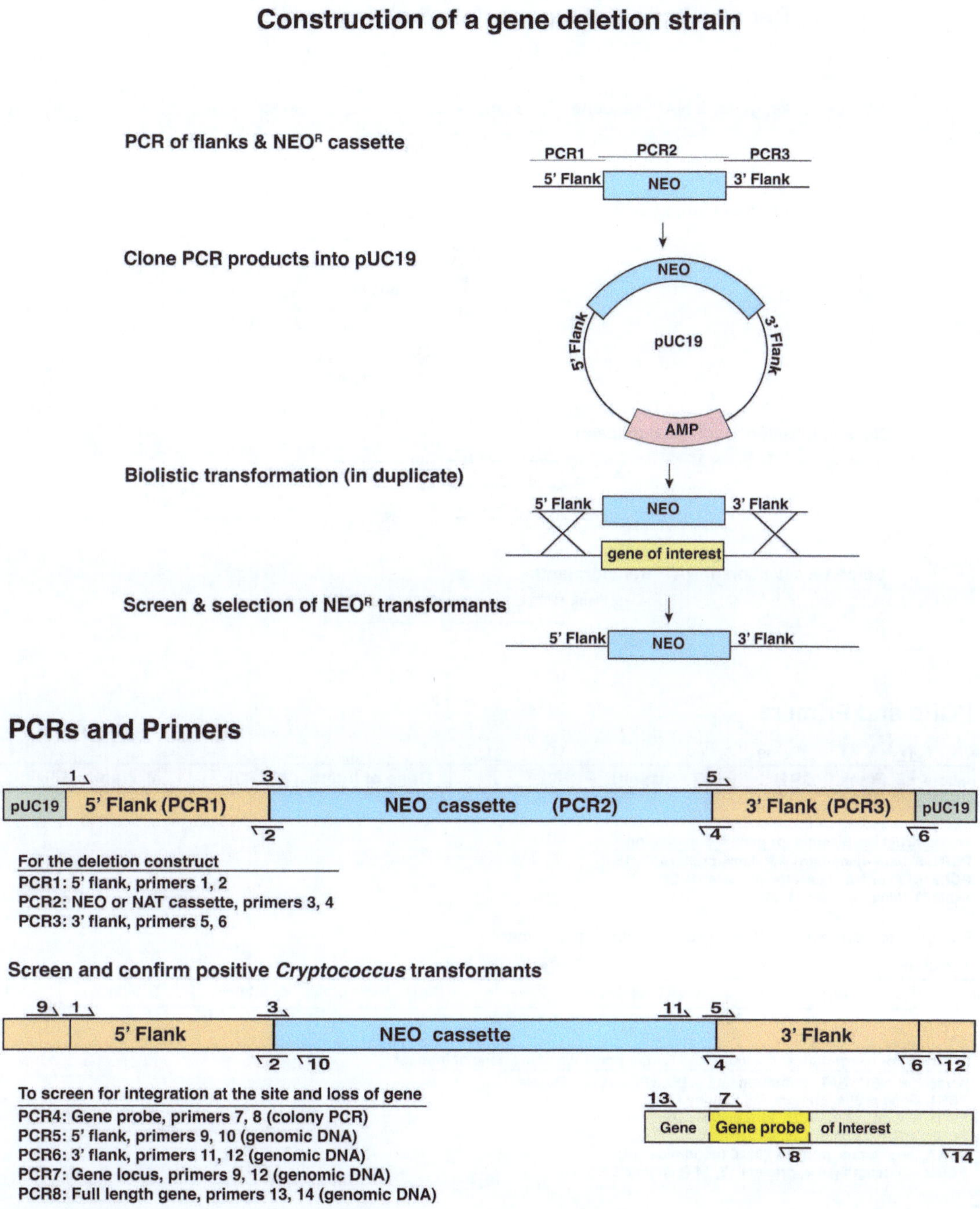

Fig. 1 Schematic diagram of the construction of a *C. neoformans* gene deletion strain

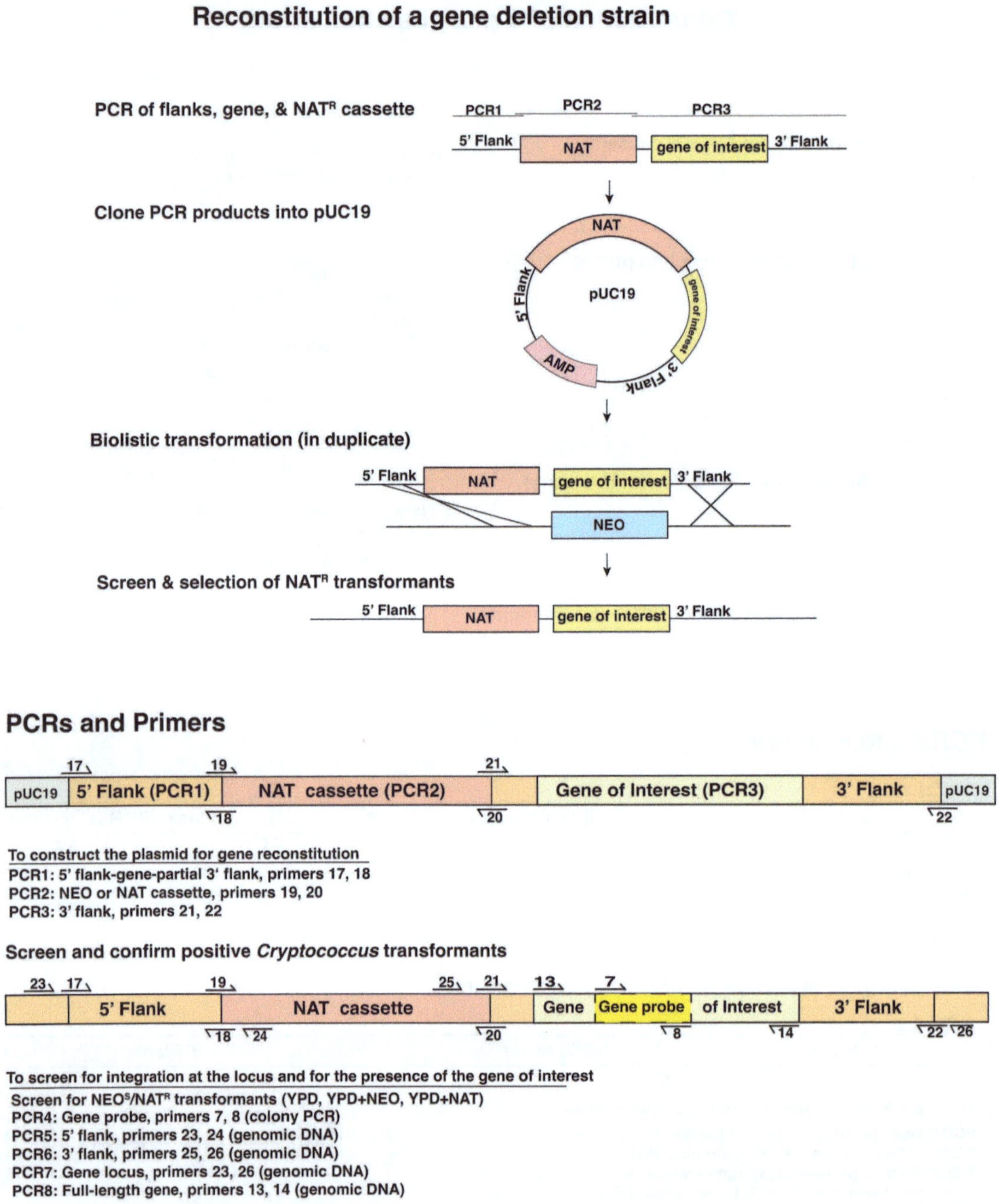

Fig. 2 Schematic diagram of the reconstitution of a *C. neoformans* gene deletion strain

8. PCR8: Assessment for loss of full-length target gene, primers 13 and 14.
9. PCR9: Assessment for loss of Cre-recombinase when applicable, primers 15 and 16 [7].

PCRs for plasmid construction to restore the gene at its native site.

10. PCR1: 5′ flank-gene-partial 3′ flank, primers 17 and 18.
11. PCR2: Drug selection cassette, primers 19 and 20.
12. PCR3: 3′ flank, primers 21 and 22.

PCRs to assess transformants for gene reconstitution.

13. PCR4: Screening for presence of gene, primers 7 and 8 (*see* **Note 1**).
14. PCR5: Assessment of 5′ integration at the target gene allele, primers 23 and 24.
15. PCR6: Assessment of 3′ integration at the target gene allele, primers 25 and 26.
16. PCR7: Assessment of target gene locus, primers 23 and 26.
17. PCR8: Assessment for loss of full-length target gene, primers 13 and 14.

2.5 Gel Extraction and Purification of PCR Products

1. Low-melting agarose.
2. Agarose gel: 1% agarose (w/v) prepared in Tris/Borate/Ethylenediaminetetraacetic acid (EDTA) (TBE) buffer.
3. 10× TBE buffer.
4. NucleoSpin Gel and PCR Clean-up Kit.

2.6 Biolistic DNA Delivery

1. Biolistics PDS-1000 He, Particle Delivery System (Bio-Rad).
2. Gold bead (also known as gold microcarriers) suspension (0.6 micron) in sterile water (125 mg/mL).
3. Rupture disks, 1350 psi.
4. Macrocarriers.
5. Stopping screens.
6. Vacuum pump.
7. Helium gas tank.
8. Plasmid DNA 2–4 μg/reaction.
9. Spent growth medium (or water) for resuspending yeast cells prior to biolistic delivery.
10. 2.5 M $CaCl_2$. Sterilized by autoclaving and stored at 4 °C.

11. 1 M spermidine in water, filter-sterilized, and stored in small aliquots at −80 °C and working stock at 4 °C.
12. Low-retention 1.7 mL microfuge tubes.

2.7 Genomic DNA Isolation

1. MasterPure Yeast DNA Purification Kit.
2. 10 mg/mL RNase A.
3. 2 mL screw cap tubes with O-rings.
4. 2.3 mm diameter silica beads.
5. Breathe-Easier membranes.
6. Qubit dsDNA BR Assay Kit.

2.8 Storage of Deletion Transformants

1. 50% (v/v) glycerol in ultrapure distilled water, filter-sterilized.
2. 1.8 mL cryotubes.
3. Freezer storage box.

2.9 Southern Blot Analysis

1. PCR DIG Probe Kit.
2. Anti-digoxigenin-AP, Fab fragments.
3. CDP-star ready to use.
4. DIG wash and block buffer set.
5. Nylon membrane.
6. Washing solutions: 2× saline–sodium citrate (SSC), 0.1% sodium dodecyl sulfate (SDS), 0.2× SSC, 0.1% SDS.

2.10 Equipment and Accessories

1. Spectrophotometer.
2. Fluorometer.
3. Lyophilizer.
4. Bead mill homogenizer.
5. GenElute linear polyacrylamide (LPA).
6. L-spreaders.
7. 32-square grid and 8-pie grid plate stickers.
8. Biolistic setup (*see* Subheading 2.6).
9. PCR machine.
10. −80 °C freezer.
11. −20 °C freezer.
12. 30 °C and 37 °C incubator.
13. 30 °C and 37 °C shaking incubator.
14. Microcentrifuge and a benchtop centrifuge.
15. 65 °C heating block.
16. Electrophoresis equipment.

17. Equipment for Southern blot analysis (*see* Subheading 2.9).
18. 50 mL conical tube.
19. 14 mL snap cap tube.
20. 5 mL snap cap tube.

2.11 General Reagents

1. Phosphate-buffered saline (PBS), pH 7.4.
2. Ultrapure distilled water for molecular procedures.
3. 70% (v/v) ethanol in ultrapure distilled water for molecular procedures.
4. 70% (v/v) ethanol in deionized water for decontaminating surface area and equipment.
5. Sani-Cloth germicidal disposable wipes.
6. Isopropanol.
7. Absolute ethanol.

3 Methods

A common approach to *Cryptococcus* genetics is to generate two strains: one strain that carries a deletion of the gene of interest and the other strain that has the gene of interest restored in the deletion strain. Figures 1 and 2 provide a basic outline for the construction of these two strains. It takes approximately 1 month to generate a strain, as highlighted in Fig. 3.

3.1 Construction of Gene Deletion Plasmid (Fig. 1)

To generate the deletion construct consisting of three DNA fragments (5′ flank, a drug selection cassette, 3′ flank), the DNA is amplified by PCR and cloned into linearized pUC19 using the HiFi DNA Assembly Cloning Kit. The 5′ and 3′ flanks should be approximately 800–1500 base pairs (bp) in length. The primers should be 39–42 nucleotides (nt) in length, which includes an overhang of 15–22 nt. Other methods such as overlap PCR can be used (*see* **Note 2**).

1. PCR1: Amplify the 5′ flank with primers 1 and 2 using genomic DNA as the template.
2. PCR2: Amplify the drug selection cassette with primers 3 and 4. The DNA template is the plasmid carrying the desired drug selection cassette.
3. PCR3: Amplify the 3′ flank with primers 5 and 6 using genomic DNA as the template.
4. Electrophorese the PCR products in a 1% agarose gel, extract band from gel, and purify the DNA using the NucleoSpin Gel and PCR Clean-up Kit.
5. Measure the concentration of the purified PCR products.

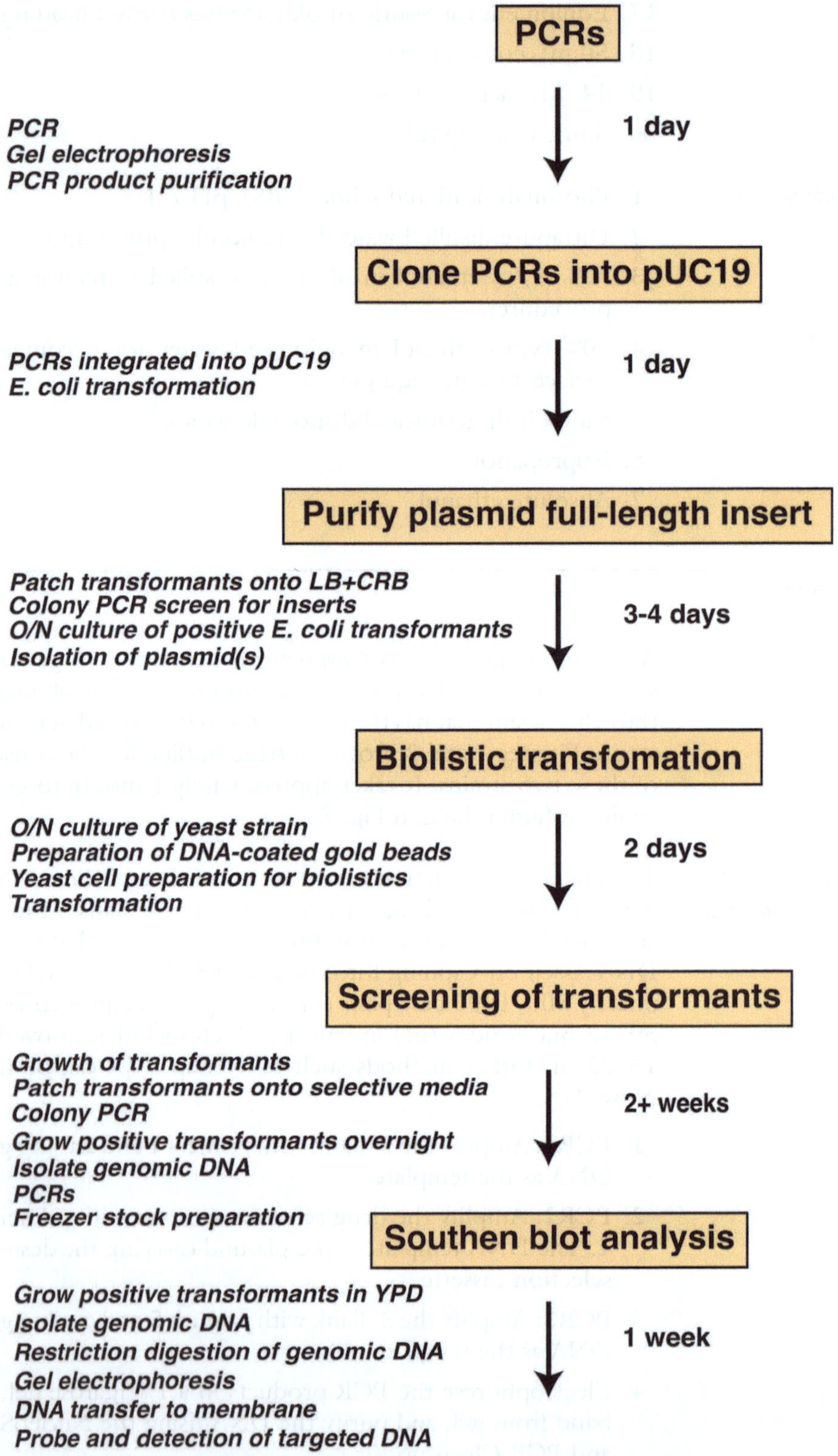

Fig. 3 Estimated timeline for the deletion or reconstitution of a gene in *C. neoformans*

6. Clone the fragments into linearized pUC19 using the HiFi DNA Assembly Kit.
7. Transform the plasmid into *E. coli* by chemical transformation following the protocol provided in the kit. Plate *E. coli* onto LB agar + CRB using a L-spreader, and incubate overnight at 37 °C.
8. Patch individual colonies onto LB agar + CRB using a 32-grid sticker, and incubate plate overnight at 37 °C.
9. Screen *E. coli* using colony PCR to detect one of the cloned PCR products. Colony PCR with *E. coli* works best when the size of the fragment is ≤1 kb.
10. Inoculate four 50 mL conical tubes containing 10 mL of LB + CRB with four *E. coli* positive transformants (one transformant per tube). Incubate cells overnight at 37 °C in a shaking incubator (225 rpm), and purify the plasmid according to the QIAprep Spin Miniprep Kit instructions.
11. Assess plasmid for individual fragments (PCR1, PCR2, and PCR3) and full-length insert (using primers 1 and 6). For the template DNA, prepare 10 μL of a 1/10 dilution of plasmid.
12. Measure plasmid concentration. You need a minimum of 2 μg DNA for each transformation. An acceptable concentration for the plasmid is ≥400 ng/μL for biolistic transformation.
13. Store the positive plasmid at −20 °C and a freezer stock of the *E. coli*, if needed.

3.2 Cell Preparation for Biolistic Transformation

1. Inoculate 10 mL of YPD with a single colony of *Cryptococcus* strain in a 50 mL tube, and incubate culture overnight at 30 °C in a shaking incubator (225 rpm).
2. Harvest cells by centrifugation (1942× *g*, 5 min).
3. Save 1–2 mL of supernatant (spent medium) and discard remaining liquid.
4. Resuspend cell pellet in 1–2 mL of the spent medium.
5. Spread 250–300 μL of cells onto YPD plus 0.75 M sorbitol +0.75 M mannitol plates using a L-spreader, and allow to air-dry on benchtop or flow hood until liquid is fully absorbed.
6. Incubate plates at 30 °C for approximately 1–2 h while preparing DNA-coated gold bead mixture.

3.3 Preparation of DNA-Coated Gold Bead Mixture and Macrocarriers

Our procedure is designed for two biolistic transformations for each DNA construct, which allows for the isolation of independent transformants. Thus, each DNA-coated gold bead mixture is designed for two bombardments per recipient strain.

1. To prepare the gold bead stock suspension: Add 2 mL of absolute ethanol to vial containing 250 mg of gold beads, vortex for 1 min, and transfer 500 μL to four low-retention

microfuge tubes. Add another 2 mL of absolute ethanol to the original vial, vortex briefly, and aliquot 500 μL to each of the microfuge tubes containing the gold bead suspension for a final volume of 1 mL. Vortex for 3 min, pellet gold beads (16,200 × *g*, 1 min), wash twice in 1 mL of sterile water, resuspend in 1 mL sterile water (*see* **Note 3**), and store at 4 °C.

2. If DNA concentration is too low, *see* **Note 4**.
3. For the preparation of two bombardments (i.e., transformations), add in the following order: 10 μL of DNA (4–8 μg), 20 μL of gold beads, 20 μL of 2.5 M $CaCl_2$, and 12 μL 1 M spermidine. It is important to vortex the tube after the addition of each component due to the rapid settling of the gold beads. Adjust volumes appropriately for less or more bombardments as needed (*see* **Note 5**).
4. Incubate DNA-coated gold bead mixture for 10 min at RT without agitation.
5. Using forceps, dip orange macrocarriers into absolute ethanol, place in sterile petri dish, and air-dry under the flow hood or benchtop.
6. Centrifuge DNA-coated gold bead mixture briefly (10 s) and remove supernatant.
7. Wash DNA-coated gold beads with 500 μL absolute ethanol (*see* **Note 6**), and resuspend beads in 14 μL absolute ethanol.
8. Spread 6 μL of DNA-coated gold beads onto the center of the macrocarrier (approximately 10 mm in diameter) in the sterile petri dish, and air-dry on the benchtop (*see* **Note** 7).
9. Using the macrocarrier insertion tool, gently press the macrocarrier into a clean metal macrocarrier holder so it is securely positioned at the bottom of the holder. Ensure that the side containing the DNA-coated gold beads is face-up in the macrocarrier holder, which will be inverted in the launch assembly.

3.4 Bombardment with DNA-Coated Gold Beads

Review the Bio-Rad instructions for proper operation of the biolistic particle delivery system. Perform two independent transformations (transformation A and transformation B).

1. Clean the inside chamber of the biolistic system with 70% ethanol or disinfectant wipes (*see* **Note 8**).
2. Open the helium tank, and adjust the regulator until the gauge reaches the maximum psi, typically in the range of 1500–2000 psi.
3. Turn on biolistic system and then the vacuum pump.
4. Using sterile forceps, dip the rupture disk into 100% isopropanol, and place the rupture disk in the retaining cap, and screw into the gas acceleration tube. Ensure cap is tightened.

5. Prepare the launch assembly by placing a stopping screen and macrocarrier with the DNA-coated gold beads in the macrocarrier holder into the assembly. Secure in place with the macrocarrier cover lid.
6. Place the launch assembly into the top position of the shelf of the biolistic chamber located just below the rupture disk cap.
7. Remove the lid from your plate containing the recipient cells, and place it in the third shelf position. Close and latch the chamber door.
8. Generate a full vacuum (> 28.5 in. Hg, 2–3 min) within the chamber, hold, and fire the biolistic system until the rupture disk bursts. Release the vacuum, and return petri dish lid to the agar plate, and remove from the chamber.
9. Wipe the internal chamber and components with either 70% ethanol or a germicidal wipe.

3.5 Post-Transformation

1. Let the cells recover for 2.5–3 h at 30 °C.
2. Harvest cells with 1.2 mL of PBS using a L-spreader, and transfer cells to a 5 mL tube. Expect to recover ~0.7–1 mL cell suspension.
3. Plate ~250–300 μL of cells onto three plates of selective medium (YPD + NEO or YPD + NAT). Place plates with lids off in a biosafety cabinet until surface liquid is absorbed into the agar. Restore lids.
4. Incubate plates for 3–5 days at 30 °C.

3.6 Screening of Transformants for Loss of Gene

1. Place a 32-grid sticker on the bottom of the selective medium. Select large colonies, as these are more likely to be positive transformants compared to the colonies that arise later. Patch 32 individual colonies onto selective medium (YPD + NEO or YPD + NAT), and incubate for 1–3 days at 30 °C. Store transformation plates at 4 °C for additional screening if needed.
2. Assess for loss of gene using colony PCR with gene probe primers 7 and 8. PCR product should be less than 600 bp.
3. Passage candidate transformants (select 4–8 colonies from transformation A and 4–8 colonies from transformation B) in 1 mL of YPD in a 5 mL tube for 1–3 days at 30 °C in a shaking incubator (225 rpm).
4. Transfer 5 μL of culture onto YPD + NEO or YPD + NAT plates, and streak for colony isolation. Incubate plates for 2–3 days at 30 °C.
5. Identify and select transformants that have similarly sized colonies for further processing (*see* **Note 9**). Typically, when testing 32 transformants, we expect up to 4 positive transformants.

3.7 Genomic DNA Isolation from Transformants to Be Used as Template DNA in PCRs

1. Select a single colony from each positive transformant plate (from Subheading 3.6, **step 5**), and circle its position on the plate.
2. Inoculate the selected colonies (circled) from transformation A and transformation B individually into 3.5 mL of YPD in a 14 mL tube, and vortex briefly.
3. Incubate cells for 1–2 days at 30 °C in a shaking incubator (225 rpm).
4. Centrifuge cells (1942× *g*, 5 min) and remove supernatant.
5. Add 8–15 silica beads to a 2 mL screw cap tube with O-ring.
6. Add 800 μL of sterile water to the cell pellet, and transfer cells to the 2 mL screw cap tube containing silica beads.
7. Centrifuge cells (16,200× *g*, 1 min), and remove supernatant with a barrier pipet tip.
8. Freeze cells at −80 °C.
9. Prepare for lyophilization by removing the caps (keep for later recapping) and covering the tubes with a breathable film. Work quickly to ensure that samples remain frozen. Lyophilize cells for several hours or overnight until pellets are desiccated. Remove film and replace cap tubes.
10. Store lyophilized cells at RT in a desiccator until pellets are processed for genomic DNA isolation.

3.8 Genomic DNA Isolation (MasterPure Yeast DNA Purification Kit)

1. Add 300 μL of yeast cell lysis solution to cell pellet.
2. Bead mill twice for 30 s at 5.5 rpm.
3. Add 5 μL of RNase A (10 mg/mL) and vortex briefly.
4. Incubate for 30–60 min at 65 °C (*see* **Note 10)** and cool on ice for at least 5 min.
5. Add 150 μL of MPC protein precipitation reagent and vortex for 10 s.
6. Centrifuge tubes (16,200× *g*, 10 min), and transfer supernatant to a new tube containing 500 μL of cold isopropanol.
7. Mix by inversion and centrifuge tubes (16,200× *g*, 15 min).
8. Remove supernatant and wash pellet with 500 μL 70% ethanol (16,200× *g*, 5 min).
9. Remove ethanol and centrifuge (16,200× *g*, 10 s).
10. Remove residual ethanol and air-dry pellet for 2–5 min.
11. Add 30–50 μL of sterile water or Tris/EDTA (TE) buffer, and place genomic DNA at 4 °C overnight.
12. Gently mix tube by flicking, and measure genomic DNA using a Qubit fluorometer (optional), and store at −20 °C (*see* **Note 11**).

3.9 PCR Amplification of Candidate Transformants

Use 2 μL of genomic DNA in a 25 μL volume for all PCRs (5′ flank, 3′ flank, gene locus, gene) to assess for homologous recombination by a double cross-over event at the target allele site. The PCRs will assess if homologous recombination has occurred in the transformant.

1. Use PCR to amplify the 5′ flank and the 3′ flank using primers that anneal to DNA outside of the construct sequence. Assess the transformant for integration of DNA at the 5′ flank and 3′ flank by PCR followed by gel electrophoresis.
2. For the 5′ flank: Use primer 9 that anneals upstream and outside of the 5′ flank and primer 10 that anneals within the drug selection cassette to generate a product ~1000–1500 bp.
3. For the 3′ flank: Use primer 11 that anneals within the drug selection cassette and primer 12 that anneals downstream and outside of the 3′ flank to generate a product ~1000–1500 bp.
4. Use PCR to amplify the locus using primers 9 and 12 that anneal to DNA outside the construct DNA sequence. Compare this PCR product to the wild-type locus PCR product using gel electrophoresis.
5. Use PCR to confirm loss of the gene of interest. Assess for loss of the entire gene with primers 13 and 14. Compare this PCR product to the wild-type locus PCR product using gel electrophoresis.
6. Prepare a glycerol stock of deletion transformant(s) that confirm the integration at the site and loss of the gene. Using the same colony selected for genomic DNA isolation, transfer a small amount of yeast cells into 2 mL of YPD in a 14 mL tube, and incubate overnight at 30 °C in a shaking incubator (225 rpm). Transfer 800 μL of culture to a freezer tube containing 800 μL sterile 50% glycerol. Store strains at −80 °C.

3.9.1 Use of the Recyclable Marker System

A recyclable marker system is available to facilitate the construction of multiple gene deletions within the same strain. To generate a deletion strain, the NEO drug selection cassette from pJL517 is used because *loxP* sites have been designed to flank the cassette [6, 7]. The *loxP* sites are essential for the later removal of this cassette from the strain. Once a deletion strain is confirmed, the NEO drug selection cassette may be flipped out. This is achieved by the transformation of the deletion strain with the plasmid pJL519 that carries the NAT drug selection cassette and the Cre-recombinase enzyme controlled by the galactose promoter. When the deletion strain is grown in the presence of galactose, the NEO drug selection cassette is removed through the action of the Cre-recombinase. Candidate transformants must then be passaged in YPD to eliminate the pJL519 plasmid. The NEO and NAT-sensitive strain is ready for the next round of gene deletion(s).

3.10 To Remove the NEO Drug Selection Cassette

1. Inoculate a single colony from the positive transformants carrying a deletion of the gene of interest (transformation A and transformation B) in 10 mL of YPG broth, and prepare cells as described in Subheading 3.2 with the exception that cells are spread onto YPG containing sorbitol and mannitol plates.
2. Prepare the DNA-labeled gold beads, and transfer to the macrocarrier as described in Subheading 3.3, **step 2–9** with the exception that 3.0 μg of pJL519 is used for two bombardments (transformation A and transformation B).
3. Transform yeast cells with the DNA-coated gold beads, as described in Subheadings 3.4 and 3.5 with the exception that the recovered cells are spread on YPG + NAT plates.

3.11 Screening of Transformants

1. Place a 32-grid sticker on the bottom of a YPG + NAT plate.
2. Patch up to 32 individual colonies onto a YPG + NAT plate, and incubate for 1–3 days at 30 °C.
3. Place a 32-grid sticker on the bottom of a YPD + NEO plate. Assess for loss of the NEO drug selection cassette: Patch the transformants from the YPG + NAT plate to the YPD + NEO plate, and incubate for 1–3 days at 30 °C. Record the transformant numbers that failed to grow on YPD + NEO. Mark these transformant numbers on the gridded YPD + NAT plate. The transformants that have lost the NEO drug selection cassette are now ready for passaging to ensure loss of pJL519.

3.12 Passaging Cells and Confirmation of NAT Sensitivity and NEO Sensitivity

1. Inoculate four transformants from transformation A and transformation B that are NEO-sensitive from the YPG + NAT plate (from Subheading 3.11, **step 2**) into 5 mL of YPD in a 50 mL tube, and incubate at 30 °C in a shaking incubator (225 rpm). Passage cells daily by transferring 5 μL of the overnight culture into a new 50 mL tube containing 5 mL of YPD. Repeat this for a total of three passages.
2. Transfer 10 μL of the third passage cultures onto a YPD plate, streak for colony isolation, and incubate for 2–3 days at 30 °C. Select and circle four colonies per transformation to screen for NAT sensitivity.
3. To identify strains that are NAT-sensitive, swipe cells from the selected colonies from each transformation A (4 colonies) and transformation B (4 colonies), and streak onto YPD + NAT plate (*see* **Note 12**). Incubate for 2–3 days at 30 °C.
4. Select transformants from transformation A and transformation B that did not grow on YPD + NAT, and inoculate these transformants from the YPD plate into 3.5 mL of YPD for genomic DNA isolation.

5. Confirm loss of the NEO drug selection cassette by PCR across the locus with primers 9 and 12. Compare this PCR product by gel electrophoresis to the wild-type locus PCR product and the PCR product from the strain with the NEO drug selection cassette as controls to confirm loss of NEO.
6. Confirm loss of pJL519 by PCR with CRE primers 15 and 16. Assess using gel electrophoresis.
7. Prepare glycerol stocks of the NEO^{s}-NAT^{s} strains as described in Subheading 3.9, **step 6**.

3.13 Reconstitution of the Wild-Type Gene at the Native Locus in a Deletion Strain (Fig. 2)

To further support that a gene confers a particular phenotype, the gene of interest is restored to its native locus, and the strain is tested for restoration/loss of the phenotype. A common approach is to transform a wild-type copy of the gene with a different drug selection cassette inserted in an adjacent intergenic region. Placement of the cassette depends on the position and orientation of the adjacent genes. Care should be taken to avoid the untranslated regions of the adjacent genes and the gene of interest (*see* **Note 13**). However, if a drug selection cassette can be placed in one of the intergenic regions (upstream or downstream of the gene), then three PCRs are performed and cloned into pUC19 using the HiFi DNA Assembly Cloning Kit. The following is an example where the drug selection cassette is inserted into the 5′ intergenic region.

1. Set up three PCR reactions.
2. PCR1: Primers 17 and 18 are used to amplify the 5′ flank region. The product size should be 700–1000 bp if possible. Genomic DNA is the template.
3. PCR2: Primers 19 and 20 are used to amplify the drug selection cassette (NAT). The plasmid carrying the drug selection cassette is the template.
4. PCR3: Primers 21 and 22 are used to amplify part of the 5′ flank, the gene, and the 3′ flank. The product size will be variable dependent on the size of the gene. Genomic DNA is the template.
5. Plasmid construct and purification: Follow the procedure as described in Subheading 3.1, **step 4–13**.
6. Biolistic transformation of deletion strain: Follow procedures as described in Subheadings 3.2–3.5. The drug selection based on our example is NAT.
7. Screening of transformants: Follow procedure as described in Subheading 3.6.
8. Patch the transformants onto a YPD + NAT plate containing a 32-grid sticker. Incubate for 1–2 days at 30 °C.

9. Patch the transformants from the YPD + NAT plate onto YPD + NEO containing a 32-grid sticker to identify those transformants with NEO sensitivity.
10. Mark the transformants on the gridded YPD + NAT plate that are NEO-sensitive. These transformants will be processed for further analyses.
11. Assess for the presence of the gene using colony PCR with gene probe primers 7 and 8.
12. Passage candidate transformants (select 4–8 colonies from transformation A and 4–8 colonies from transformation B) in 1 mL of YPD in a 5 mL tube for 1–3 days at 30 °C in a shaking incubator (225 rpm).
13. Transfer 5 μL of culture onto YPD + NAT plates and streak for colony isolation. Incubate plates for 2–3 days at 30 °C.
14. Follow Subheadings 3.7 and 3.8 to isolate genomic DNA.
15. Two PCRs are performed to confirm the presence of the gene at its native locus. To confirm the presence of the entire gene, primers 13 and 14 are used. To confirm the integration of the gene at its native locus, primers 23 and 26 are used.
16. Prepare glycerol stocks of reconstituted strains as described in Subheading 3.9, **step 6**.

3.14 Southern Blot Analysis

Southern blot analysis is used to confirm that the transformants have the desired genetic change(s), such as a deletion or restoration of the gene of interest at the native site. PCR is performed to generate a DIG-labeled DNA probe of approximately ~500–1000 bp. This PCR product is designed to hybridize with the gene of interest or the drug selectable marker (NEO or NAT). Genomic DNA (~10 μg) is treated with restriction enzyme(s) that will result in a 3500–7500 bp fragment that contains the DNA region of interest. The digested genomic DNA is then electrophoresed through a 0.9% agarose gel and transferred to a nylon membrane overnight following standard methods. The membrane is processed by immunological detection via an anti-digoxigenin antibody conjugated to alkaline phosphatase and a chemiluminescent substrate.

4 Notes

1. The PCR product for colony PCR should be less than 600 bp for reliable screening of transformants. Primers are designed to amplify a 250–600 bp region of the gene.

2. Alternatively, other labs have used overlap PCR to generate cassettes carrying selectable markers flanked by regions of targeted gene homology for biolistic transformation [10, 11].
3. A gold bead suspension concentration of 62.5 mg/mL is recommended by Bio-Rad; however, we typically use 125 mg/mL. This change in the gold bead concentration has resulted in increased numbers of transformants. The 125 mg/mL gold bead suspension is prepared by resuspending the gold beads in 0.5 mL of sterile water per microfuge tube.
4. To concentrate DNA there are two approaches: The first approach precipitates the DNA, and the second simply removes water from the gold bead suspension to offset the volume of DNA needed. To precipitate the DNA, transfer the volume of DNA needed to reach 4–8 μg/mL to perform two transformations. Add 1.5 μL of LPA, 3 M sodium acetate pH 5.2 (1/10 the volume of DNA), and absolute ethanol (3 times the volume of the DNA). Mix by inversion, centrifuge at 16,200× *g* for 15 min, remove the liquid, wash the pellet with 500 μL 70% ethanol, centrifuge at 16,200× *g* for 5 min, remove all liquid, air-dry pellet 2–5 min, and resuspend DNA in 10 μL of sterile water (for two bombardments). The second approach can be done if the volume of DNA needed is between 21 and 30 μL. Since the gold beads are stored in water, one can remove water simply by centrifuging the beads (20 μL) for 10 s at full speed. Once the beads are pelleted, the water can be removed and offset by the volume of DNA.
5. The order of adding DNA and gold beads can be reversed.
6. To wash the gold beads coated with DNA, first add 50 μL of absolute ethanol, and vortex until most of beads are in suspension. Then, add the remaining 450 μL of absolute ethanol and vortex again.
7. Resuspension in 14 μL of ethanol ensures that 6 μL of beads can be spread onto each macrocarrier.
8. Decontaminate the biolistic chamber thoroughly with 70% ethanol before and after performing all bombardments. Germicidal disposable wipes can also be used and are particularly useful if you want to remove microbial contaminants.
9. After passaging in YPD broth and reselection on drug selective medium, some transformants may have differently sized colonies arise. Unless associated with the strain phenotype, select only the transformants with colonies that are similarly sized to each other.
10. Vortex cells 2–4 times during incubation period.
11. Typically, genomic DNA is not measured for routine PCRs.

12. Place an 8-pie grid sticker on the YPD + NAT plate for assessment of transformation A colonies and transformation B colonies. Passaging four transformants will ensure that at least two transformants are NAT-sensitive. This approach allows reduction of the overall passaging time needed to obtain a NAT-sensitive transformant.
13. When a drug selection cassette cannot be placed in the intergenic region, different strategies can be taken: (1) The drug selection cassette can be inserted into the safe haven site, which is specific site within the genome of H99 [19]. The safe haven site is a site where DNA can be inserted without impacting neighboring genes. (2) The drug selection cassette and the DNA that spans the locus for the gene of interest are co-transformed where the DNA fragments are not on the same strand of DNA. (3) A more recently accepted approach is to generate two independent deletion strains and confirm that the phenotypes are the same between the two strains.

References

1. Arras SD, Chua SM, Wizrah MS et al (2016) Targeted genome dditing via CRISPR in the pathogen *Cryptococcus neoformans*. PLoS One 11:e0164322. https://doi.org/10.1371/journal.pone.0164322
2. Wang Y, Wei D, Zhu X et al (2016) A 'suicide' CRISPR-Cas9 system to promote gene deletion and restoration by electroporation in *Cryptococcus neoformans*. Sci Rep 6:31145. https://doi.org/10.1038/srep31145
3. Wang P (2018) Two distinct approaches for CRISPR-Cas9-mediated gene editing in *Cryptococcus neoformans* and related species. mSphere 3:e00208-18. https://doi.org/10.1128/mSphereDirect.00208-18
4. Fan Y, Lin X (2018) Multiple applications of a transient CRISPR-Cas9 coupled with electroporation (TRACE) system in the *Cryptococcus neoformans* species complex. Genetics 208:1357–1372. https://doi.org/10.1534/genetics.117.300656
5. Zhang P, Wang Y, Li C et al (2021) Simplified all-in-one CRISPR-Cas9 construction for efficient genome editing in *Cryptococcus* species. J Fungi (Basel) 7:505. https://doi.org/10.3390/jof7070505
6. Patel RD, Lodge JK, Baker LG (2010) Going green in *Cryptococcus neoformans*: the recycling of a selectable drug marker. Fungal Genet Biol 47:191–198. https://doi.org/10.1016/j.fgb.2009.11.007
7. Baker LG, Lodge JK (2012) Multiple gene deletion in *Cryptococcus neoformans* using the Cre-lox system. Methods Mol Biol 845:85–98. https://doi.org/10.1007/978-1-61779-539-8_6
8. Erpf PE, Stephenson CJ, Fraser JA (2019) amdS as a dominant recyclable marker in *Cryptococcus neoformans*. Fungal Genet Biol 131:103241. https://doi.org/10.1016/j.fgb.2019.103241
9. Toffaletti DL, Rude TH, Johnston SA et al (1993) Gene transfer in *Cryptococcus neoformans* by use of biolistic delivery of DNA. J Bacteriol 175:1405–1411
10. Davidson RC, Blankenship JR, Kraus PR et al (2002) A PCR-based strategy to generate integrative targeting alleles with large regions of homology. Microbiology (Reading) 148(Pt 8):2607–2615. https://doi.org/10.1099/00221287-148-8-2607
11. Kim MS, Kim SY, Yoon JK et al (2009) An efficient gene-disruption method in *Cryptococcus neoformans* by double-joint PCR with NAT-split markers. Biochem Biophys Res Commun 390:983–988. https://doi.org/10.1016/j.bbrc.2009.10.089
12. Chang YC, Kwon-Chung KJ (1994) Complementation of a capsule-deficient mutation of *Cryptococcus neoformans* restores its virulence. Mol Cell Biol 14:4912–4919. https://doi.org/10.1128/mcb.14.7.4912
13. Lodge JK, Jackson-Machelski E, Toffaletti DL et al (1994) Targeted gene replacement demonstrates that myristoyl-CoA: protein N-myristoyltransferase is essential for viability

of *Cryptococcus neoformans*. Proc Natl Acad Sci U S A 91:12008–12012. https://doi.org/10.1073/pnas.91.25.12008

14. Chang YC, Penoyer LA, Kwon-Chung KJ (1996) The second capsule gene of *Cryptococcus neoformans*, CAP64, is essential for virulence. Infect Immun 64:1977–1983. https://doi.org/10.1128/iai.64.6.1977-1983.1996
15. Salas SD, Bennett JE, Kwon-Chung KJ et al (1996) Effect of the laccase gene CNLAC1, on virulence of *Cryptococcus neoformans*. J Exp Med 184:377–386
16. Odom A, Muir S, Lim E et al (1997) Calcineurin is required for virulence of *Cryptococcus neoformans*. EMBO J 16:2576–2589
17. Idnurm A, Reedy JL, Nussbaum JC et al (2004) *Cryptococcus neoformans* virulence gene discovery through insertional mutagenesis. Eukaryot Cell 3:420–429
18. Fraser JA, Subaran RL, Nichols CB et al (2003) Recapitulation of the sexual cycle of the primary fungal pathogen *Cryptococcus neoformans* var. *gattii*: implications for an outbreak on Vancouver Island, Canada. Eukaryot Cell 2:1036–1045. https://doi.org/10.1128/EC.2.5.1036-1045.2003
19. Arras SD, Chitty JL, Blake KL et al (2015) A genomic safe haven for mutant complementation in *Cryptococcus neoformans*. PLoS One 10:e0122916. https://doi.org/10.1371/journal.pone.0122916

Chapter 6

Genetic Transformation of *Cryptococcus* Species with *Agrobacterium* Transfer DNA

Nicholas F. Chong, Alexander Idnurm, and Bridgit C. Nugent

Abstract

Transformation of foreign DNA into *Cryptococcus* species is a powerful tool for exploring gene functions in these human pathogens. *Agrobacterium tumefaciens*-mediated transformation (*At*MT) has been used for the stable introduction of exogenous DNA into *Cryptococcus* for over two decades, being particularly impactful for insertional mutagenesis screens to discover new genes involved in fungal biology. A detailed protocol to conduct this transformation method is provided in the chapter. Scope for modifications and the benefits and disadvantages of using *At*MT in *Cryptococcus* species are also presented.

Key words *Agrobacterium*, Complementation, *Cryptococcus*, Forward genetics, Genetic modification, Mutant screens

1 Introduction

The ability to insert foreign DNA into the genome of an organism is a powerful tool for altering its biology, with a suite of potential applications ranging from making fundamental discoveries about gene functions through to the generation of commercial lines featuring genetic modifications. One method used extensively by plant molecular biologists is *Agrobacterium tumefaciens*-mediated transformation (*At*MT), a system adopted from a natural transconjugation process that occurs during the crown gall disease caused by this bacterium infecting plants [1]. The technique was subsequently adopted for use in numerous fungal species, as described in detail in review articles [2, 3].

*At*MT was developed for *Cryptococcus* species two decades ago [4, 5]. Since then the method has been particularly useful for insertional mutagenesis screens, that is, making strains with transfer DNA (T-DNA) integrated at random positions in the genome, screening the strains for phenotypes of interest, and then identifying the mutated genes based upon knowledge about the inserted

Erin E. McClelland (ed.), *Cryptococcus neoformans: Methods and Protocols*, Methods in Molecular Biology, vol. 2775, https://doi.org/10.1007/978-1-0716-3722-7_6,

DNA (*see* **Note 1**). Just one example of this application of insertional mutagenesis is the extensive screening for genes that impact the transition between yeast and filamentous forms [6, 7]. Another use is in generating complementation strains, in which mutants that were created by other means receive a wild-type copy of the gene, creating control strains to confirm the mutation, and therefore the gene is associated with the phenotype of interest.

The *At*MT method attracted attention when it was developed because it was the first transformation technique that reliably resulted in the stable integration of DNA into *Cryptococcus* species. Prior to that time, the other methods for DNA delivery, by electroporation or biolistics, yielded a high proportion of colonies that were not stable, i.e., the DNA had entered the cells, but had not integrated into the fungus' genomic DNA, so it would be subsequently lost when selection for transformation was removed from cultures [8, 9]. A more recent technique for gene manipulation is gene editing, e.g., using CRISPR-Cas9, whose components do not require integration into the genome, although the editing components may also be delivered using *At*MT [10]. Nonetheless, *At*MT still has potential uses in the routine creation of transgenic *Cryptococcus* strains.

2 Materials

2.1 Microbial Strains

1. A *Cryptococcus* strain, likely a wild type or a mutant that requires complementation.
2. An *Agrobacterium tumefaciens* "helper" strain. These are designed to carry a plasmid encoding virulence proteins, in the sense of virulence being the ability to transfer DNA into a host cell, and are then transformed with a second plasmid containing the DNA to be integrated into the fungus between 25 bp repeats, known as the "left" and "right" borders (*see* **Note 2**).

2.2 Chemicals or Other Reagents

1. For media for culturing bacteria and *Cryptococcus*: yeast extract, peptone, tryptone, glucose, NaCl, and agar.
2. For induction medium: KH_2PO_4, K_2HPO_4, NaCl, $MgSO_4{\cdot}7H_2O$, $CaCl_2{\cdot}2H_2O$, $FeSO_4{\cdot}7H_2O$, and $(NH_4)_2SO_4$ and agar.
3. Glycerol (100% and 10% glycerol v/v in water sterilized by autoclaving).
4. Dimethyl sulfoxide (DMSO).
5. Glucose.
6. Acetosyringone (3′,5′-dimethoxy-4-hydroxyacetophenone).
7. 5 M KOH: 14.02 g dissolved in 50 mL H_2O.

8. 2-(*N*-morpholino)ethanesulfonic acid (MES).
9. Antibiotics for the selection of bacteria containing plasmids and transformed *Cryptococcus*: kanamycin (50 mg/mL 1000× stock), cefotaxime (100 mg/mL 500× stock), *Cryptococcus* selection (e.g., nourseothricin, G418, or hygromycin; 100 mg/mL 1000× stock). Weigh the chemicals, dissolve in a suitable proportion of water (e.g., 5 g kanamycin in 10 mL water or 10 g of the other chemicals in 10 mL water), allow to dissolve, then sterilize through a 0.22 μm filter using a syringe, and aliquot as 1 mL volumes in sterile Eppendorf tubes. Store at −20 °C until needed. Some of these antimicrobial agents are available commercially as sterile solutions.

2.3 Media (see Note 3)

1. Lysogeny broth (LB): Dissolve 5 g yeast extract, 10 g tryptone, and 10 g NaCl in 1000 mL water. If making agar plates, add 20 g agar. Autoclave and allow to cool to 50 °C. For the selection of either *E. coli* or *A. tumefaciens* carrying plasmids, add an antibiotic to select for the plasmid, e.g., 50 μg/mL kanamycin. Store LB containing antibiotics at 4 °C.
2. Yeast extract–peptone–dextrose (YPD): Dissolve 10 g yeast extract and 20 g peptone in 900 mL water. If making agar plates, add 20 g agar. Dissolve 20 g glucose in 100 mL water. Autoclave both solutions, allow to cool to 50 °C, and then add the glucose solution to the larger volume. For agar plates, pour approximately 25 mL per 9 cm diameter Petri dish. To modify for the selection of transformants, after mixing the two solutions, add cefotaxime to a final concentration of 200 μg/mL to select against *Agrobacterium* and either G418, hygromycin, or nourseothricin (100 μg/mL) to select for transformed *Cryptococcus*.
3. Bacterial recovery medium (SOC, for super optimal broth with catabolite repression). Purchase or make as described as in reference [11]. That is, add 20 g tryptone, 5 g yeast extract, 0.5 g NaCl, and 10 mL of 250 mM KCl into a 1 L Schott bottle. Dissolve in a final volume of 975 mL water, and then autoclave. Cool to room temperature. Then, add 5 mL of 2 M $MgCl_2$ and 20 mL of 1 M glucose, both of which may be sterilized either by autoclaving or passing through a 0.22 μm filter.
4. Induction medium: First make a 2.5× stock of salts, made in 2 L H_2O: KH_2PO_4 (7.25 g), K_2HPO_4 (10.25 g), NaCl (0.75 g), $MgSO_4{\cdot}7H_2O$ (2.5 g), $CaCl_2{\cdot}2H_2O$ (0.33 g), $FeSO_4{\cdot}7H_2O$ (0.0124 g), and $(NH_4)_2SO_4$ (2.5 g). Use a magnetic stir bar to keep the salts in suspension as the solution will be cloudy. It is stable at room temperature. For the induction medium, mix 400 mL 2.5× salts, 0.9 g glucose, 5 mL

glycerol, and 20 g agar into 525 mL water. Autoclave and cool to 50 °C. Make the pH buffer, with 7.7 g MES in 40 mL water, adjust pH to 5.3 with drops of 5 M KOH, and then make up to 50 mL volume with water. Dissolve 19 mg acetosyringone in 250 μL DMSO. Add the acetosyringone in DMSO to the MES solution. Filter sterilize through a 0.22 μm filter directly into the molten agar. Mix and pour into 9 cm diameter Petri dishes.

2.4 Plasticware and Glassware

1. Disposable pipette tips.
2. 9 cm Petri dishes.
3. 14 mL tubes.
4. 50 mL Falcon tubes.
5. 0.2 mm gap electroporation cuvette.
6. Syringes (10 mL and 50 mL).
7. 0.22 μm filters.
8. 100 mL and 1 L Schott bottles.
9. Sterilized Eppendorf (1.5 mL) tubes.
10. Plastic spatula/cell spreading tool or 4 mm glass beads.
11. Cuvettes for spectrometer (optional).

2.5 Equipment

1. Electroporator.
2. Pipettes.
3. Incubators, static, and shaking, set between 22 °C and 30 °C.
4. Centrifuges (micro, ideally chilled to 4 °C, and benchtop for 50 mL tubes).
5. pH meter (alternatively use pH indicator paper).
6. Spectrometer (optional).

3 Methods

This protocol uses microbiology techniques, so it is assumed that the laboratory will have standard microbiology equipment and regulatory approvals for working under aseptic conditions and with a human pathogenic microbe (*see* **Note 1**).

3.1 Transform *Agrobacterium* Strain with a Plasmid Carrying a Selectable Marker Between the Left and Right Borders of T-DNA

1. Culture the *Agrobacterium* helper strain (e.g., EHA105) on a 9 cm diameter LB agar plate for 2 days.
2. Pre-chill an electroporation cuvette on ice.
3. Scrape the bacterial cells from approximately 1/8 of the plate, and place into a pre-chilled Eppendorf tube containing 1 mL ice-cold 10% glycerol. Centrifuge for 1 min at maximum speed (13,000 rpm/ ~15,000 × *g* to pellet cells), remove the

supernatant, resuspend in 1 mL ice-cold 10% glycerol, and centrifuge again. Remove the supernatant and resuspend in 250 μL ice-cold glycerol. Keep cells on ice.

4. Add 50 μL *Agrobacterium* cells and 1 μL plasmid between the cuvette electrodes.
5. Electroporate as for *Escherichia coli*: 2500 V, 200 Ω, and 25 μF. The time constant will be 5.6–5.9 ms.
6. Recover the *Agrobacterium* for 2 h in 1 mL SOC, shaking at 250 rpm at 22 °C–28 °C.
7. Plate on LB + antibiotic medium (e.g., 50 μg/mL kanamycin).
8. Incubate at 22–28 °C, with colonies appearing in 2 days (*see* **Note 4**).

3.2 Culture Agrobacterium and Cryptococcus Strains

1. Starting with multiple colonies, culture the *Agrobacterium* carrying the plasmid with the T-DNA for transformation in 2 mL of LB medium plus antibiotic (e.g., kanamycin) in a 14 mL tube at 20–28 °C in a shaking incubator or equivalent device for agitation.
2. Culture the *Cryptococcus* strain overnight in 2 mL of YPD in a 14 mL tube at 22–30 °C in a shaking incubator or equivalent device for agitation.
3. Dilute the overnight cultures of both organisms with sterile water until opaque, yet it remains possible to see the test tube volume measurement lines through the suspension, or use a spectrometer to obtain an optical density between 0.6 and 1.0 at 600 nm.
4. Co-culture the *Agrobacterium* and *Cryptococcus* on induction medium, using four combinations of concentrations, as below, mixed first in a sterile Eppendorf tube:
   - 100 μL *Agrobacterium* cells and 100 μL *Cryptococcus*.
   - 10 μL *Agrobacterium* cells and 100 μL *Cryptococcus* + 90 μL H_2O.
   - 100 μL *Agrobacterium* cells and 10 μL *Cryptococcus* + 90 μL H_2O.
   - 10 μL *Agrobacterium* cells and 10 μL *Cryptococcus* + 180 μL H_2O.

   Pipette all 200 μL of the mixtures onto an induction medium plate, without spreading (Fig. 1). Allow the inoculum to evaporate.
5. Incubate the plates, without inverting them and unsealed by Parafilm® M, at 20–24 °C for 2 or 3 days in a dark incubator (*see* **Note 5**).

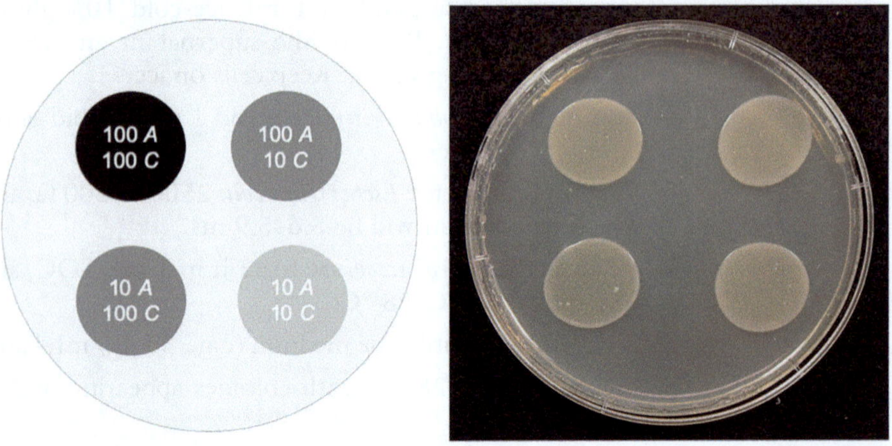

Fig. 1 Plating strategy to create different ratios of *Cryptococcus* (*C*) and *Agrobacterium* (*A*) cells, as described in Subheading 3.2, **step 4**. Four combinations are inoculated onto a single plate, illustrated in the diagram on the left where the intensity of shading reflects the overall cell density and a photograph of a plate on the right after 3 days coincubation on induction medium

3.3 Selection of Transgenic Cryptococcus *Strains*

The selection of transformed strains is illustrated in Fig. 2, with the following section describing the processes involved.

1. Add 10 mL of YPD to the induction medium plate, and lightly scrape the surface with the four bacterial and fungal co-cultures with a sterile disposable plastic tool or spatula. Transfer the liquid, approximately 9 mL, into a 50 mL Falcon tube.
2. Centrifuge the cell mixture (2000 rpm/ ~900× *g*, 5 min), which will pellet the *Cryptococcus* cells while leaving much of the *Agrobacterium* cells in suspension. Pour off the supernatant, and resuspend the pelleted cells in 1 mL of YPD.
3. The cell suspension may be serially diluted (e.g., 1×, 0.1×, and 0.01× concentrations), and then 300 μL should be plated and spread onto YPD medium supplemented with cefotaxime (200 μg/mL) to inhibit the *Agrobacterium* growth and an agent to select for the transformed *Cryptococcus* (e.g., 100 μg/mL nourseothricin, G418, or hygromycin) (*see* **Notes 6** and 7).
4. Transformants will appear in 2–3 days after incubation at 30 °C.
5. Using a microbiology inoculation loop or toothpicks, streak transformants onto fresh selection media to isolate colonies derived from a single cell and isolated away from any residual *Agrobacterium*. After this point, antibiotic selection is no longer needed as the T-DNA will have integrated into the genome (*see* **Notes 8** and **9**).

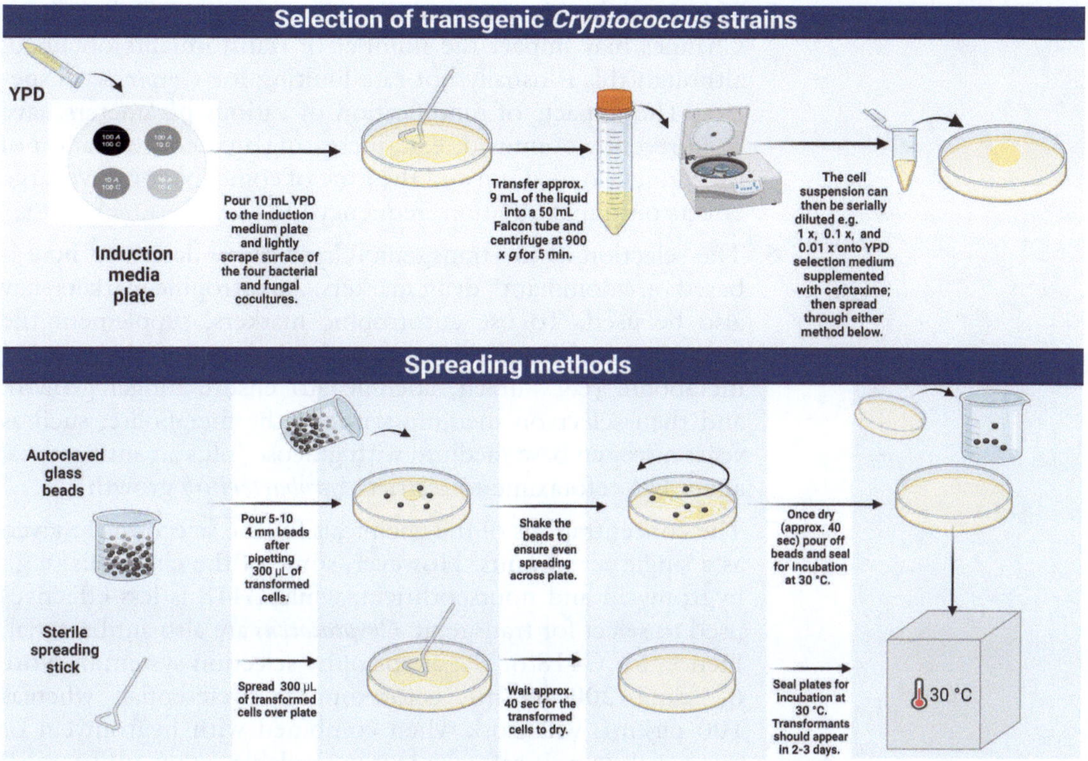

Fig. 2 Strategy to generate transformed *Cryptococcus* strains, with two methods for spreading the cells onto medium for the selection of transformed cells after the coincubation step. (Figure created with BioRender.com)

4 Notes

1. *At*MT results in the generation of transgenic organisms. The regulations associated with this technique vary around the world. It is essential to ensure that the necessary approvals, facilities, and equipment are acquired prior to performing the method.
2. There are different *Agrobacterium* strains and plasmids that may be used for transformation; one review of the plasmids available is reference [12]. This chapter does not cover aspects about these vectors; however, several are available from the Fungal Genetics Stock Center (Kansas State University, USA).
3. The volumes for media preparation are given for liters. For small numbers of transformants, a liter of media is more than needed, so the proportions may be reduced accordingly.
4. *Agrobacterium*-carrying plasmids are generally less stable than the equivalent *E. coli* with the same plasmid. Hence, if the *Agrobacterium* strain will be used again in the future, save an aliquot mixed with an equal volume of sterile 30% glycerol in a cryotube at −80 °C.

5. This is a flexible protocol that is amenable to alterations. Changes may impact the number of transformants obtained, although this is usually not rate limiting for *Cryptococcus* species. The impacts of modification of various parameters have been recently examined, e.g., increasing agar concentration of the induction medium and the time of coincubation have large effects on transformation frequency [13].
6. The selection of the transgenic *Cryptococcus* described here is based on "dominant" drug markers. Auxotrophic markers may also be used. To use autotrophic markers, supplement the induction medium with a suitable amount of the missing metabolite (e.g., uracil, adenine) to ensure fungal growth, and then select on medium without the metabolite, such as yeast nitrogen base medium with glucose (plus an antibacterial agent like cefotaxime to restrict *Agrobacterium* growth).
7. The concentrations of the chemicals for the selection are given as a single set of units. However, some of the chemicals (e.g., hygromycin and nourseothricin, while G418 is less effective) used to select for transgenic *Cryptococcus* are also antibacterial. Hence, for G418 or the auxotrophy selection system in **Note 6**, using 200 μg/mL cefotaxime is preferential, whereas 100 μg/mL will work when combined with hygromycin or nourseothricin. If cefotaxime is unavailable, other antibacterial chemicals may be used, being best if first tested on the *Agrobacterium* helper strain to make sure they are inhibitory and at which concentration.
8. To date, the integration of the T-DNA in *Cryptococcus* species appears exclusively through the nonhomologous end-joining pathway, i.e., targeted gene replacements are not possible. This contrasts to other fungal species where *At*MT may be used for both targeted and random integration of DNA [2]. While this is beneficial for insertional mutagenesis, for complementation there is the risk of random integration of T-DNA into the genome in that it may inadvertently "hit" a gene. A solution to this concern is with transformation using other methods wherein targeted integration occurs, such as targeting constructs to "save havens" within the genome, i.e., regions where it is established that integration of DNA has no impact on fungal growth, on development in vitro, or on virulence in animals [14–16]. For complementation experiments using *At*MT, a less laborious approach, especially for mutants with in vitro phenotypes, is to test multiple transformants, as the probability of insertion into the same region is negligible.
9. The integration of T-DNA is not completely random, with often a bias in inserting outside of coding regions. As illustrated in Idnurm et al. 2017 [2], of T-DNA insertions identified in

the *LAC1* gene encoding laccase for melanin synthesis, all insertions were within the promoter or 5′ untranslated region 1 kb upstream of the gene start codon rather than in the 2.8 kb coding region. This could result in mutations yielding alleles with reduced gene expression rather than a complete loss of function. However, this property has been leveraged to identify genes, using *At*MT insertional mutagenesis, with essential gene function that could not otherwise be identified, with the insertion of a promoter region regulated by carbon source (galactose vs. glucose) as part of the T-DNA [17].

Author Contributions

The authors are listed alphabetically by surname and contributed equally to the chapter. The first draft was prepared by AI. The protocol was tested by NFC and BCN, who edited the text and prepared the figures.

References

1. Gelvin SB (2003) *Agrobacterium*-mediated plant transformation: the biology behind the "gene-jockeying" tool. Microbiol Mol Biol Rev 67:16–37. https://doi.org/10.1128/MMBR.67.1.16-37.2003
2. Idnurm A, Bailey AM, Cairns TC et al (2017) A silver bullet in a golden age of functional genomics: the impact of *Agrobacterium*-mediated transformation of fungi. Fungal Biol Biotechnol 4:6. https://doi.org/10.1186/s40694-017-0035-0
3. Michielse CB, Hooykaas PJJ, van den Hondel CAMJJ et al (2005) *Agrobacterium*-mediated transformation as a tool for functional genomics in fungi. Curr Genet 48:1–17. https://doi.org/10.1007/s00294-005-0578-0
4. Idnurm A, Reedy JL, Nussbaum JC et al (2004) *Cryptococcus neoformans* virulence gene discovery through insertional mutagenesis. Eukaryot Cell 3:420–429. https://doi.org/10.1128/EC.3.2.420-429.2004
5. McClelland CM, Chang YC, Kwon-Chung KJ (2005) High frequency transformation of *Cryptococcus neoformans* and *Cryptococcus gattii* by *Agrobacterium tumefaciens.* Fungal Genet Biol 42:904–913. https://doi.org/10.1016/j.fgb.2005.07.003
6. Feretzaki M, Heitman J (2013) Genetic circuits that govern bisexual and unisexual reproduction in *Cryptococcus neoformans.* PLoS Genet 9:e1003688. https://doi.org/10.1371/journal.pgen.1003688
7. Lin X, Jackson JC, Feretzaki M et al (2010) Transcription factors Mat2 and Znf2 operate cellular circuits orchestrating opposite- and same-sex mating in *Cryptococcus neoformans.* PLoS Genet 6:e1000953. https://doi.org/10.1371/journal.pgen.1000953
8. Toffaletti DL, Rude TH, Johnston SA et al (1993) Gene transfer in *Cryptococcus neoformans* by use of biolistic delivery of DNA. J Bacteriol 175:1405–1411. https://doi.org/10.1128/jb.175.5.1405-1411.1993
9. Edman JC, Kwon-Chung KJ (1990) Isolation of the *URA5* gene from *Cryptococcus neoformans* var. *neoformans* and its use as a selective marker for transformation. Mol Cell Biol 10 (9):4538–4544:4538. https://doi.org/10.1128/mcb.10.9.4538-4544.1990
10. Li Z, Kim KS (2020) RELATe enables genome-scale engineering in fungal genomics. Sci Adv 6:eabb8783. https://doi.org/10.1126/sciadv.abb8783
11. Sambrook J, Russell DW (2001) Molecular cloning. A laboratory manual, vol 3, 3rd edn. Cold Spring Harbor Laboratory Press, New York
12. Frandsen RJN (2011) A guide to binary vectors and strategies for targeted genome modification in fungi using *Agrobacterium tumefaciens*-mediated transformation. J Microbiol Methods 87:247–262. https://doi.org/10.1016/j.mimet.2011.09.004

13. Fu J, Brockman NE, Wickes BL (2021) Optimizing transformation frequency of *Cryptococcus neoformans* and *Cryptococcus gattii* using *Agrobacterium tumefaciens.* J Fungi 7:520. https://doi.org/10.3390/jof7070520
14. Arras SDM, Chitty JL, Blake KL et al (2015) A genomic safe haven for mutant complementation in *Cryptococcus neoformans.* PLoS One 10(4):e0122916. https://doi.org/10.1371/journal.pone.0122916
15. Fan Y, Lin X (2020) An intergenic "safe haven" region in *Cryptococcus neoformans* serotype D genomes. Fungal Genet Biol 144:103464. https://doi.org/10.1016/j.fgb.2020.103464
16. Li Y, Pham T, Xie X et al (2022) Identification and characterization of an intergenic "safe haven" region in human fungal pathogen *Cryptococcus gattii.* J Fungi 8:178. https://doi.org/10.3390/jof8020178
17. Ianiri G, Boyce KJ, Idnurm A (2017) Isolation of conditional mutations in genes essential for viability of *Cryptococcus neoformans.* Curr Genet 63:519–530. https://doi.org/10.1007/s00294-016-0659-2

Chapter 7

Gene Silencing via RNA Interference in *Cryptococcus*

Indrani Bose

Abstract

RNA interference (RNAi) is a molecular biology technique for silencing specific eukaryotic genes without altering the DNA sequence in the genome. The silencing effect occurs because of decreased levels of mRNA that then result in decreased protein levels for the gene. The specificity of the silencing is dependent upon the presence of sequence-specific double-stranded RNA (dsRNA) that activates the cellular RNAi machinery. This chapter describes the process of silencing a specific target gene in *Cryptococcus* using a dual promoter vector. The plasmid, pIBB103, was designed with two convergent *GAL7* promoters flanking a *ura5* fragment that acts as a reporter for efficient RNAi. The target gene fragment is inserted between the promoters to be transcribed from both directions leading to the production of dsRNA in cells that activate the RNAi pathway.

Key words RNA interference, Gene silencing, Convergent promoters, Dual promoters, *Cryptococcus*

1 Introduction

RNA interference (RNAi) is a conserved mechanism that modulates gene expression through posttranscriptional silencing of target genes by small noncoding RNAs. Although the mechanism was first described in the nematode *Caenorhabditis elegans* [1], this pathway has now been shown to be present in diverse eukaryotic organisms, including a wide variety of fungi [2–5]. RNAi was initially thought to provide protection against viruses and transposons [6], but is now known to affect many cellular pathways. In *Cryptococcus* sp., it has been shown to affect processes as diverse as genome defense [7], centromere maintenance [8], and fluconazole resistance [9].

The RNAi process is activated by the introduction of double-stranded RNA (dsRNA) in cells, either naturally by synthesized oligonucleotides, or by expression from an introduced construct. dsRNA is recognized and cleaved by the RNase III family ribonuclease, Dicer, into short 20–25 base pair (bp) fragments called "small interfering RNAs" or siRNAs [10]. The siRNAs are incorporated into the RNA-induced silencing complex (RISC),

Erin E. McClelland (ed.), *Cryptococcus neoformans: Methods and Protocols*, Methods in Molecular Biology, vol. 2775, https://doi.org/10.1007/978-1-0716-3722-7_7,

which includes the Argonaute protein [11, 12]. This complex then uses the siRNA to identify cognate mRNA, targeting them for degradation or transcriptional repression leading to "silencing" of genes in a sequence-specific manner [13, 14]. Many fungi have the core components of the RNAi pathway [5]. *Cryptococcus* spp. have two Dicer, one to two Argonaute, and a single RNA-dependent RNA polymerase (*RDP1*) gene that are essential to the RNAi process [15, 16] (*see* **Note 1**); deletion of these genes in cryptococcal cells abrogates silencing of target genes [15, 17, 18]. The RNA-dependent RNA polymerase (RdRP) is thought to amplify the siRNAs in a positive feedback loop to prolong the RNAi effect and probably helps in the heritability of this mechanism [19–21].

Over the years, this pathway has been developed as an effective laboratory tool and utilized for targeting genes for repression in a sequence-dependent manner in many organisms, including in many fungi and mushrooms [22–32]. Liu et al. (2002) first performed RNAi in *C. deneoformans* to repress the *CAP59* and *ADE2* genes using a single promoter construct. An approximately 500 bp fragment of the gene to be silenced was cloned into the construct twice, in opposite orientation, around a small spacer that led to the transcription of a short hairpin RNA (shRNA) [22]. Later, Panepinto et al. (2005) modified this technique by using an intron from the *LAC1* gene as the spacer [25]. This allowed the spacer to be maintained within the plasmid in bacterial cells but was subsequently spliced out of the hairpin RNA to produce a dsRNA in cryptococcal cells. Although these systems worked well for a variety of genes, they required two cloning steps, with the fragments inserted in a specific configuration, making the preparation of the RNAi vector somewhat time-consuming. To simplify the construction of the silencing vector for a target gene in *Cryptococcus*, this system was updated with the discovery that a dual promoter construct could be used to generate the dsRNA required for RNAi [18, 23]. Although shRNA constructs have mainly been used for RNAi in a number of fungi [24–28], there are now examples where a convergent dual promoter vector has been effectively used for gene silencing [29–32].

The advantage of using RNAi in *Cryptococcus* is that it is effective as a plasmid-borne system [18, 22, 26, 32]; genomic integration of the RNAi plasmid is not mandatory, although it may stabilize heritability of the phenotype. Depending upon the promoter used to drive transcription, the process may be constitutive (*GPD1*p or *ACT1*p) [33, 34] or inducible (*CTR4*p or *GAL7*p) [35–37]. As shown in other fungi [38, 39], this system is amenable for designing inserts that simultaneously target multiple genes of a family. In addition, the dual convergent promoter vector may be used in a high-throughput manner for screening the genome and to identify genes involved in a pathway [40].

Silencing by RNAi leaves the DNA sequence of a gene intact. A plasmid-based RNAi system, therefore, has inherent variability and is not as completely repressive as the deletion of a gene (a knockdown vs. a knockout). Differentially activated RNAi pathways in independent transformants lead to inconsistent levels of silencing, often resulting in incomplete phenotypic penetrance depending upon the levels and stability of the targeted mRNA [41]. To select for colonies with active RNAi, a "sentinel" gene system was developed where a second gene was repressed simultaneously to provide a read-out of the level of silencing obtained in a transformant [18, 22–25]. The sentinel gene was one whose expression could be readily determined, either visually as for *GFP* or *ADE2* or by viability as with the *URA5* gene [18, 42].

The use of long dsRNA for RNAi has the disadvantage of possible "off-target" silencing [43]. To mitigate this, a number of strategies should be adopted. Homology searches should be performed to ensure that unique exonic sequences are used to create the targeting vectors. Sequences with longer than 15 bases of continuous homology to a non-targeted region of the genome and low-complexity regions with homopolymer stretches of five or more should be avoided [44]. When possible, it should be determined whether different fragments from the same gene give the same mutant phenotype. As a control, either a plasmid with a nonspecific sequence or the RNAi vector without the target gene fragment should be used to verify the specificity of the phenotype [45].

This chapter details the protocol (Fig. 1b) for using the dual convergent promoter plasmid, pIBB103, for RNAi in *Cryptococcus.*

2 Materials

2.1 The pIBB103 Cn*RNAi* Plasmid

The pIBB103 plasmid is 8505 bp in length and contains an RNAi cassette consisting of two convergent *GAL7* promoters, each flanked by a *GAL7* terminator [36, 37] (*see* **Note 2**). This cassette is placed within *Kpn*I sites, making it transferable to other plasmids with different selectable markers, if necessary. The *GAL7* promoters flank a 252 bp fragment of the *URA5* gene from *C. deneoformans* [44] and unique *Spe*I and *Nde*I sites, either of which may be used to insert a polymerase chain reaction (PCR)-amplified fragment of the target gene (Fig. 1a). This feature allows for the simultaneous regulated transcription of sense and anti-sense RNA from the *ura5* fragment, as well as the target gene fragment in the presence of galactose, allowing for concomitant silencing of both the *URA5* gene and the target gene in cryptococcal cells [18]. Cells eliciting a strong RNAi response silence the *URA5* gene and are able to survive on media containing 5-fluoroorotic acid (5-FOA) [46] (Fig. 2), whereas those that have a weak response are

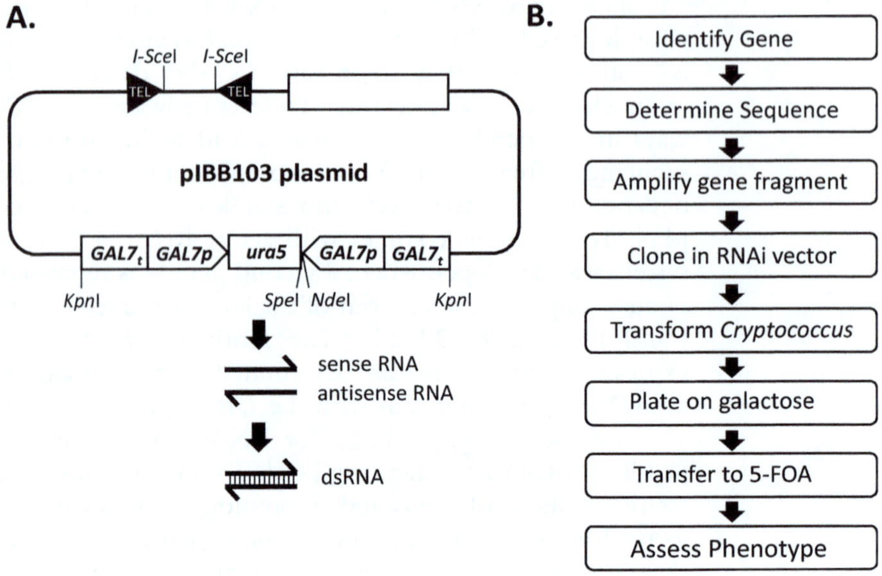

Fig. 1 A plasmid-based RNAi system for *Cryptococcus*. (**a**) Schematic of the RNAi vector, pIBB103, and the transcripts produced by two convergent *GAL7* promoters (*GAL7*p) flanked by the *GAL7* terminators ($GAL7_t$) to form dsRNA that activates the RNAi process. The telomeric sequences (TEL) are exposed upon digestion with the *I-Sce*I endonuclease, which linearizes the vector. The target gene fragment may be inserted in the unique *Spe*I or *Nde*I sites. (**b**) Overview of the steps of the RNAi process using the pIBB103 plasmid and a *ura5* gene fragment as a reporter. 5-FOA; 5-fluoroorotic acid

A.

Media	pIBB103	Essential	Non-essential
- 5-FOA	+++	++ / -	+++
+ 5-FOA	+++ no phenotype	-	+++ with phenotype

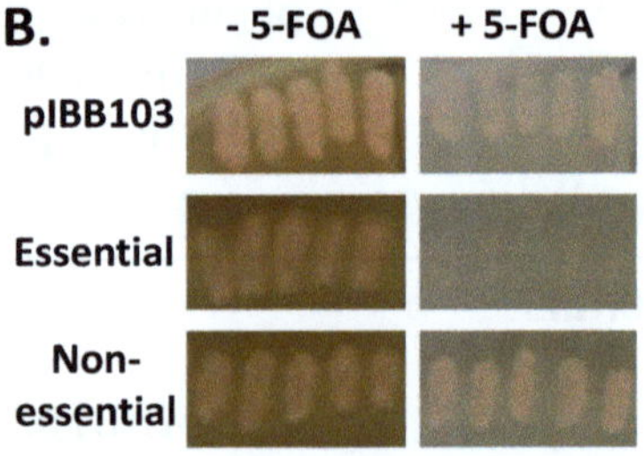

Fig. 2 Assessment of RNAi phenotype. (**a**) Summary of growth characteristics of transformant colonies containing the empty vector (pIBB103), the vector carrying a fragment of an "essential" gene, or the vector with a fragment of a "nonessential" gene on media without or with 5-fluoroorotic acid (5-FOA). The colonies with empty vector transcribe dsRNA from the *ura5* fragment and grow well on 5-FOA media. These colonies do not show a second phenotype. Transformants that contain a fragment from the *ura5* gene, as well as an essential gene, silence the *URA5* gene, but do not survive the knockdown of the target gene. If the target gene is nonessential for viability, the cells survive on 5-FOA and may be assessed for a second phenotype associated with the gene knockdown. [+++ = good growth, ++ = some growth, — = no growth]. (**b**) Example of the growth of colonies on media with and without 5-FOA

eliminated. Therefore, transformants with high levels of repression are usually selected for, thereby decreasing the amount of screening that is needed. A Geneticin (G418) resistance gene is present for selection in *Cryptococcus* sp. [47], and the stability of the plasmid in cryptococcal cells is enhanced by the fact that it contains telomeric repeats that can be exposed by digestion with the *I-Sce*I enzyme, converting the circular plasmid into a mini-chromosome [48]. This increases transformation efficiency and heritability of the plasmid.

2.2 General Instructions

1. Maintain bacteria and *Cryptococcus* cultures according to microbiological methods.
2. Prepare all microbial media using distilled water. Media without added antibiotics may be stored at room temperature (RT, 22–25 °C), while media containing antibiotics should be stored at 4 °C.
3. All molecular biology reactions should be performed using sterile deionized ultrapure water, such as that from a Milli-Q® Ultrapure System.

2.3 Inserting the Target Gene Fragment into the pIBB103 RNAi Plasmid

1. Access to the NCBI website (ncbi.nlm.nih.gov/) or FungiDB (fungidb.org).
2. A DNA sequence editor, such as SnapGene Viewer or ApE plasmid, is desirable.
3. Autoclaved sterile distilled deionized water, such as from a Milli-Q® Ultrapure System, for all molecular biology solutions and reactions.
4. 1.5 mL microfuge tubes.
5. 0.2 mL PCR tubes.
6. 13 mL culture tubes.
7. 90–100 mm petri dishes.
8. PCR reagents. Use thermostable proofreading polymerase and buffer to avoid the incorporation of unintended mutations in the sequence.
9. Thermocycler.
10. PCR purification kit (e.g., QIAquick® PCR Purification Kit from Qiagen or Wizard® SV Gel and PCR Clean-up System from Promega).
11. Restriction enzymes *Nde*I or *Spe*I, with their corresponding buffers (*see* **Note 3**).
12. *I-Sce*I with its corresponding buffer.
13. A ligation amount calculator, such as NEBioCalculator at New England Biolabs (https://nebiocalculator.neb.com/#!/ligation) or Ligation Calculator from Omni Calculator (https://www.omnicalculator.com/biology/ligation), is optional but helpful.
14. Ligation reagents: T4 DNA ligase and its corresponding buffer.

15. DNA gel electrophoresis apparatus.
16. Low-melt agarose.
17. Tris–acetate–EDTA (TAE) or Tris–borate–EDTA (TBE) buffer. 10× TAE buffer: 2 M Tris base, 1 M acetic acid, and 50 mM EDTA disodium hydrate. 10× TBE buffer: 890 mM Tris–borate, 890 mM boric acid, and 20 mM EDTA. The pH of the buffers is adjusted to 8.3.
18. A nucleic acid stain, such as ethidium bromide, SYBR® Green, or GelRed® (Biotium).
19. DNA gel extraction kit, such as QIAquick (Qiagen) or Wizard® SV Gel extraction Kit (Promega).
20. *E. coli* competent cells for transformation, such as NEB® 5-alpha competent cells, either high efficiency or sub-cloning efficiency.
21. 1000× ampicillin stock solution: Dissolve 100 mg/mL ampicillin in water and filter sterilize. Prepare 250 or 500 μL aliquots in sterile microfuge tubes, and store at −20 °C. Aliquots may be refrozen and reused if kept on ice after thawing.
22. Luria Bertani (LB) media plates with filter-sterilized 100 μg/mL ampicillin solution: Mix 1% tryptone, 0.5% yeast extract, and 1% NaCl in distilled water. Add 2% (w/v) Bacto agar and autoclave for 20 min. Cool to approximately 55 °C, and add the filter-sterilized ampicillin solution to a final concentration of 100 μg/mL. Store at 4 °C.
23. Refrigerated chilled 70% and 100% (v/v) ethanol prepared with distilled water.

2.4 Transforming Cryptococcal Cells by Electroporation

1. 1.5 mL microfuge tubes.
2. 13 mL culture tubes.
3. Chilled 50 mL conical tubes.
4. Chilled 2 mm electroporation cuvettes, one per transformation.
5. Chilled, autoclaved ultrapure sterile distilled water. Store at 4 °C.
6. Chilled, filter-sterilized electroporation buffer: 10 mM Tris–HCl pH 7.5, 270 mM sucrose, 1 mM $MgCl_2$ in ultrapure water. Store at 4 °C.
7. 1 M dithiothreitol (DTT) prepared in ultrapure water and filter-sterilized. Store in 250 μL aliquots at −20 °C (*see* **Note 4**).
8. 1000× stock solution of G418: Add 100 mg/mL G418 Sulfate (Geneticin) in ultrapure distilled water and filter sterilize. Divide into 500 μL aliquots and store at −20 °C. Aliquots may be refrozen and reused if kept on ice after thawing.

9. Yeast extract–peptone–dextrose (YPD) broth: Add 1% yeast extract (w/v), 2% peptone (w/v), and 2% dextrose (w/v) to distilled water in flasks or bottles. Stir to mix and autoclave for 20 min. Store at RT or 4 °C.
10. Yeast extract–peptone–dextrose (YPD) plates with 100 μg/mL G418: Add 2% Bacto agar (w/v) to the above medium before autoclaving for 20 min. Cool to approximately 55 °C, and add the filter-sterilized G418 solution to a final concentration of 100 μg/mL before adding media to plates. Store at 4 °C.

2.5 Determining the RNAi Phenotype of Transformants

1. Yeast extract–peptone–galactose (YPG) liquid broth with 100 μg/mL G418: Add 1% yeast extract (w/v), 2% peptone (w/v), and 2% galactose (w/v) to distilled water in flasks or bottles. Stir to mix and autoclave for 20 min. Cool the autoclaved solution to approximately 55 °C, and add filter-sterilized G418 solution to a final concentration of 100 μg/mL. Store at 4 °C.
2. YPG agar plates with 100 μg/mL G418: Stir in 2% Bacto agar (w/v) to the YPG liquid broth above *before* autoclaving. Add the filter-sterilized G418 solution as above, and stir slowly to mix before pouring the plates. Store at 4 °C.
3. 20% galactose solution: Add 20 g galactose to 100 mL of ultrapure water. Stir to mix. Filter sterilize and store at RT.
4. 5-Fluoroorotic acid hydrate (5-FOA) solution: For a 10× solution, add 100 mg 5-FOA in 10 mL of distilled water with constant stirring, and warm to dissolve. Filter sterilize and use immediately.
5. Uracil dropout mix, such as the yeast synthetic dropout media supplement without uracil (e.g., from Sigma-Aldrich).
6. 100× uracil solution (2 mg/mL) in sterile distilled water. Warm solution to dissolve, if necessary. Store at RT.
7. Synthetic media plates containing 5-FOA and 100 μg/mL G418. To prepare 100 mL of media, add 0.67 g of yeast nitrogen base without amino acids, 80 mg of the uracil dropout mix, 600 μL 100× uracil solution, and 2 g Bacto agar to 75 mL of deionized water. Stir to combine. Autoclave the mixture for 20 min with the stir bar in the flask. After autoclaving, add 10 mL of filter-sterilized 20% galactose solution and 10 mL of filter-sterilized 5-FOA solution. Stir slowly to combine without creating bubbles. Cool to approximately 55 °C, and then add G418 solution to a final concentration of 100 μg/mL. Bring up to a final volume of 100 mL with autoclaved deionized water. Pour the plates immediately. Protect from light and store at 4 °C.
8. Autoclaved flat-tipped toothpicks.

2.6 Analysis of mRNA Levels of the Target Gene

1. TRIzol® reagent.
2. Chloroform.
3. Isopropanol.
4. RNase-free ultrapure water.
5. 70% (v/v) ethanol prepared with RNase-free water, chilled at 4 °C.
6. 0.4–0.6 mm glass beads and bead beater.
7. First-strand cDNA kit.

3 Methods

Inserting the Target Gene Fragment into the RNAi Plasmid, pIBB103.

3.1 Selection of the Sequence of the Gene to Be Silenced

1. Obtain the sequence of the gene to be silenced from a database, such as FungiDB or NCBI (*see* **Note 5**). Either gene name or accession number may be used to query these databases. In case of sequence homology, the BLAST function at NCBI may be used to determine the ortholog (*see* **Note 6**).
2. Choose a <500 bp exonic region (preferably 250–350 bp) to use in the pIBB103 plasmid (*see* **Note 7**). The sequence should not be either AT- or GC-rich and have few homopolymer stretches greater than 3 bp in length (*see* **Note 8**).
3. The sequence should be determined to have as few off-target matches as possible. Use the sequence of the fragment in the BLAST tool in NCBI or FungiDB (https://fungidb.org/fungidb/app/workspace/blast/new) to search the *Cryptococcus* genome for possible matches (*see* **Note 9**).
4. The sequence should not contain recognition sites for the restriction enzymes to be used for cloning or linearizing the finished plasmid, such as *Spe*I or *Nde*I, and *I-Sce*I (*see* **Note 10**).

3.2 Designing Primers to Amplify the Gene Fragment

1. The sense primer (S) should include 20–25 bases of the 5′ end of the top strand of the fragment, whereas the antisense primer (AS) should contain 20–25 bases of the 5′ end of the complementary strand.
2. Both primers should include a *Spe*I or *Nde*I overhang and approximately six random nucleotides at the 5′ end. The primers should read as follows:

   5′-(6 random nucleotides)-(*Spe*I or *Nde*I restriction site)-(20–25 bases of the 5′-end of the sequence to be amplified)-3′ (*see* **Note 11**).
3. The melting temperatures (T_m) of the two primers should be compatible, preferably within a couple of degrees of each other.

3.3 Amplification of the Sequence by PCR

1. Amplify the target sequence to be cloned into pIBB103 using primers S and AS in a 50–100 μL PCR reaction. The template may be either genomic DNA or cDNA for primers that fall within a single exon. cDNA should be used as the template when amplifying fragments that overlap two exons.
2. Run 2–5 μL of the PCR reaction (approximately 25–50 ng) on a 1–1.5% agarose gel to confirm a DNA band of the correct size (*see* **Note 12**).
3. Purify the PCR-amplified DNA using a PCR purification kit, according to the manufacturer's instructions. Elute in sterile ultrapure water, and quantify the purified DNA using a NanoDrop or a similar instrument.

3.4 Cloning the Target Sequence into the pIBB103 RNAi Plasmid

1. Prepare a miniprep of the RNAi vector, pIBB103, using any plasmid miniprep kit, according to the manufacturer's instructions. Plasmid DNA may also be prepared using the lab protocol in any molecular biology protocol book.
2. Set up digests with approximately 1 μg of vector and 1–2 μg of the purified PCR product using 1–2 μL (10–15 U) of either the *Spe*I or *Nde*I restriction enzymes (as incorporated in the primers) for 5 h to overnight at 37 °C.
3. Add 1 μL of calf-intestinal phosphatase or 2 μL of shrimp alkaline phosphatase to the vector reaction mix only, and incubate at 37 °C for an additional 30 min. *[Do not add phosphatase to the reaction with the PCR fragment.]*
4. Purify the restriction enzyme-digested and phosphatase-treated vector, as well as the enzyme-digested PCR fragment, using any PCR purification kit, according to the manufacturer's instructions. Elute the DNA in 30–35 μL nuclease-free or ultrapure water.
5. Quantify the purified and digested vector, and insert using a NanoDrop or similar methods.
6. Set up several ligation reactions with different vector DNA to insert DNA ratios, such as 1:0, 1:3, and 1:5 vector DNA/insert DNA (*see* **Note 13**). The ligation reaction should contain the appropriate amount of vector and insert fragments with ligation buffer, 1 μL T4 DNA ligase, and water up to a final volume of 20 μL. Incubate at RT for 30–60 min or overnight at 16 °C.
7. Store the ligation reactions on ice if it is used immediately for transformation or at −20 °C if it is to be used later.

3.5 Selecting Transformant Colonies with the Insert DNA

1. Transform 2–6 μL of the above ligations into *E. coli* competent cells, according to the manufacturer's protocol (*see* **Note 14**).
2. Plate 20% and 80% of the transformed cells onto LB media containing ampicillin (50–100 μg/μL), and incubate at 37 °C for approximately 16 h (*see* **Note 15**).

3. Select six to eight single colonies to be screened for the insert. Screening may be performed either (a) by digesting 2–3 μL of plasmids obtained from broth cultures of each colony with *Spe*I/*Nde*I or (b) by colony PCR from cells taken from the plate (*see* **Note 16**). The digest or PCR should be run on a 1–1.5% agarose gel for visualization.
4. To prepare the plasmid for electroporation into cryptococcal cells, digest approximately 2 μg of plasmid with 10–15 U of the *I-Sce*I restriction enzyme for 5–6 h at 37 °C. The presence of an 801 bp fragment in addition to a larger vector fragment is evidence of complete digestion. This helps to linearize the plasmid and expose the telomeric (*TEL*) sequences.

3.6 Transform into *Cryptococcus* by Electroporation

1. Grow the strain to be transformed on a fresh YPD agar plate at 30 °C for 2 days.
2. Inoculate a single colony from the above plate into 50 mL of YPD liquid broth in a 250 mL conical flask, and grow at 30 °C overnight with shaking at 200 rpm until the culture reaches a density of 3–5 × 10^7 cells/mL.
3. Dilute the culture with fresh YPD to 2 × 10^6 cells/mL. The amount of total culture needed should be determined by the fact that 40–50 mL of freshly diluted culture will be needed for each transformation.
4. Allow the above culture to reach a density between 6 and 10 × 10^6 cells/mL. Transfer to 50 mL conical tubes, and centrifuge at 4 °C for 5 min at 2000 ×*g*.
5. Discard the supernatant, and wash each pellet twice with 50 mL of ice-cold ultrapure distilled water by centrifuging as above. Make sure to resuspend the cell pellet thoroughly each time.
6. Resuspend the cell pellets after the two water washes in 5–10 mL of ice-cold electroporation buffer (EB), and pool them into one centrifuge tube. Add 200 μL of 1 M DTT. Invert 5–6 times to mix, bring up the volume to 50 mL with ice-cold EB, and incubate on ice for 10–15 min. Centrifuge as above and discard the supernatant. Keep the pellet on ice.
7. Wash the pellet by resuspending it in ice-cold EB (with no DTT), and centrifuge as above.
8. Decant the supernatant, and resuspend the pellet in ice-cold EB to a final volume of 1–2 mL. Keep the tube on ice.
9. Label one sterile 1.5 mL microcentrifuge tube per transformation and chill on ice. Include a tube for a water-only and a vector-only (pIBB103 without a target gene insert) control. Add enough of the above cell suspension for 3 × 10^8 cells/transformation. Centrifuge at 4 °C for 1 min at 2000 ×*g* and aspirate the supernatant.

10. For each tube, resuspend the pellet in 20 μL ice-cold EB. Add 1–5 μg of the linearized plasmid to be transformed in a final volume of 10 μL. Add 10 μL of sterile water for the water-only control tube. Mix the cells and DNA by flicking the tube.
11. Transfer the cell-DNA mixture to a chilled 0.2 cm electroporation cuvette.
12. Electroporation conditions are as follows: voltage 500 V, capacitance 25 μF, resistance 1000 Ω or more. Note the time constant (*see* **Note 17**).
13. Add 1 mL of room temperature YPD broth to the cuvette immediately after the pulse, and transfer to a 13–15 mL culture tube.
14. Incubate the tubes with shaking for 2 h at 30 °C. Pre-warm selective plates at 30 °C.
15. Centrifuge the tubes at RT for 1 min at 2000 ×*g*. Aspirate off most of the supernatant, leaving approximately 200 μL in the tube.
16. Resuspend the pellet in the remaining media, and plate differing quantities onto two plates (such as 10%/90%, 20%/80%, or 25%/75%). Incubate at 30 °C till colonies appear. This may take 2–7 days.

3.7 Determine the RNAi Phenotype of the Transformant Colonies

1. Pick eight to ten single colonies from the transformation plate, transfer onto a fresh YPG plate containing 100 μg/mL of G418, and incubate at 30 °C for 2 days or until colonies are sufficiently grown. [If replica plating is to be performed, then make small lines or dots for each colony that are efficiently transferable.]
2. Transfer or replica-plate onto (a) YPG plate containing 100 μg/mL G418 or (b) 5-FOA plate containing galactose and 100 μg/mL G418.
3. Incubate at 30 °C for 2 days or until growth is observed on the plates (Fig. 2).

3.8 Analysis of mRNA Levels of the Target Gene

1. Streak the transformants to be tested for single colonies onto YPG media supplemented with 5-FOA and G418 (as above).
2. Pick a single colony from each transformant, and inoculate into 5 mL of the above liquid medium. Let the culture grow overnight at 30 °C with shaking at 200–225 rpm.
3. Pellet 3 mL of the culture (1.5 mL × 2) in a 2 mL sterile screw cap tube at 500 ×*g* at 4 °C.
4. Resuspend the pelleted cells in 750 μL of TRIzol®, and add approximately 750 μL of sterile 0.4–0.6 mm glass beads to each tube.
5. Bead beat the tubes for eight intervals consisting of 1 min, followed by a 2 min incubation on ice.

6. Transfer the lysate to a fresh sterile microfuge tube. Wash the beads in the original tube with 750 μL of TRIzol®, and add to the second tube, bringing the pooled volume to approximately 1 mL.
7. Follow the manufacturer's instructions to complete the extraction of total RNA. Quantify using a NanoDrop or a similar method. The samples should be kept at −80 °C for long-term storage (*see* **Note 18**).
8. Prepare cDNA from the RNA samples using a first-strand synthesis kit following the manufacturer's instructions.
9. Use 1 μL of cDNA from each sample and the S and AS primers used for cloning the insert to perform a PCR reaction to determine the level of expression of the target gene (*see* **Note 19**).

4 Notes

1. The funRNA website (http://funrna.riceblast.snu.ac.kr/) is a platform that curates the putative Argonaute, Dicer, and RdRP genes from the fungal kingdom.
2. pIBB103 plasmid: GenBank accession number HQ455038.
3. Use either *Spe*I or *Nde*I to insert the target gene fragment into the pIBB103 plasmid. Directional cloning is not required since the fragment is transcribed from both directions by the convergent *GAL7* promoters.
4. For optimal transformation efficiency, DTT aliquots should not be refrozen and reused after they are thawed.
5. Sequences of several strains of *C. neoformans* may be found at fungidb.org or through NCBI at http://www.ncbi.nlm.nih.gov/genomes/leuks.cgi.
6. Most of the resources for designing RNAi sequences are based on short siRNAs [43, 49]. However, some resources, such as the E-RNAi site https://www.dkfz.de/signaling/e-rnai3/idseq.php, have applications for designing long dsRNA for RNAi [50].
7. The total length of the dsRNA transcribed from the promoters should be kept to approximately 500 bp, if possible. Fragments longer than 500 bp lead to loss of silencing efficiency (Bose, unpublished data). Since the *ura5* fragment is 252 bp long, the target gene fragment should be approximately 250–300 bp in length for an optimal RNAi response.
8. Homopolymer stretches may lead to off-target effects, possibly due to misalignment.

9. Sequences that have at least 15–18 bp homology to an off-target sequence with few mismatches (approximately 1–3 bases) should be avoided, if possible.
10. It is preferable to choose more than one fragment from different parts of the gene to test since there may be variability in their effectiveness. If possible, at least one of the fragments should be from the 5′ end of the gene, avoiding the first 50–100 bases. If a unique sequence of the desired length is not readily available within an exon, partial exonic sequences may be used, with the caveat that the insert fragment should be amplified from cDNA for cloning.
11. The extra nucleotides at the 5′ end of the primers ensure efficient digest of the PCR fragments by restriction enzymes. See "Cleavage Close to the End of DNA Fragments" at neb.com.
12. To make a 1% agarose gel, add 1 g molecular grade agarose to 100 mL of 1× TAE or TBE buffer. Boil for 1–2 min until the solution is clear, with no flakes or lumps detectable when held up to the light. Add a DNA visualization dye, according to the manufacturer's instructions.
13. The NEBioCalculator tool at neb.com or a similar tool, such as the Ligation Calculator at Omni Calculator, may be used for convenience.
14. The amount of ligation reaction to be transformed into competent cells, as well as the volume of transformed competent cells to be plated, depends both upon the efficiency of the ligation reaction and that of the uptake of DNA during transformation. High-efficiency competent cells will require smaller amounts of ligation reaction (~2 μL) for transformation and a smaller amount of the transformation to be plated (~50 μL of the transformation). Sub-cloning efficiency competent cells may need a higher amount of ligation reaction (5–6 μL) for the transformation and the entire transformation to be plated on a single plate.
15. Ampicillin is degraded by beta-lactamase enzymes produced by plasmid-containing cells. Therefore, longer incubations at 37 °C will lead to the formation of satellite colonies and may make the identification of the original transformant difficult. The satellite colonies do not contain a plasmid and will not grow in liquid media containing ampicillin.
16. Plasmids from broth cultures often provide cleaner and more reliable results than colony PCR, and the remaining liquid culture for the colony (or colonies) with an insert may be frozen for stock cultures.

17. Time constants of over 20 ms result in better transformation efficiencies.
18. RNA samples often degrade after a few months, even at −80 °C. The best results are obtained with samples that have been prepared within a few weeks.
19. Northern blots or qPCRs may also be performed, as necessary.

References

1. Fire A, Xu S, Montgomery MK et al (1998) Potent and specific genetic interference by double-stranded RNA in *Caenorhabditis elegans*. Nature 391(6669):806–811. https://doi.org/10.1038/35888
2. Ghildiyal M, Zamore PD (2009) Small silencing RNAs: an expanding universe. Nat Rev Genet 10(2):94–108. https://doi.org/10.1038/nrg2504
3. Lax C, Tahiri G, Patiño- Medina JA et al (2020) The evolutionary significance of RNAi in the fungal kingdom. Int J Mol Sci 21(24): 9348. https://doi.org/10.3390/ijms21249348
4. Nicolás FE, Garre V (2016) RNA interference in fungi: retention and loss. Microbiol Spectr 4(6). https://doi.org/10.1128/microbiolspec.FUNK-0008-2016
5. Nakayashiki H, Kadotani N, Mayama S (2006) Evolution and diversification of RNA silencing proteins in fungi. J Mol Evol 63(1):127–135. https://doi.org/10.1007/s00239-005-0257-2
6. Malone CD, Hannon GJ (2009) Small RNAs as guardians of the genome. Cell 136(4): 656–668. https://doi.org/10.1016/j.cell.2009.01.045
7. Dumesic PA, Natarajan P, Chen C et al (2013) Stalled spliceosomes are a signal for RNAi-mediated genome defense. Cell 152(5): 957–968. https://doi.org/10.1016/j.cell.2013.01.046
8. Yadav V, Sun S, Billmyre RB et al (2018) RNAi is a critical determinant of centromere evolution in closely related fungi. Proc Natl Acad Sci U S A 115(12):3108–3113. https://doi.org/10.1073/pnas.1713725115
9. Ma X, Li C, Ma L et al (2023) RNAi machinery regulates nutrient metabolism and fluconazole resistance in the pathogenic fungus *Cryptococcus deneoformans*. Med Mycol 61(2):myac095. https://doi.org/10.1093/mmy/myac095
10. Zamore PD, Tuschl T, Sharp PA et al (2000) RNAi: double-stranded RNA directs the ATP-dependent cleavage of mRNA at 21 to 23 nucleotide intervals. Cell 101(1): 25–33. https://doi.org/10.1016/S0092-8674(00)80620-0
11. Hammond SM, Bernstein E, Beach D et al (2000) An RNA-directed nuclease mediates post-transcriptional silencing in *Drosophila* cells. Nature 404(6775):293–296. https://doi.org/10.1038/35005107
12. Rivas FV, Tolia NH, Song JJ et al (2005) Purified Argonaute2 and an siRNA form recombinant human RISC. Nat Struct Mol Biol 12(4): 340–349. https://doi.org/10.1038/nsmb918
13. Martinez J, Patkaniowska A, Urlaub H et al (2002) Single-stranded antisense siRNAs guide target RNA cleavage in RNAi. Cell 110(5):563–574. https://doi.org/10.1016/s0092-8674(02)00908-x
14. Meister G, Landthaler M, Patkaniowska A et al (2004) Human Argonaute2 mediates RNA cleavage targeted by miRNAs and siRNAs. Mol Cell 15:185–197. https://doi.org/10.1016/j.molcel.2004.07.007
15. Burke JE, Longhurst AD, Natarajan P et al (2019) A non-dicer RNase III and four other novel factors required for RNAi-mediated transposon suppression in the human pathogenic yeast *Cryptococcus neoformans*. G3 9(7): 2235–2244. https://doi.org/10.1534/g3.119.400330
16. Choi J, Kim K, Jeon J et al (2014) funRNA: a fungi-centered genomic platform for genes encoding key components of RNAi. BMC Genomics 15(Suppl 9):S14. https://doi.org/10.1186/1471-2164-15-S9-S14
17. Janbon G, Maeng S, Yang DH et al (2010) Characterizing the role of RNA silencing components in *Cryptococcus neoformans*. Fungal Genet Biol 47(12):1070–1080. https://doi.org/10.1016/j.fgb.2010.10.005
18. Bose I, Doering TL (2011) Efficient implementation of RNA interference in the pathogenic yeast *Cryptococcus neoformans*. J Microbiol Methods 86:156–159. https://doi.org/10.1016/j.mimet.2011.04.014
19. Zhang H, Gao J, Chen J et al (2022) RNA-dependent RNA polymerase could

extend the lasting validity period of exogenous dsRNA. Pest Manag Sci 78(11):4569–4578. https://doi.org/10.1002/ps.7076

20. Sijen T, Fleenor J, Simmer F et al (2001) On the role of RNA amplification in dsRNA-triggered gene silencing. Cell 107(4): 465–476. https://doi.org/10.1016/s0092-8674(01)00576-1
21. Lipardi C, Wei Q, Paterson BM (2001) RNAi as random degenerative PCR: siRNA primers convert mRNA into dsRNAs that are degraded to generate new siRNAs. Cell 107(3): 297–307. https://doi.org/10.1016/s0092-8674(01)00537-2
22. Liu H, Cottrell T, Pierini L et al (2002) RNA interference in the pathogenic fungus *Cryptococcus neoformans*. Genetics 160:463–470. https://doi.org/10.1093/genetics/160.2.463
23. Skowyra ML, Doering TL (2012) RNA interference in *Cryptococcus neoformans*. Methods Mol Biol 845:165–186. https://doi.org/10.1007/978-1-61779-539-8_11
24. Rappleye CA, Engle JT, Goldman WE (2004) RNA interference in *Histoplasma capsulatum* demonstrates a role for alpha-(1,3)-glucan in virulence. Mol Microbiol 53:153–165. https://doi.org/10.1111/j.1365-2958.2004.04131.x
25. Krajaejun T, Gauthier GM, Rappleye CA et al (2007) Development and application of a green fluorescent protein sentinel system for identification of RNA interference in *Blastomyces dermatitidis* illuminates the role of septin in morphogenesis and sporulation. Eukaryot Cell 6:1299–1309. https://doi.org/10.1128/EC.00401-06
26. Panepinto J, Liu L, Ramos J et al (2005) The DEAD-box RNA helicase Vad1 regulates multiple virulence-associated genes in *Cryptococcus neoformans*. J Clin Invest 115(3):632–641. https://doi.org/10.1172/JCI200523048
27. Liu C, Kang L, Lin M et al (2022) Molecular mechanisms by which the GATA transcription factor CcNsdD2 regulates the developmental fate of *Coprinopsis cinerea* under dark or light conditions. MBio 13(1):e03626-21. https://doi.org/10.1128/mbio.03626-21
28. Liu Z, Deng B, Yuan H et al (2022) Transcription factor FfMYB15 regulates the expression of cellulase gene FfCEL6B during mycelial growth of *Flammulina filiformis*. Microb Cell Factories 21(1):216. https://doi.org/10.1186/s12934-022-01932-z
29. Nguyen QB, Kadotani N, Kasahara S et al (2008) Systematic functional analysis of calcium-signalling proteins in the genome of the rice-blast fungus, *Magnaporthe oryzae*, using a high-throughput RNA-silencing system. Mol Microbiol 68(6):1348–1365. https://doi.org/10.1111/j.1365-2958.2008.06242.x
30. Mu D, Shi L, Ren A et al (2012) The development and application of a multiple gene co-silencing system using endogenous *URA3* as a reporter gene in *Ganoderma lucidum*. PLoS One 7(8):e43737. https://doi.org/10.1371/journal.pone.0043737
31. Zhang J, Chen H, Chen M et al (2017) Construction and application of a gene silencing system using a dual promoter silencing vector in *Hypsizygus marmoreus*. J Basic Microbiol 57(1):78–86. https://doi.org/10.1002/jobm.201600291
32. Reilly MC, Levery SB, Castle SA et al (2009) A novel xylosylphosphotransferase activity discovered in *Cryptococcus neoformans*. J Biol Chem 284(52):36118–36127. https://doi.org/10.1074/jbc.M109.056226
33. Varma A, Kwon-Chung KJ (1999) Characterization of the glyceraldehyde-3-phosphate dehydrogenase gene [correction of glyceraldehyde-3-phosphate gene] and the use of its promoter for heterologous expression in *Cryptococcus neoformans* a human pathogen. Gene 232:155–163. https://doi.org/10.1016/s0378-1119(99)00132-8
34. Cox GM, Rude TH, Dykstra CC et al (1995) The actin gene from *Cryptococcus neoformans*: structure and phylogenetic analysis. J Med Vet Mycol 33(4):261–266. https://doi.org/10.1080/02681219580000521
35. Ory JJ, Griffith CL, Doering TL (2004) An efficiently regulated promoter system for *Cryptococcus neoformans* utilizing the CTR4 promoter. Yeast 21(11):919–926. https://doi.org/10.1002/yea.1139
36. Wickes BL, Edman JC (1995) The *Cryptococcus neoformans* GAL7 gene and its use as an inducible promoter. Mol Microbiol 16(6): 1099–1109. https://doi.org/10.1111/j.1365-2958.1995.tb02335.x
37. Ruff JA, Lodge JK, Baker LG (2009) Three galactose inducible promoters for use in *C. neoformans var. grubii*. Fungal Genet Biol 46(1):9–16. https://doi.org/10.1016/j.fgb.2008.10.003
38. Lacroix H, Spanu PD (2009) Silencing of six hydrophobins in *Cladosporium fulvum*: complexities of simultaneously targeting multiple genes. Appl Environ Microbiol 75(2): 542–546. https://doi.org/10.1128/AEM.01816-08

39. Néstor G, González MA, González C et al (2017) Simultaneous silencing of xylanase genes in *Botrytis cinerea*. Front Plant Sci 8: 2174. https://doi.org/10.3389/fpls.2017.02174
40. Morris JC, Wang Z, Motyka S et al (2004) An RNAi-based genomic library for forward genetics in the African Trypanosome. In: Sohail M (ed) Gene silencing by RNA interference. CRC Press, Boca Raton. https://doi.org/10.1201/9780203489253.ch16
41. Espino J, González MA, González C et al (2014) Efficiency of different strategies for gene silencing in *Botrytis cinerea*. Appl Miccrobiol Biotechnol 98(22):9413–9423. https://doi.org/10.1007/s00253-014-6087-7
42. Edman JC, Kwon-Chung KJ (1990) Isolation of the *URA5* gene from *Cryptococcus neoformans* var. *neoformans* and its use as a selective marker for transformation. Mol Cell Biol 10(9):4538–4544. https://doi.org/10.1128/mcb.10.9.4538-4544.1990
43. Chen J, Peng Y, Zhang H et al (2021) Off-target effects of RNAi correlate with the mismatch rate between dsRNA and non-target mRNA. RNA Biol 18(11):1747–1759. https://doi.org/10.1080/15476286.2020.1868680
44. Jain CK, Wadhwa G (2017) Computational tools: RNA interference in fungal therapeutics. Curr Trends Bioinformat Insight 2017:207–225. https://doi.org/10.1007/978-981-10-7483-7_12
45. Cottrell TR, Doering TL (2003) Silence of the strands: RNA interference in eukaryotic pathogens. Trends Microbiol 11:37–43. https://doi.org/10.1016/s0966-842x(02)00004-5
46. Boeke JD, Trueheart J, Natsoulis G et al (1987) 5-Fluoroorotic acid as a selective agent in yeast molecular genetics. Methods Enzymol 154:164–175. https://doi.org/10.1016/0076-6879(87)54076-9
47. Hua J, Meyer JD, Lodge JK (2000) Development of positive selectable markers for the fungal pathogen *Cryptococcus neoformans*. Clin Diagn Lab Immunol 7:125–128. https://doi.org/10.1128/cdli.7.1.125-128.2000
48. Edman JC (1992) Isolation of telomere like sequences from *Cryptococcus neoformans* and their use in high-efficiency transformation. Mol Cell Biol 12:2777–2783. https://doi.org/10.1128/mcb.12.6.2777-2783.1992
49. Park YK, Park SM, Choi YC et al (2008) Asi-Designer: exon-based siRNA design server considering alternative splicing. Nucleic Acids Res 36:W97–W103. https://doi.org/10.1093/nar/gkn280
50. Horn T, Boutros M (2010) E-RNAi: a web application for multi-species design of RNAi reagents – 2010 update. Nucleic Acids Res 38 (Web Server Issue):W332–W339. https://doi.org/10.1093/nar/gkq317

Part III

Omics

Chapter 8

Cryptococcus neoformans Transcriptome Analysis to Identify Differentially Expressed Genes Using the STAR Pipeline in CyVerse Discovery Environment

Rebecca L. Seipelt-Thiemann

Abstract

RNA sequencing is a next-generation sequencing approach that may be used to investigate many aspects of gene expression changes between cells. Analysis of the data is typically a multistep process using several bioinformatics tools. The following protocol utilizes a reliable pipeline for identifying differentially expressed genes among samples of *Cryptococcus neoformans* that is approachable for the adventurous beginner.

Key words RNA sequencing analysis, Transcriptome analysis, Differential expression

1 Introduction

Many external and internal stimuli initiate changes in gene expression that are of interest to biologists. RNA sequencing followed by data analysis provides a clearer picture of which genes and which biological pathways are affected by a particular treatment. Different tools and pipelines have been developed to address questions of differential expression of genes, alternative splicing, and quantification of quantitative trait loci (see [1] for a broad overview). This tutorial/protocol follows a typical workflow of quality checks, read alignment, annotation generation, differential expression analysis, and functional categorization (Fig. 1) to identify differentially expressed genes (DEGs) between treatment groups.

Erin E. McClelland (ed.), *Cryptococcus neoformans: Methods and Protocols*, Methods in Molecular Biology, vol. 2775,
https://doi.org/10.1007/978-1-0716-3722-7_8,

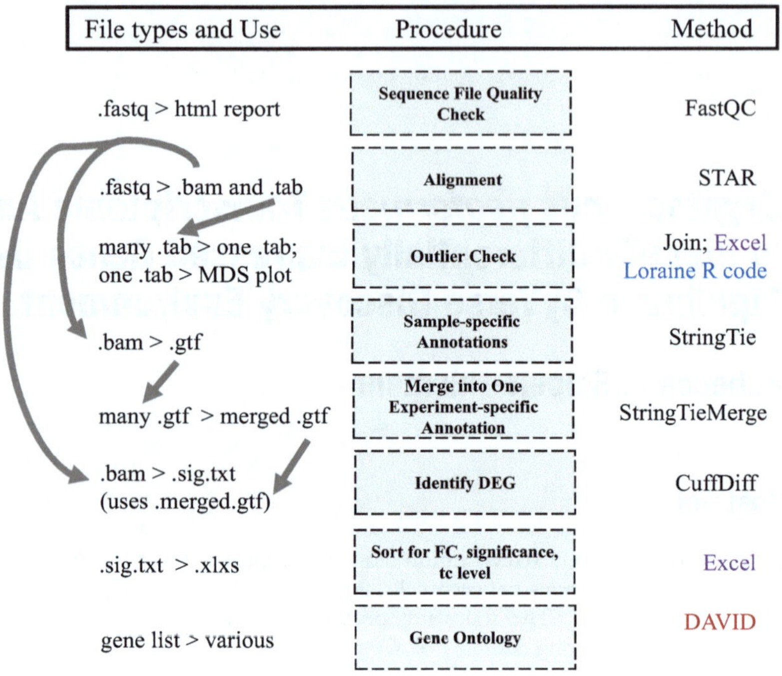

Fig. 1 Transcriptomics workflow. The center column is the analysis step or procedure. The left-most column identifies the file types of input and output for each analysis step. The right-most column is the tool, method, or program name used in the analysis step. Black text in the right-most column indicates a tool used in the CyVerse Discovery Environment, while blue, purple, and red texts indicate use of R code, Microsoft Excel, or a different web-based tool (DAVID)

2 Materials

1. Paired-end fastq files from an RNA sequencing experiment with three biological replicates per treatment with adapters removed.
2. A "regular" or "pro" CyVerse Discovery Environment subscription: https://de.cyverse.org/.
3. Internet access.
4. An external hard drive for backing up files.
5. The following applications should be downloaded and installed:
6. Cyberduck: https://cyberduck.io/.
7. Excel: https://www.microsoft.com/en-us/microsoft-365/excel.
8. R: https://www.r-project.org/.
9. R Studio/Posit: https://posit.co/download/rstudio-desktop/.

3 Methods

3.1 A Workbench for Beginners

Several workbench platforms are free or low cost and easy to use for beginners. In this tutorial, we will use the CyVerse Discovery Environment (DE; [2]) for this pipeline (Fig. 1), but Galaxy [3] also has many of the same tools. In CyVerse DE, we will mainly use three buttons: Home, Data, and Apps/Tools (Fig. 2). I have provided annotated screenshots in the earlier steps but use more text directions in the later steps since you will have become more familiar with the CyVerse DE structure. Please *see* **Notes 1–4** for good data practices.

3.2 Genome File Source and Formats

Large databases, such as the National Center for Biotechnology Information and the European Bioinformatics Institute, as well as species-specific databases provide reference genome (.fasta, .fa) and annotation (.gtf) files, but I have generally had the smoothest workflow using genome files from ENSEMBL (EBI). You may download many other types of genome-related files at ENSEMBL fungi: http://fungi.ensembl.org/info/data/ftp/index.html [4].

1. Use the search tool to find your species (Fig. 3a, purple rectangle). For this workflow, we will use the genome (DNA, yellow rectangle) in FASTA format and the annotation (called Gene Sets) in GTF format (green rectangle). It is important that you use the same reference files for the entire pipeline, as errors may occur if you switch to a new reference mid-pipeline.

Fig. 2 CyVerse Discovery workbench icons. Workbench icons are labeled under the icon in yellow text (HOME, DATA, APPS)

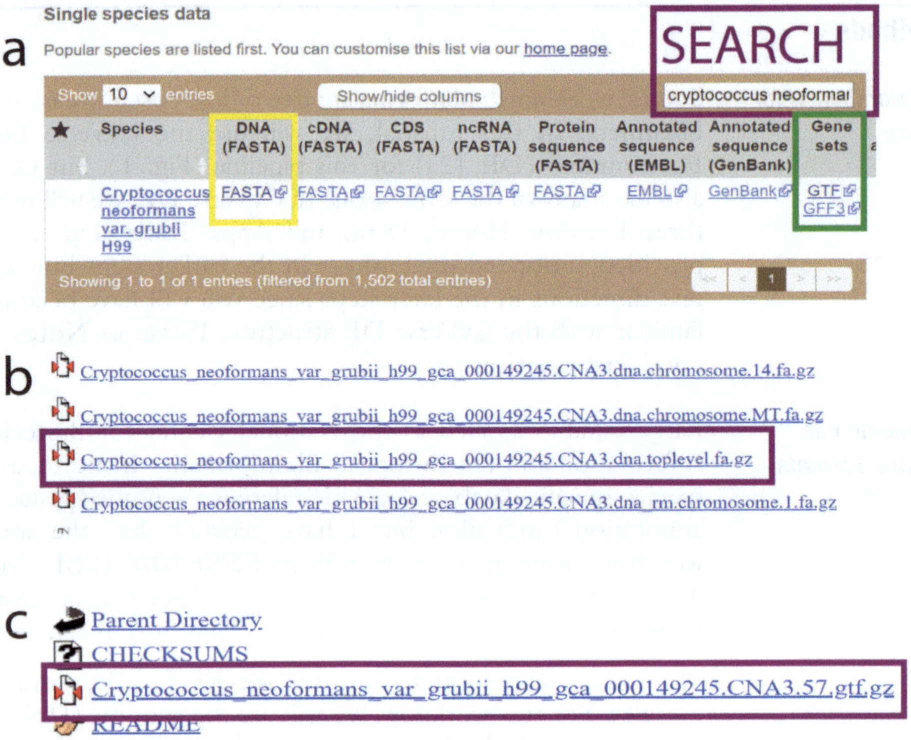

Fig. 3 Reference database and downloads. (**a**) ENSEMBL fungi download search page with search bar (purple rectangle), reference genome column (yellow rectangle, column header is DNA (FASTA)), and reference annotation column (green rectangle, column header is Gene sets) highlighted. (**b**) Naming style for reference genome file with correct file highlighted in the purple rectangle (file name contains words dna.toplevel). (**c**) Naming style for reference annotation file with correct file highlighted in the purple rectangle (file name ends in .gtf.gz)

2. When you click on these links, a new window will open with a list of files you can download. We want the full genome listed as "dna.toplevel" (Fig. 3b, purple rectangle).
3. Right-click on this file to obtain and copy the URL for import. It should look something like this: https://ftp.ebi.ac.uk/ensemblgenomes/pub/release-56/fungi/fasta/fungi_basidiomycota1_collection/cryptococcus_neoformans_var_grubii_h99_gca_000149245/dna/Cryptococcus_neoformans_var_grubii_h99_gca_000149245.CNA3.dna.toplevel.fa.gz.
4. For the annotation (GTF), there is only one choice for our species (Fig. 3c, purple rectangle). Right-click on this to obtain and copy the URL, which should look something like this: https://ftp.ebi.ac.uk/ensemblgenomes/pub/release-56/fungi/gtf/fungi_basidiomycota1_collection/cryptococcus_neoformans_var_grubii_h99_gca_000149245/Cryptococcus_neoformans_var_grubii_h99_gca_000149245.CNA3.56.gtf.gz.
5. *See* **Note 5** if an analysis fails.

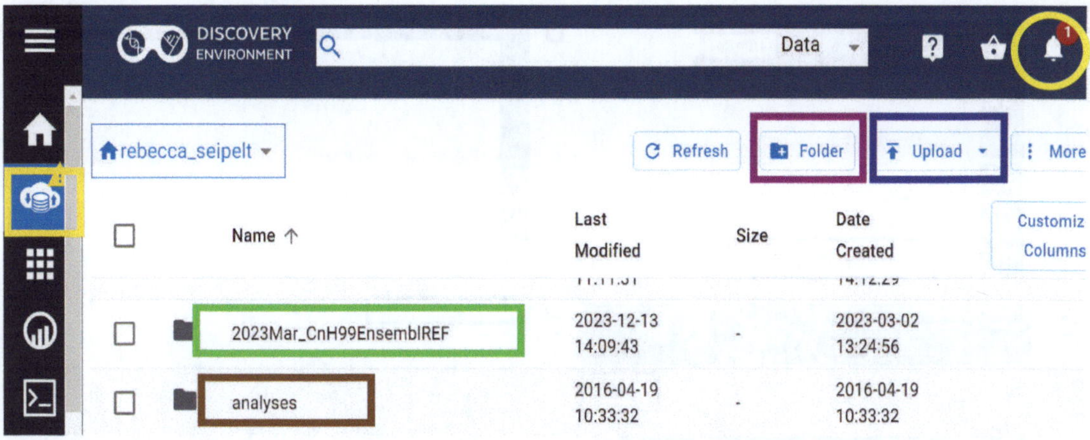

Fig. 4 CyVerse Discovery Environment organization. The data icon is highlighted in a yellow rectangle. The button to add folders is noted by a purple rectangle, while the button to upload a file or import from a URL is shown in the blue rectangle. The green rectangle highlights a folder containing the genome and annotation reference files, while the brown rectangle shows the location of the default analyses folder. The yellow circle is where notifications can be viewed

3.3 Getting Reference Data into the Platform

We will use both the genome and annotation reference files from ENSEMBL but also your own sequencing files. It is best to organize your reference files, analyses, and raw sequencing data into folders in your DE data area (Fig. 4, yellow rectangle).

1. Add folders (purple rectangle) for your raw data and reference files that are appropriately named, for example, 2023Mar_CnH99EnsemblREF for the reference files (green rectangle) and RawData_2023MarCnStrainEvol for the fastq read files (not shown). For organization purposes, I will direct you to place the analyses on these data inside the default analyses folder (brown rectangle); more on this later.
2. Using the URLs you copied earlier, directly import both your genome and annotation files from ENSEMBL into your reference folder using a URL import under the upload button (blue rectangle). You may check the status of any job using the bell icon (yellow circle). These files are currently zipped/compressed files (.gz file extension), but the tools we will be using require these files to be uncompressed.
3. To unzip both reference files, use the gunzip application in CyVerse. To set this up, select the App button (Fig. 5a, yellow rectangle), then use the search bar to search for unzip (gunzip 1.6-2, purple rectangle), and select the name of the application (blue rectangle).
4. In the new window, use Browse (Fig. 5b, blue rectangle) to navigate to your genome reference folder (e.g., 2023Mar_CnH99EnsemblREF), and then click Next, which is just below the Browse button (not shown).

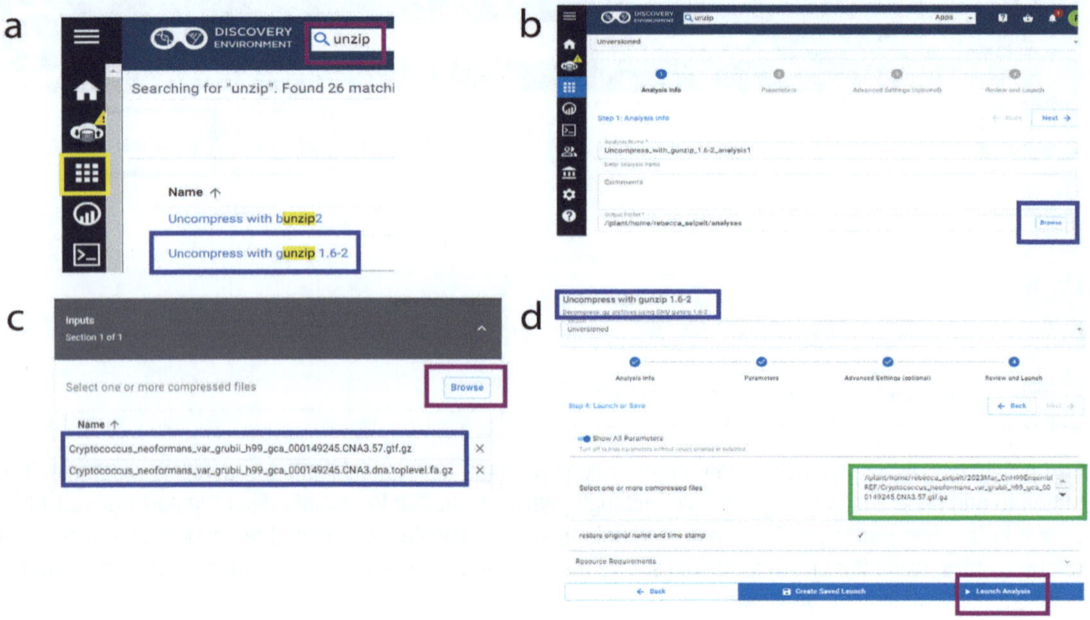

Fig. 5 Setting up a CyVerse application run. (**a**) Use the search bar (purple rectangle) to search for an application (blue rectangle). (**b**) Use the browse button (blue rectangle) to navigate to your preferred output folder. (**c**) Use the browse button (purple rectangle) to navigate to the folder containing your input files, which are then displayed as input (blue rectangle). (**d**) Launch the analysis (blue rectangle) with the desired inputs (green rectangle) using the Launch button (purple rectangle)

5. The next page is for Inputs (Fig. 5c; AFTER inputs selected). Use the Browse button (purple rectangle) to navigate to the genome reference folder, and select both zipped files. Once selected, they will show up in the input section (blue rectangle).
6. Click the Next button at the bottom right (not shown) for this page and the next (Resource Parameters – no changes).
7. The last page is the launch page (Fig. 5d). It is a good idea to take a screenshot of this entire page for your notes, as it has the tool (blue rectangle), version number (blue rectangle), and inputs (green rectangle) listed. Click Launch (purple rectangle). *See* **Note 5** if an analysis fails.

3.4 Getting Your Data into the Platform

Uploading your own data (your sequencing reads) to these platforms is specific to the platform. Directions from CyVerse for uploading files using various methods are available at their datastore page (https://cyverse.org/data-store), and I find using the "Drag and Drop" via Cyberduck to be the easiest to use: https://learning.cyverse.org/ds/cyberduck/.

1. You should upload all of your sequencing files into your raw data folder using the "drag and drop" feature in Cyberduck (Fig. 6a).

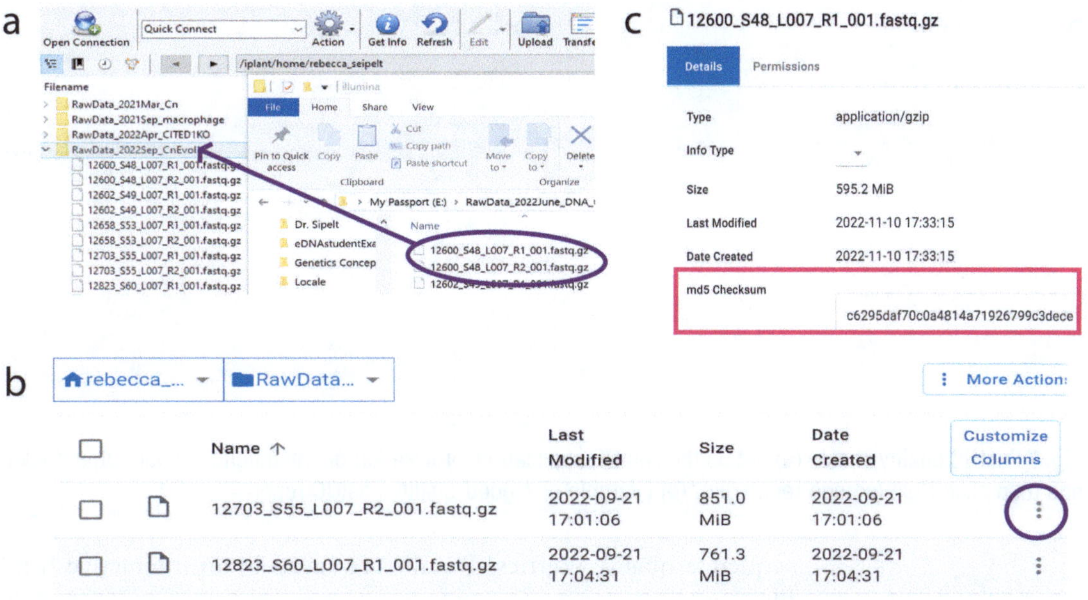

Fig. 6 Data upload and check. (**a**) Use the drag-and-drop feature in Cyberduck to move your data files (circle) to your CyVerse folders (arrow). (**b**) Use the "more" dots (purple circle) to get to the file details, so you can check the file completeness. (**c**) md5 (pink rectangle)

3.5 Checking for File Completeness for Your Data

Large files may appear to transfer completely, but actually do not. You should check the completeness by confirming that the checksum or md5 provided by your sequencing provider matches that of the uploaded files.

1. CyVerse checksums are found by navigating to the file, selecting the three vertical dots (Fig. 6b, purple circle), and selecting details (Fig. 6c) to find the checksum.
2. Ensure this series of numbers and letters matches exactly so that you know the file is complete. Incomplete files will cause errors in the pipeline. Clues that point to this being a problem will be found in the error logs and likely note something about end of file not being found.

3.6 Read Quality Checks

The quality of the sequencing reads is critical to having the reads align to the genome with confidence, so the first step is to check the quality of each read file. You will have two fastq files for each sample if you requested paired-end reads (PE reads, files usually named R1/R2 or 1/2), but only a single fastq file if you requested single-end reads (SE reads). To check the quality, we will use FastQC (https://www.bioinformatics.babraham.ac.uk/projects/fastqc/), which is a java application that provides a quality control report and spots problems originating either in the sequencer or in the starting library material. FastQC uses plots to display several different

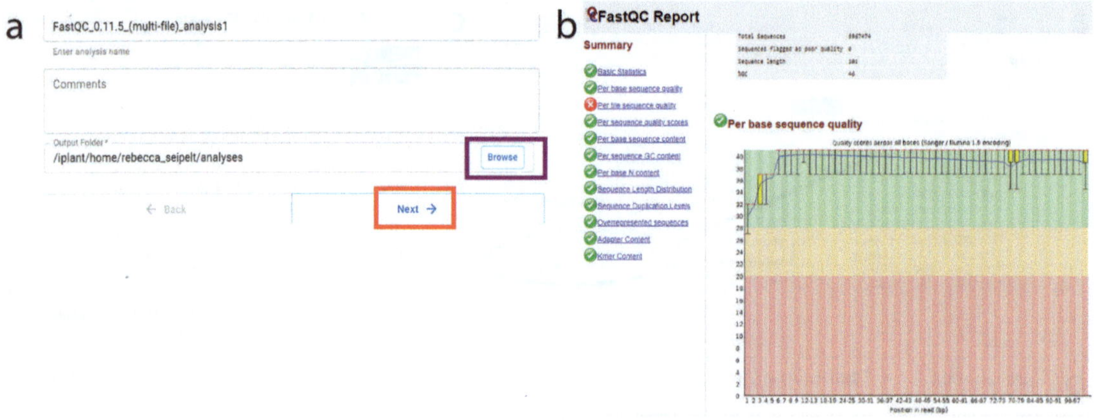

Fig. 7 FastQC quality check. (**a**) Set up the FastQC application, browse (purple rectangle) to your output folder, and then click Next (orange rectangle). (**b**) Example of a good-quality FastQC report

sequence quality metrics. Results are saved in an interactive html file.

1. First, make a folder where all of the analyses for this project will be stored ("Data" button > analyses folder > "new folder" button; *see* Fig. 4 for a refresher of where these are located).
2. Name your project folder (e.g., 2023Mar_CnEvol_RNAresults).
3. It will be best for you to keep all analysis step results (e.g., FastQC, STAR, and StringTie) in sub-folders of one project folder, so make a new folder inside your project folder called FastQC.
4. To launch the application and check the quality of all fastq files, use the app search bar to find the FastQC 0.11.5 (multi-file) app. *See* Fig. 5a for a refresher on where the app search bar is located.
5. On this first app page, choose your Output Folder by clicking the Browse button (Fig. 7a, purple rectangle) and navigating to your new FastQC folder. Then, select the Next button (Fig. 7a, orange rectangle).
6. The next page has a Browse button to select the input, which are all the fastq files in your raw data folder. Select all your fastq files, and then click Next at the bottom of this page and the following Resources page.
7. The final page is the launch page that has a summary of the analysis. Click Launch. Typically, if an analysis fails, it will fail in the first 45 min. You can check the status using the bell icon at the top right (Fig. 4, yellow circle).

8. Once your analysis completes, you will see two output files per fastq file in your output folder (an html file and a zip file). Click on the html file, and it will open a new browser tab. The report shows a graph that visualizes the quality (y axis) at each base position (x axis) for all reads (Fig. 7b, a good quality FastQC report image). The blue line/curve is the average phred quality score for that base across all reads. Good quality is in the green range (phred scores of 28 and higher); questionable quality is in the peach range (phred scores of 20–28); bad quality is in the red range (phred scores of 0–20). The links in the left column may also clue you in on overrepresented sequences and adapter content. If the sequence is of poor quality at the ends, you may trim the reads using a tool called Trimmomatic [5] or CutAdapt [6]; however, be aware that all reads will be trimmed. Trimmomatic has many options for trimming and masking if you have poor-quality sequence. You can view a poor-quality report image in the FastQC manual (https://www.bioinformatics.babraham.ac.uk/projects/fastqc/Help/3%20Analysis%20Modules/2%20Per%20Base%20Sequence%20Quality.html. Generally, a little dip near the ends into the 28–20 phred score range is OK because the reads will still align well, but if the quality goes lower and is a longer portion, you will probably want to consult someone more experienced to help you make decisions about whether trimming and/or masking is even worth doing.

3.7 Aligning Reads to the Genome

Once you have confirmed your fastq files have good-quality sequence, you can align the reads to the genome using STAR [7].

1. Make a new folder inside your project folder called STAR where your aligner output will be stored.
2. Search for the STAR application (STAR 2.5.3a -index-align). For the inputs, use browse to navigate to and select the following:
3. Your genome reference. The top-level DNA file you imported and unzipped; it will have the file extension .fa.
4. Your reference annotation file. The file you imported and unzipped and has the file extension .gtf.
5. Your first read files (usually named Rl or simply 1); yes, all of them.
6. Your second read files (usually named R2 or simply 2); yes, all of them.
7. File type is PE (paired end reads).
8. Output quantification method is GeneCounts.

   For example, the genome and annotation files I am using are named Cryptococcus_neoformans_var_grubii_h99_gca_

000149245.CNA3.dna.toplevel.fa and Cryptococcus_neoformans_var_grubii_h99_gca_000149245.CNA3.56.gtf, respectively.

9. All other parameters should stay the same. Click next, then next again to get to the launch page, and launch the analysis. It would be a good time to take a screenshot for your notes.
10. The output for this analysis will be folders containing read counts per gene (output folder > sampleID folder > sampleID.out.tab) and .bam files (bam_output folder > sampleID.Aligned.sortedByCoord.out.bam).

3.8 Identifying Outlier Samples

The high variability among the replicates makes it difficult to identify DEGs, so one step that is widely accepted to reduce noise is to check for and eliminate distinctly outlier replicates. This method uses R and R Studio/Posit. Excellent directions for performing this step are found in Loraine et al. [8, 9], Subheadings 3.1 and 3.2. Also, *see* **Note 6**. The GeneCounts output from STAR should be joined into one table where the genes are listed in column 1 and gene counts are listed in the remaining columns.

1. Use CyVerse application: Join_multiple_tab-delimited_files_analysis1) to join the single data files (now one file per sample) into one tabular file called ccombine_result.txt.
2. Download this file (data icon > navigate to and select your joined file > more actions button > select download from the menu).
3. The file needs some editing before it can be imported for the outlier check. The file has three problems we need to resolve: (1) There are no sample names, (2) there are four rows we need to remove, and (3) there are three columns for every sample.
4. Open the file in excel and insert a row at the very top (Fig. 8a, row 1 inserted).
5. We need to label which columns belong to which samples. Generally, these are in the order you joined them, but it is a good idea to check. You can tell using the N_counts currently in row 2. Compare this number to the number in your CyVerse individual GeneCounts files (output from STAR that you joined to make the table) to confirm which samples are which.
6. Label your sample sets of three—I find it easy to color-code sets of three columns and then label the sample ID in the first column (Fig. 8a, rows B–D are for sample 10). These three columns are (1) counts for the combined R1 and R2 files (Fig. 8a, column B), (2) counts for the R1 file only (Fig. 8a, column C), and (3) counts for the R2 file only (Fig. 8a, column D).

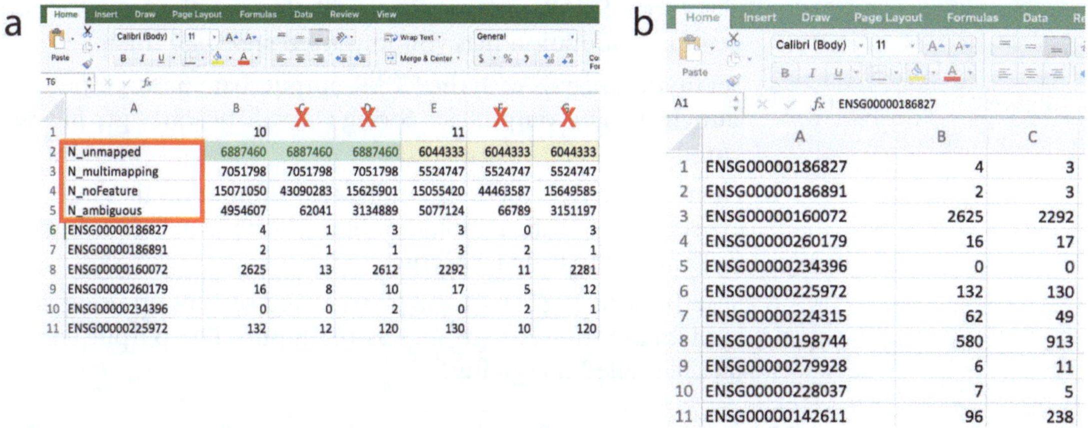

Fig. 8 Reorganization of the joined gene count file. (**a**) View of gene count file in excel after labeling, but before reorganization. Row 1 is added to accommodate the sample name before deleting any rows or columns. After labeling, individual (R1 and R2) read columns (red Xs), rows labeled with N_ (orange rectangle), and the inserted header row (row 1) are deleted to produce the R-ready file for the outlier check. (**b**) View of gene count file in Excel after reorganization

7. Once you have labeled and confirmed your sample IDs, delete the R1 only and R2 only columns for each sample, e.g., columns C and D for sample 10 (Fig. 8a, red Xs).
8. Next, delete the four "N_" rows (Fig. 8a, orange rectangle). Save this file as GeneCounts_reorganized.xlxs to ensure you have the sample identification.
9. Next, delete the header row with the sample names. I know this sounds crazy, but you know the order of your samples from part e and the header row will complicate the outlier check R code. It should now look like Fig. 8b where column B has sample 10's counts and column C has sample 11's counts.
10. Outlier samples should not be included in any further analysis.

3.9 Making an Experiment-Specific Genome Annotation File

Differential expression for well-annotated genomes, such as human or mouse, may be performed without making an experiment-specific annotation. If you are not interested in identifying transcripts for unannotated genes, you can skip Subheading 3.8, and use the genome annotation file you downloaded instead of the "merged.gtf." If you are interested in identifying all transcripts, we will use the StringTie [10] and StringTie merge [11] tools to make our own annotation.

1. To set this up, make two new folders inside your project folder called StringTie and StringTieMerge.
2. Search for the application StringTie (StringTie – 1.3.3).
3. Use browse to navigate to and select the StringTie output folder you just made. The inputs for this will be your .bam

files (your STAR output). You will also use the genome and genome annotation files you imported and then unzipped.

4. Set up the analysis with the STAR output .bam files as input and launch it. The output for StringTie will be one .gtf file per sample.
5. Next, we need to merge all individual annotation files into one experiment-specific annotation file, which we will do using the CyVerse tool StringTieMerge.
6. Search for StringTie – 1.3.3_merge, and set this up using your individual StringTie .gtf files as the input. The output will be one file called merged.gtf.

3.10 Identifying DEGs

Using the .bam files from STAR and the merged.gtf from StringTie merge, we are now ready to identify DEGs between our treatment groups using a tool called CuffDiff [12]. Also, *see* **Note 7**.

1. Make a new folder in your project folder called CuffDiff. Search for and set up the "CuffDiff2 16 way max -2.2.1" application.
2. For inputs, name each treatment group something simple (Fig. 9), and use Browse to select all the .bam files (STAR output) for that treatment group. Due to the alphabetical

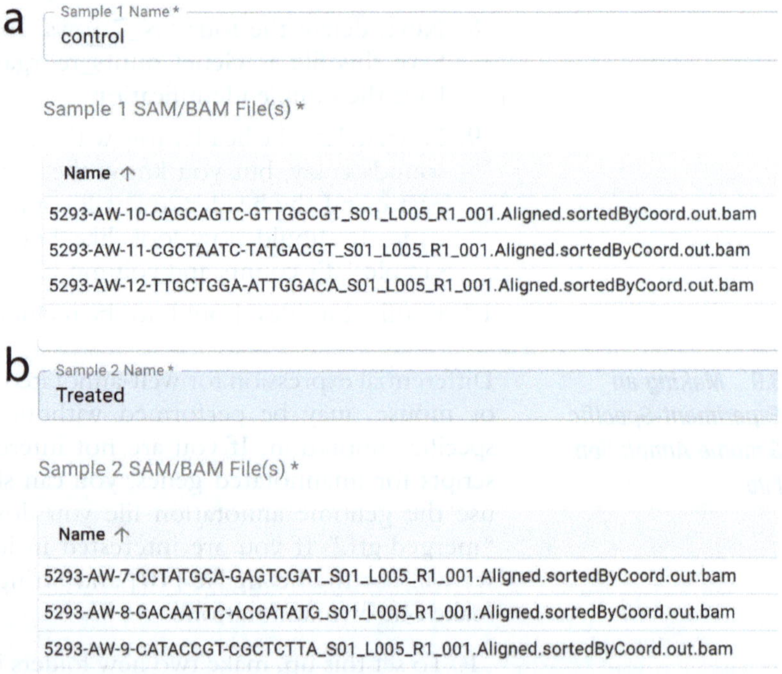

Fig. 9 CuffDiff sample organization. All replicates that belong to a sample group are listed together for analysis in CuffDiff. (**a**) Example of Control group biological replicates used for CuffDiff input. (**b**) Example of Treated group biological replicates used for CuffDiff input

nature of sample name sorting in CuffDiff, you will want your control sample labeled with something nearer the beginning of the alphabet, such as control.

3. In Subheading 3 of inputs, use the experiment-specific annotation file (merged.gtf output file from StringTieMerge).
4. In Subheading 4 of inputs, use your reference genome file (FASTA) that you imported and unpacked.
5. No other parameters need to be changed, so keep clicking Next and Launch the analysis.
6. The output for this analysis will be several folders, but the folder you want is called sorted_data. This folder has the data sorted by: (1) feature type (gene or transcript), (2) an expression value (fold change or expression level), and (3) statistical significance ("sig" = statistically significant genes only or all genes). The statistical metric is the q-value, which was set at 0.05 by default and is a modified p-value. The file we will work with is called gene.sorted_by_fold.sig.txt. Download this file and open it in excel.

3.11 Sorting for Genes with Statistical and Biological Significance

First, we only want statistically significant genes that have a fold change (either up or down) of 2 or greater. The reason is that two-fold is a level of change that is generally agreed upon to be a change in level that would significantly affect the cell—we call it a "biologically relevant" change.

1. Starting with your excel file, save the file under a new name: genes.sorted_by_fold_sig_2FC.xlxs. The columns should be as follows:
2. Column A is the gene_id, which is an ENSEMBL transcript ID (or something similar).
3. Column B is the common gene name.
4. Column C is the sample 1 identifier.
5. Column D is the sample 2 identifier.
6. Column E is the fold change.
7. Column F is the direction of change.
8. Column G is a measure of the read count (fragments per kilo).
9. Column H is the q-value (which is a corrected p-value for statistical significance).
10. All genes should have q-values of 0.05 or less. They should also be sorted with the largest fold change at the top.
11. We next need to delete genes that have fold change of less than 2.0, so scroll down in the list, highlight all genes with a fold change of less than 2, and delete these rows.

12. If you have multiple comparisons to make, you can sort on the sample identifier columns (first D, then A) and copy-paste comparison sets to new tabs. For example, control vs. drug1, control vs. drug2, and drug1 vs drug2. These are your DEG lists.
13. One issue that you will want to keep track of is that low-expressed genes may show an artificially high fold change. So, it is a good idea to be wary of any genes whose RPKM value is </= 0.5. They may be valuable and real, but it is good to be aware.
14. XLOC "genes" are genomic locations that encode transcripts in your data set, but are not yet annotated genes in the genome annotation. Depending on your specific species, they may be new genes or may be genes that have been identified in a similar species. If you want to find out more about them, you can use genome coordinates [locus column], which are located in a CuffDiff output folder (CuffDiff > cuffdiff_out folder > gene_exp.diff) to search the ENSEMBL genome for your species/strain to identify any orthologs.
15. Additionally, if you would like to see how different comparison groups overlap and are unique, you can build a Venn diagram. Simply copy-paste your comparison gene lists into the text boxes at: https://bioinformatics.psb.ugent.be/webtools/Venn/.

3.12 Identifying Functional Aspects/Gene Ontology of DEG Lists

The final step is to give some meaning of the DEG list(s) by identifying the functions of the genes and the biological pathways in which they work. We will be using a web-based tool called Database for Annotation, Visualization, and Integrated Discovery (DAVID; [12]).

1. Go to DAVID (https://david.ncifcrf.gov/tools.jsp), and click on the tab called Start Analysis (Fig. 10a, orange rectangle).
2. Open your excel file with the gene lists and sort by gene names (column A).
3. Copy all the gene names that are not XLOC, and paste them into the Gene List box (Fig. 10a, purple rectangle). XLOC genes are unannotated genes, so you will not want to copy them as they will not be in DAVID. If there are any noncoding RNA genes in your gene list, they may be listed with the gene ID, such as CNAG_13146-1.
4. DAVID is very particular in its "Step 2 Select Identifier" for the gene/protein name. It is better to pick "unsure" from the very bottom of the list and let DAVID identify it for you. You can then go backwards one step and pick the correct identifier, such as ENSEMBL_PROTEIN_ID (Fig. 10a, green rectangle).

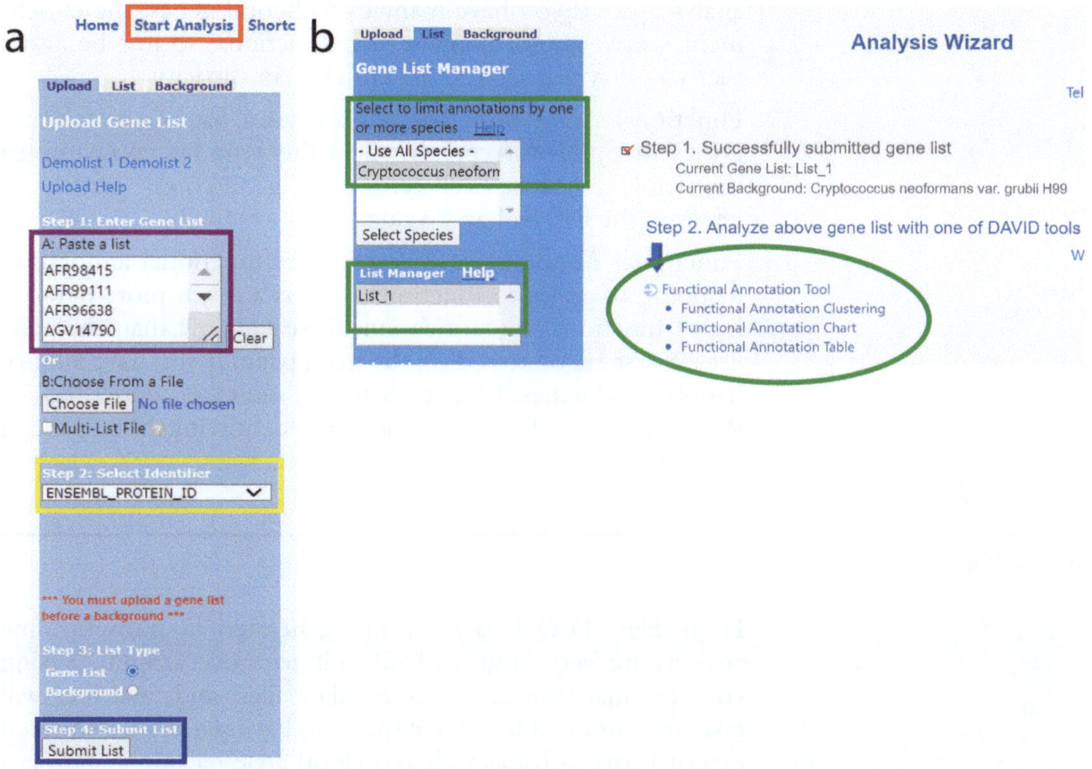

Fig. 10 DAVID input and analysis. (**a**) Start the DAVID analysis (orange rectangle), copy-paste your gene/protein list (purple rectangle), select the correct identifier for your gene/protein list (green rectangle), select the correct list type (yellow rectangle), and then click the submit button (blue). (**b**) Set up the analysis of the gene list by selecting all species, list 1 (green rectangles), and then use each of the DAVID toolds (green circle) to analyze the gene/protein list

5. Select Gene List as your List Type (Fig. 10a, yellow rectangle).
6. Click submit (Fig. 10a, blue circle).
7. The gene list with its corresponding gene names now shows on the main page. You can download the list (right top of table).
8. We are interested in which pathways are enriched, which we can find using the functional annotation tools in DAVID. From the List Manager column, select "all species," list _1, and Use (Fig. 10b, yellow rectangles).
9. In the Shortcut to DAVID Tools tab, select Functional Annotation (Fig. 10b, yellow circle).
10. Click on each of the buttons to examine the details:
11. Functional Annotation Clustering provides a clustering or relationship of genes within the list. Each clustered group has an enrichment score, gene count, and e-value. CAVEATS: (1) Some genes will not have any functional annotations, so take note of how many genes were "not found" in each

analysis. (2) If you have many hypothetical genes, the enrichment scores cannot include their functions, so just be aware that you may not see the full enrichment capacity.

12. Functional Annotation Chart provides a list of terms/categories that are overrepresented in the gene list compared to the whole genome. Each term has a gene count, percent of genes in the list, and an e-value.
13. Functional Annotation Table provides functional annotation mapping of genes to functions. This is a much more detailed list of functional categorizations for each gene than the gene list we first encountered on the main page. If you have a lot of hypothetical genes, here is where you can take a closer look at the functions of these genes by identifying the domain functions.

4 Notes

1. Large files. Data files used and generated in transcriptome analysis are very large and difficult to move around. Saving your original data files, intermediate files, and final files will take up considerable storage space, so having a large, dedicated external hard drive as well as a cloud-style backup system is a good idea. You will not generally be able to send them by email, and uploads/downloads can take hours, if not days, to complete. Moving files will be easier using ftp transfer, and many free, user-friendly, drag-and-drop programs exist, such as Cyberduck (recommended by CyVerse Discovery Environment) and FileZilla. Sequencing providers generally provide a narrow window of time (7–10 days) for you to download your data, so be sure to be prompt. Also, be sure to download your files to two distinct places, such as your computer and an external hard drive, so you have an independent backup set of files if one is or becomes corrupted.
2. Quality checking file completeness. Because files are very large and not human-interpretable, you will need to rely on different quality checks than the visual inspection of files. First, your sequencing provider and genome data source will provide you checksum/md5 identifiers to ensure the data file you have is complete. A checksum program will calculate a checksum based on the file contents and will be a unique combination of letters and numbers per file. Once you download or upload a file, you should calculate the checksum for each file and compare it to the checksum provided by the source. If they do not match, then your file is corrupted or incomplete and will not be usable in your pipeline. Some sequencing providers, such as Novogene, provide a script that will check all the files and highlight

files with errors. The platform we use in this chapter, CyVerse Discovery Environment, also calculates checksums on each file.

3. File names and renaming. When working with these large files, it can be tempting to rename them, but do not. The naming system is used with various bioinformatics tools, and you do not want to take the chance of a misnaming error. It is better to make a table with the file name, sample ID, and replicate number to keep with you. For any files or folders that you do name, I recommend using only numbers, letters, and underscores as unusual characters sometimes cause errors. Additionally, many text and number editors, such as Microsoft Word or Excel, add invisible characters to the files, so do not open or save sequence, annotation, or DEG list files in this format until you are at the end of the pipeline.
4. Taking notes. It is important to take copious notes and screenshots as notes. This will ensure that you can follow the pipeline again for your next experiment and writing the methods section of manuscripts but also will allow you to recheck any questions you might have regarding tool versions or settings. Google Docs is a good note-taking platform because it is accessible from any computer and you can easily share with colleagues or students.
5. Failed analyses. If an analysis fails, you may try to identify why by looking at the error log. This will be inside your output folder for the analysis. It is called condor-stderr-out. Look for where the output mentions "failing" or "exiting" due to an error in a particular file. Whichever file is mentioned is the likely cause. Most times, the file in question is corrupted and you replace it and re-run. If I cannot immediately figure out what has gone wrong, I usually look in the tool's manual and google the exact text of the error and figure out a way to fix it from there.
6. Outlier check R code. I have pared down the Loraine et al. [9] code to use in my classes. If you would like this pared down version of their code, please email me.
7. Interpretation. Please be aware that this analysis is undertaken on steady-state levels of RNA material. This means that transcriptional regulation or mRNA decay may be responsible for affecting transcript levels. There may also be alternative splicing differences that can be elucidated using the Salmon-IsoformSwitchAnalyzeR pipeline [13, 14].

References

1. Kukurba KR, Montgomery SB (2015) (2015) RNA sequencing and analysis. Cold Spring Harb Protoc 11:951–969. https://doi.org/10.1101/pdb.top084970
2. Merchant N, Lyons E, Goff S et al (2016) The iPlant collaborative: cyberinfrastructure for enabling data to discovery for the life sciences. PLoS Biol 14(1):e1002342. https://doi.org/10.1371/journal.pbio.1002342
3. Afgan E, Baker D, van den Beek M et al (2016) The Galaxy platform for accessible, reproducible and collaborative biomedical analyses: 2016 update. Nucleic Acids Res 44(W1):W3–W10. https://doi.org/10.1093/nar/gkw343
4. Howe KL, Contreras-Moreira B, De Silva N et al (2020) Ensembl Genomes 2020-enabling non-vertebrate genomic research. Nucleic Acids Res 48(D1):D689–D695. https://doi.org/10.1093/nar/gkz890
5. Bolger AM, Lohse M, Usadel B (2014) Trimmomatic: a flexible trimmer for Illumina sequence data. Bioinformatics 30(15): 2114–2120. https://doi.org/10.1093/bioinformatics/btu170
6. Kechin A, Boyarskikh U, Kel A et al (2017) CutPrimers: a new tool for accurate cutting of primers from reads of targeted next generation sequencing. J Comput Biol 24(11): 1138–1143. https://doi.org/10.1089/cmb.2017.0096
7. Dobin A, Davis CA, Schlesinger F et al (2013) STAR: Ultrafast universal RNA-seq aligner. Bioinformatics 29(1):15–21. https://doi.org/10.1093/bioinformatics/bts635
8. Loraine AE, Blakley IC, Jagadeesan S et al (2015) Analysis and visualization of RNA-Seq expression data using RStudio, Bioconductor, and Integrated Genome Browser. Methods Mol Biol 1284:481–501. https://doi.org/10.1007/978-1-4939-2444-8_24
9. Robinson MD, McCarthy DJ, Smyth GK (2010) EdgeR: a bioconductor package for differential expression analysis of digital gene expression data. Bioinformatics 26(1): 139–140. https://doi.org/10.1093/bioinformatics/btp616
10. Pertea M, Pertea GM, Antonescu CM et al (2015) StringTie enables improved reconstruction of a transcriptome from RNA-seq reads. Nat Biotechnol 33(3):290–295. https://doi.org/10.1038/nbt.3122
11. Trapnell C, Hendrickson DG, Sauvageau M et al (2013) Differential analysis of gene regulation at transcript resolution with RNA-seq. Nat Biotechnol 31(1):46–53. https://doi.org/10.1038/nbt.2450
12. Sherman BT, Hao M, Qiu J et al (2022) DAVID: a web server for functional enrichment analysis and functional annotation of gene lists (2021 update). Nucleic Acids Res 50(W1):W216–W221. https://doi.org/10.1093/nar/gkac194
13. Patro R, Duggal G, Love MI et al (2017) Salmon provides fast and bias-aware quantification of transcript expression. Nat Methods 14(4):417–419. https://doi.org/10.1038/nmeth.4197
14. Vitting-Seerup K, Sandelin A (2019) IsoformSwitchAnalyzeR: analysis of changes in genome-wide patterns of alternative splicing and its functional consequences. Bioinformatics 35(21):4469–4471. https://doi.org/10.1093/bioinformatics/btz247

Chapter 9

Proteomic Profiling of Samples Derived from a Murine Model Following *Cryptococcus neoformans* Infection

Ben Muselius, Antoine Bodein, Florence Roux-Dalvai, Arnaud Droit, and Jennifer Geddes-McAlister

Abstract

Proteomic profiling provides in-depth information about the regulation of diverse biological processes, activation of and communication across signaling networks, and alterations to protein production, modifications, and interactions. For infectious disease research, mass spectrometry-based proteomics enables detection of host defenses against infection and mechanisms used by the pathogen to evade such responses. In this chapter, we outline protein extraction from organs, tissues, and fluids collected following intranasal inoculation of a murine model with the human fungal pathogen *Cryptococcus neoformans*. We describe sample preparation, followed by purification, processing on the mass spectrometer, and a robust bioinformatics analysis. The information gleaned from proteomic profiling of fungal infections supports the detection of novel biomarkers for diagnostic and prognostic purposes.

Key words *Cryptococcus neoformans*, Murine model, Cryptococcosis, Mass spectrometry-based proteomics, Bioinformatics

1 Introduction

The field of mass spectrometry-based proteomics encompasses a diverse set of scientific disciplines, including biochemistry, computer science, mathematics, analytical chemistry, biology, and engineering with broad applicability and adaptability. Proteomics provides global information about protein production across a biological system, identifies proteins released into an extracellular environment, defines modifications regulating protein function, and detects protein–protein interactions and complexes that influence cellular signaling [1]. Proteomics is classically defined as top-down (i.e., analyze intact proteins for identification of protein complexes and proteoforms) [2, 3] or bottom-up (i.e., discovery-

Erin E. McClelland (ed.), *Cryptococcus neoformans: Methods and Protocols*, Methods in Molecular Biology, vol. 2775, https://doi.org/10.1007/978-1-0716-3722-7_9,

driven method that digests proteins into peptides followed by liquid chromatography separation and mass spectrometry) [4]. Targeted proteomics also analyzes digested proteins for detection and accurate quantification of predefined peptides in complex mixtures, such as for biomarker discovery [5]. For this chapter, we focus on the application of bottom-up proteomics within infectious disease research.

Protein-level analysis of infected in vivo samples provides a dual perspective on disease, which is ideal for studying the complex relationship between a host and pathogen [6, 7]. Specifically, the application of mass spectrometry-based proteomics provides qualitative information on the presence of condition-specific proteins of both the host and pathogen and quantitative measurements of relative protein abundance and changes driven by infection, as well as novel drug discovery [1, 8–10]. Such information may be leveraged for diverse purposes. For instance, comparison of infected versus uninfected host samples may confirm immune system activation and propose novel biomarkers of infection [11, 12]. Additionally, proteomic profiling of infection may identify putative novel infection-associated proteins or virulence factors produced by a pathogen to overcome host defenses, which may reveal new druggable targets for the design of novel therapeutics [13–15]. Moreover, comparison of samples collected across a temporal axis of infection or from distinct compartmental host sites may highlight unique strategies used by both host and pathogen during infection [16].

In this chapter, we discuss proteome preparation of organ, tissue, and fluid samples collected from a murine model of infection (Fig. 1). We focus on infection by the human fungal pathogen, *Cryptococcus neoformans*, which colonizes the lungs of immunocompromised individuals and disseminates throughout the host in the wake of a partial immune response [17]. Notably, the detailed protein extraction process is relatively pathogen-agnostic and provides equally important information about the host response to infection as the pathogenic drivers of disease, and it is applicable to diverse pathogens and host systems (Fig. 2). For technological prowess, we also cover the use of isobaric mass tagging to increase protein quantification and reduce variance combined with high-pH fractionation to increase resolution of complex multiplexed samples, as well as a general framework for robust data processing and analysis using freely available bioinformatics tools.

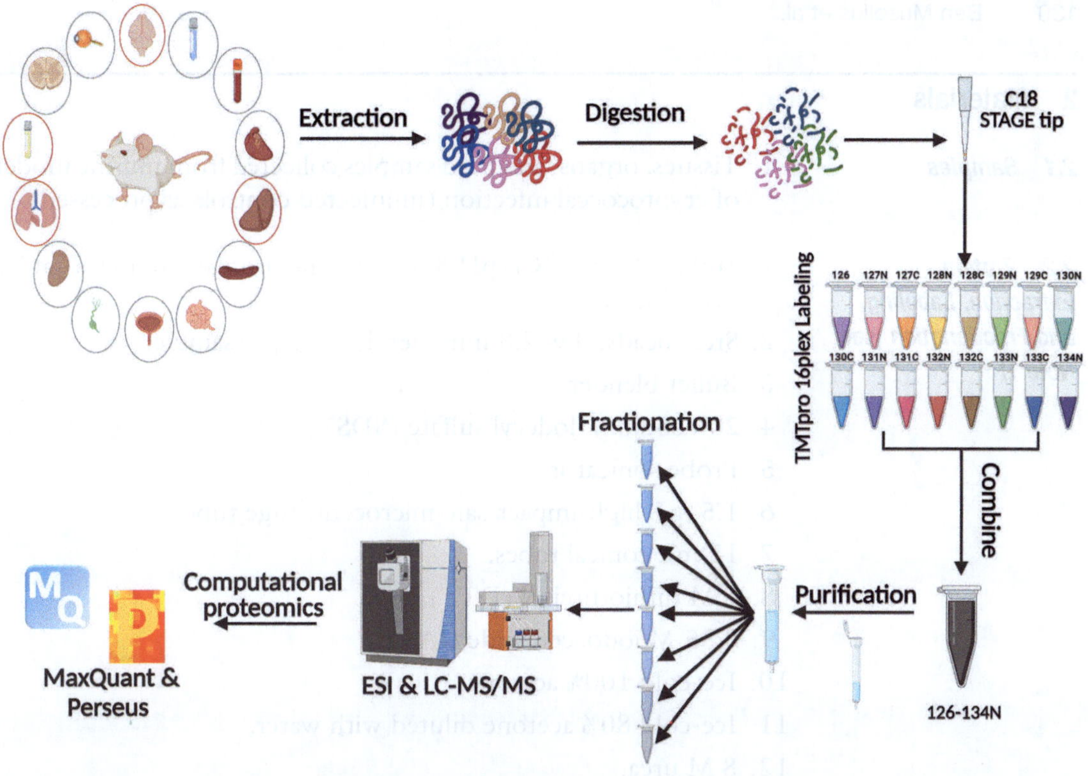

Fig. 1 Workflow of sample preparation for mass spectrometry-based proteomics. Proteins are extracted by chemical and mechanical disruption, digested by proteases, and purified on a C18 resin. Next, samples are labeled with isobaric mass tags, multiplexed, and re-purified. Samples are then fractionated (high pH) before separation on a high-performance liquid chromatography system and measured on a high-resolution mass spectrometer. Resulting data is processed using appropriate software. (Figure generated with Biorender.com)

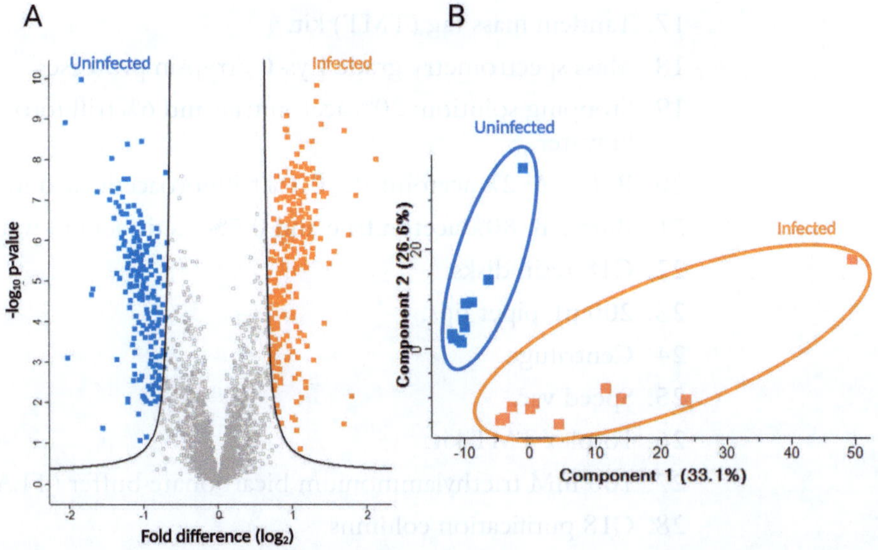

Fig. 2 Preliminary proteomics data analysis of cryptococcosis from murine samples. (**a**) Volcano plot comparing host and pathogen protein abundance between splenic infected (orange) and uninfected (blue) samples collected 3 days postinfection. Significant changes in abundance were determined using Student's *t*-test ($p < 0.05$, false discovery rate = 5%, $S_0 = 1$). (**b**) Principal component analysis showing clustering of the ten infected and ten uninfected samples collected 3 days postinfection

2 Materials

2.1 Samples

1. Tissues, organs, and fluid samples collected from murine model of cryptococcal infection (uninfected controls as processed).

2.2 Protein Extraction, Labeling, and Fractionation (see Note 1)

1. 100 mM Tris–HCl (pH 8.5) with one protease inhibitor tablet per 10 mL.
2. Steel beads, 0.9–2.0 mm blend, 0.8 g per sample.
3. Bullet blender.
4. 20% sodium dodecyl sulfate (SDS).
5. Probe sonicator.
6. 1.5 mL high-impact-safe microcentrifuge tubes.
7. 15 mL conical tubes.
8. 1 M dithiothreitol (DTT).
9. 0.55 M iodoacetamide (IAA).
10. Ice-cold 100% acetone.
11. Ice-cold 80% acetone diluted with water.
12. 8 M urea.
13. 40 mM N-2-hydroxyethylpiperazine-N′-2-ethanesulfonic acid (HEPES).
14. Water bath sonicator.
15. 50 mM ammonium bicarbonate (ABC).
16. 0.22 μm filter (one per sample).
17. Tandem mass tag (TMT) kit.
18. Mass spectrometry grade Lys-C/trypsin proteases.
19. Stopping solution: 20% acetonitrile and 6% trifluoroacetic acid in water.
20. Buffer A: 2% acetonitrile, 0.1% trifluoroacetic acid in water.
21. Buffer B: 80% acetonitrile and 0.5% acetic acid in water.
22. C18 resin disks.
23. 200 μL pipet tips.
24. Centrifuge.
25. Speed vac.
26. 16-plex TMT kit.
27. 100 mM triethylammonium bicarbonate buffer (TEAB).
28. C18 purification columns.
29. 0.1% trifluoroacetic acid (TFA).
30. Hydroxylamine.

31. High-pH fractionation columns.
32. High-performance liquid chromatography system.
33. High-resolution Orbitrap mass spectrometer.

2.3 Bioinformatics

1. Mass spectrometry analysis software suitable for proteomics data.
2. Bioinformatics and visualization software.

3 Methods

3.1 Protein Extraction from Murine Organ and Tissue Samples (see Note 2)

1. Transfer collected tissue samples to 1.5 mL high-impact microcentrifuge tubes containing 600 μL of ice-cold 100 mM Tris–HCl (pH 8.5) combined with a protease inhibitor cocktail tablet.
2. Add one scoop of steel beads.
3. Place samples in a bullet blender for 5 min at power 5 or until sample is homogenized.
4. Using a 1000 μL pipet, transfer sample to a 15 mL conical tube, ensuring that no steel beads transfer with the sample.
5. Add 1:10 volume of 20% SDS to each sample to a final concentration of 2% SDS.
6. Further lyse samples using a probe sonicator set for three cycles of 30 s on/30 s off at an amplitude of 20% in an ice bath.
7. Pulse-centrifuge samples (approximately 10 s) to collect spray from the sides of the tube, but avoid pelleting the sample.
8. Add 1:100 volume of 1 M DTT to each sample to a final concentration of 10 mM DTT.
9. Incubate for 10 min at 95 °C, shaking at 800 rpm.
10. Cool samples to room temperature.
11. Add 1:10 volume of 0.55 M IAA to each sample to a final concentration of 55 mM IAA (*see* **Note 3**).
12. Incubate at room temperature for 20 min, avoiding exposure to light.
13. Centrifuge sample at 10,000× *g* for 5 min at 4 °C, and discard pellet to remove tissue debris (*see* **Note 4**).
14. Add ice-cold 100% acetone to a final concentration of 80%.
15. Incubate overnight at −20 °C.
16. Centrifuge samples at 10,000× *g* at 4 °C for 10 min.
17. Remove supernatant and wash pellet with 1 mL of 80% ice-cold acetone.
18. Repeat steps 16 and 17 above.

19. Air-dry pellet.
20. Add 100 μL of 8 M urea/40 mM HEPES.
21. Using a pipet, disrupt the pellet from the bottom of the tube.
22. Using a water bath sonicator, sonicate samples in an ice bath for eight cycles of 1 min on/1 min off to completely dissolve the pellet.
23. Centrifuge samples for 5 min at 10,000× *g* and 4 °C, and collect supernatant in a 2 mL low-bind microcentrifuge tube (*see* **Note 4**).
24. Quantify sample protein concentrations using a suitable assay (e.g., bicinchoninic acid protein assay, bovine serum albumin tryptophan fluorescence assay).
25. Add ABC to each sample to reach a final concentration of 2 M urea.
26. Collect the appropriate volume to obtain 100 μg of protein (values determined in step 24).
27. Add Lys-C/trypsin protease mixture 2:50 (v/w) enzyme to protein.
28. Tap gently to mix.
29. Incubate overnight at room temperature.
30. Add 1:10 stopping solution.

3.2 Protein Extraction for Murine Fluid Samples

1. Thaw samples and filter using a 0.22 μm filter. Expel filtered sample into a 1.5 mL low-bind microcentrifuge tube.
2. Heat sample to 95 °C for 10 min, and then cool to room temperature.
3. Add 1/3 sample volume of 8 M urea/40 mM HEPES.
4. Using a water bath sonicator, sonicate samples in an ice bath for eight cycles of 1 min on/1 min off.
5. Add 1:100 volume of 1 M DTT to each sample to a final concentration of 10 mM DTT.
6. Incubate at room temperature for 30 min.
7. Add 1:10 volume of 0.55 M IAA to each sample to a final concentration of 55 mM IAA.
8. Incubate at room temperature for 20 min, avoiding exposure to light.
9. Add 4 μL of 0.5 μg/μL Lys-C/trypsin protease mixture.
10. Tap gently to mix.
11. Incubate overnight at room temperature.
12. Add 1:10 stopping solution.

3.3 Peptide Purification

1. Prepare an equal number of STAGE tips [18] to samples by packing six layers of C18 into a p200 pipet tip.
2. Wash STAGE tips with 100 μL of 100% ACN.
3. Wash STAGE tips with 50 μL of Buffer B.
4. Wash STAGE tips with 200 μL of Buffer A (*see* **Note 5**).
5. Centrifuge samples at 10,000× *g* at 4 °C for 10 min.
6. Load supernatant from each sample into a separate STAGE tip (*see* **Note 6**).
7. Centrifuge the tip for 1–2 min. Discard the flow-thru.
8. Wash tip with 200 μL of Buffer A.
9. Add and elute peptides from STAGE tip using 50 μL of Buffer B, collecting flow-thru in a 0.2 mL microcentrifuge tube (*see* **Note 7**).
10. Speed vac samples at 45 °C for 60–80 min or until no liquid remains.

3.4 TMT Labeling (See Note 8)

1. Add 100 μL of 100 mM TEAB to all samples.
2. Vortex samples for 1.5 min.
3. Using a water bath sonicator, sonicate samples in an ice water bath for four cycles (120 s on/60 s off).
4. Repeat steps 2 and 3 above an additional three times.
5. Pulse-centrifuge samples (approximately 10 s) to collect spray from sides, but avoid pellet formation.
6. Quantify protein content of each sample (e.g., measure absorbance at 280 nm) (*see* **Note 9**).
7. Based on the protein concentration determined, add the appropriate volume of sample to obtain 10 μg of peptide into a separate 0.2 mL microcentrifuge tube for each sample.
8. Add the corresponding selected TMT reagent to each sample at 1:10 (sample/label) ratio.
9. Incubate at room temperature for 1 h.
10. Add 5 μL of 5% hydroxylamine to quench the reaction.
11. Incubate for 15 min at room temperature.
12. Combine samples into desired multiplexing combinations.
13. Speed vac for 3–5 h at 45 °C or until no liquid remains.
14. Remove excess unincorporated label from the sample using a desalting spin column, according to the manufacturer's instructions.

3.5 High-pH Fractionation

1. Resuspend dried TMT-labeled samples in 300 μL of 0.1% TFA.
2. Vortex samples for 1 min.
3. Use a high-pH fractionation kit, as directed by the manufacturer to fractionate samples.
4. Speed vac for ~1 h at 45 °C until no liquid remains.

3.6 Mass Spectrometry (MS)

1. Reconstitute dried fractionated peptides in 10 μL of Buffer A.
2. Inject the liquid volume equivalent of 1 μg peptide onto a high-performance liquid chromatography column.
3. Separate peptides over an ACN gradient (*see* **Note 10**).
4. Acquire full MS scans using an Orbitrap mass spectrometer operated in data-dependent acquisition mode (*m/z* 375–1575) with Orbitrap resolution of 60,000.

3.7 Data Analysis

1. Import RAW mass spectrometer output files into processing software (*see* **Note 11**).
2. Set all fractions (1–8) from each combined TMT sample as an "experiment."
3. If the experimental design includes a standardized channel to normalize comparison between TMT samples, set the associated isobaric mass tag as the reference channel.
4. Set reporter ion to MS2, and import isobaric mass tag information and batch corrections (*see* **Note 12**).
5. Set normalization as "weighted ratio to reference channel."
6. Set carbamidomethylation of cysteine as a fixed modification and N-acetylation and oxidation of methionine as variable modifications.
7. Enable quantification and set the isobaric weight exponent to 0.75.
8. Enable match between runs.
9. Set minimum peptides for protein identification to two.
10. Search raw MS files against species-specific databases (e.g., FASTA files from UniProt).
11. Import the processed MS file to data analysis and visualization software (*see* **Note 13**).
12. Import the "Reporter Intensity Corrected [...]" columns for all samples.
13. Filter rows for reverse hits and potential contaminants.
14. Log_2 transform all intensity values.
15. Add labels to each column corresponding to the sample name.
16. Remove columns corresponding to the reference channels.

17. Group columns by "experiment."
18. Filter rows by valid values (>50%, in at least one group).
19. Annotate rows with taxonomic identification.
20. Replace missing values drawing from a normal distribution originating at the median of each column with a downshift of 1.8 and width of 0.3.
21. Proceed with statistical analysis and data visualization in an experiment-oriented manner (*see* **Note 14**).
22. Best practices for proteomics experiments supports deposition of raw and processed MS files in a publicly available repository (*see* **Note 15**).

4 Notes

1. Stock solutions of IAA, DTT, and 80% acetone may be prepared ahead of time and stored at −20 °C until required. For DTT and IAA, after use, discard remaining thawed stock.
2. During protein extraction, samples should be kept on ice whenever possible.
3. IAA should be prepared and stored in the dark, when possible, with minimal exposure to light during use.
4. Steps designed to remove debris during protein extraction may be performed discretionally, depending on the amount of debris present.
5. During STAGE tipping, ensure all washes flow through easily, but not too quickly (ideally ~30 s). If samples flow through too quickly, discard and make a new tip as it is improperly packed.
6. It is normal for very minimal or no debris to be present.
7. Liquid can be eluted using either a syringe or by centrifugation.
8. During TMT labeling, multiple standards may be used, including organ-specific and global standards, allowing for better normalization and comparison across samples.
9. All samples within a sample must be standardized to the same protein concentration, as quantification is relative.
10. Liquid chromatography gradient length and ACN concentrations will depend on the system used; data from this protocol used a 24-min gradient.
11. Examples of software with this functionality include MaxQuant, Proteome Discoverer, and FragPipe [19, 20].
12. Batch corrections are supplied with the TMT kit by the manufacturer to account for any variation between batches of the product.

13. Examples of software with this functionality include Perseus and Proteome Discoverer [21].
14. Statistical interpretation used a Benjamini–Hochberg false discovery rate (FDR) correction with a cutoff of 0.05 [22].
15. PRoteomics IDEntifications (PRIDE) Database supports the upload of proteomics data sets [23].

Acknowledgments

The authors thank the members of the Geddes–McAlister lab for their informative and constructive feedback on project design and manuscript preparation, as well as Samiha Kazi and Kiyan Kheradvar for technical assistance.

Author Contributions B.M. & J.G.-M. conceptualized and designed the study. B.M. & F.R.-D. performed the experiments and data analysis. B.M. designed and developed the figures and wrote the first draft of the manuscript. B.M., F.R.-D., A.D., & J. G.-M. edited the manuscript and wrote the final version. All authors have read and approved the submitted manuscript.

Funding

This project is supported, in part, by the University of Guelph, the Canada Foundation for Innovation (JELF 38798), and a Canadian Institutes of Health Research (Project Grant) for J.G.-M. and by Université Laval (Québec, QC), CHU de Québec–Université Laval Research Center, Génome Québec (Genomic Integration Program), and Fonds de Recherche du Québec–Nature et Technologies (Team Grant) for A.D.

References

1. Aebersold R, Mann M (2016) Mass-spectrometric exploration of proteome structure and function. Nature 537(7620): 347–355. https://doi.org/10.1038/nature19949
2. Toby TK, Fornelli L, Kelleher NL (2016) Progress in top-down proteomics and the analysis of proteoforms. Annu Rev Anal Chem (Palo Alto, Calif.) 9(1):499–519. https://doi.org/10.1146/annurev-anchem-071015-041550
3. Tran JC, Zamdborg L, Ahlf DR et al (2011) Mapping intact protein isoforms in discovery mode using top-down proteomics. Nature 480(7376):254–258. https://doi.org/10.1038/nature10575
4. Gillet LC, Leitner A, Aebersold R (2016) Mass spectrometry applied to bottom-up proteomics: Entering the high-throughput era for hypothesis testing. Annu Rev Anal Chem 9(1):449–472. https://doi.org/10.1146/annurev-anchem-071015-041535
5. Aebersold R, Bensimon A, Collins BC et al (2016) Applications and developments in targeted proteomics: From SRM to

DIA/SWATH. Proteomics 16(15–16): 2065–2067. https://doi.org/10.1002/pmic.201600203

6. Hartlova A, Krocova Z, Cerveny L et al (2011) A proteomic view of the host-pathogen interaction: The host perspective. Proteomics 11(15):3212–3220. https://doi.org/10.1002/PMIC.201000767
7. Sukumaran A, Woroszchuk E, Ross T et al (2021) Proteomics of host–bacterial interactions: new insights from dual perspectives. Can J Microbiol 67(3):213–225. https://doi.org/10.1139/cjm-2020-0324
8. Ball B, Bermas A, Carruthers-Lay D et al (2019) Mass spectrometry-based proteomics of fungal pathogenesis, host–fungal interactions, and antifungal development. J Fungi 5(2):52. https://doi.org/10.3390/jof5020052
9. Meissner F, Geddes-McAlister J, Mann M et al (2022) The emerging role of mass spectrometry-based proteomics in drug discovery. Nat Rev Drug Discov 21(9):637–654. https://www.nature.com/articles/s41573-022-00409-3
10. Sukumaran A, Ball B, Krieger JR et al (2022) Cross-kingdom infection of macrophages reveals pathogen- and immune-specific global reprogramming and adaptation. mBio 13(4): e0168722. https://doi.org/10.1128/mbio.01687-22
11. Rieckmann JC, Geiger R, Hornburg D et al (2017) Social network architecture of human immune cells unveiled by quantitative proteomics. Nat Immunol 18(5):583–593. https://doi.org/10.1038/ni.3693
12. Sukumaran A, Coish J, Yeung J et al (2019) Decoding communication patterns of the innate immune system by quantitative proteomics. J Leukoc Biol 106(6):1221–1232. https://doi.org/10.1002/JLB.2RI0919-302R
13. Ball B, Woroszchuk E, Sukumaran A et al (2021) Proteome and secretome profiling of zinc availability in *Cryptococcus neoformans* identifies Wos2 as a subtle influencer of fungal virulence determinants. BMC Microbiol 21(1): 341. https://doi.org/10.1186/s12866-021-02410-z
14. Muselius B, Sukumaran A, Yeung J et al (2020) Iron limitation in *Klebsiella pneumoniae* defines new roles for Lon protease in homeostasis and degradation by quantitative proteomics. Front Microbiol 11:546. https://doi.org/10.3389/fmicb.2020.00546
15. Sukumaran A, Pladwig S, Geddes-McAlister J (2021) Zinc limitation in *Klebsiella pneumoniae* profiled by quantitative proteomics influences transcriptional regulation and cation transporter-associated capsule production. BMC Microbiol 21(1):1–15. https://doi.org/10.1186/s12866-021-02091-8
16. Muselius B, Dalvai-Roux F, Droit A et al (2023) Resolving the temporal splenic proteome during fungal infection for discovery of putative dual perspective biomarker signatures. J Am Soc Mass Spectrom 34(9):1928–1940. https://doi.org/10.1021/jasms.3c00114
17. Kronstad JW, Attarian R, Cadieux B et al (2011) Expanding fungal pathogenesis: *Cryptococcus* breaks out of the opportunistic box. Nat Rev Microbiol 9(3):193–203. https://doi.org/10.1038/nrmicro2522
18. Rappsilber J, Mann M, Ishihama Y (2007) Protocol for micro-purification, enrichment, pre-fractionation and storage of peptides for proteomics using StageTips. Nat Protoc 2(8): 1896–1906. https://doi.org/10.1038/nprot.2007.261
19. Cox J, Mann M (2008) MaxQuant enables high peptide identification rates, individualized p.p.b.-range mass accuracies and proteome-wide protein quantification. Nat Biotechnol 26(12):1367–1372. https://doi.org/10.1038/nbt.1511
20. Kong AT, Leprevost FV, Avtonomov DM et al (2017) MSFragger: Ultrafast and comprehensive peptide identification in mass spectrometry–based proteomics. Nat Methods 14(5):513–520. https://doi.org/10.1038/nmeth.4256
21. Tyanova S, Temu T, Sinitcyn P et al (2016) The Perseus computational platform for comprehensive analysis of (prote) omics data. Nat Methods 13(9):731–740. https://doi.org/10.1038/nmeth.3901
22. Benjamini Y, Hochberg Y (1995) Controlling the false discovery rate: A practical and powerful approach to multiple testing. J Royal Stat Soc Ser B (Methodological) 57(1):289–300. https://doi.org/10.1111/j.2517-6161.1995.tb02031.x
23. Perez-Riverol Y, Csordas A, Bai J et al (2019) The PRIDE database and related tools and resources in 2019: Improving support for quantification data. Nucleic Acids Res 47 (D1):D442–D450. https://doi.org/10.1093/nar/gky1106

Part IV

Microscopy

Chapter 10

Electron Microscopy of *Cryptococcus neoformans*: Processing Challenges to Avoid Artifacts

Glauber R. de S. Araújo, Bruno Pontes, and Susana Frases

Abstract

This chapter describes methodological details for preparing specimens of *Cryptococcus neoformans* (although it can be applied to any species of the genus) and their subsequent analysis by scanning and transmission electron microscopy. Adaptations to conventional protocols for better preservation of the sample, as well as to avoid artifacts, are presented. The protocols may be used to examine both the surface ultrastructure and the interior of this pathogenic fungus in detail.

Key words Fungi, *Cryptococcus neoformans*, Scanning electron microscopy, Transmission electron microscopy, Polysaccharide capsule, Cell wall

1 Introduction

Scanning and transmission electron microscopies have been extensively used to examine the morphology and structural organization of several biological materials [1, 2]. Despite sample variation, ranging from soft to hard materials, from molecules to cells or tissues, or even from nanosized to macroscopic structures, electron microscopy allows an "in-depth" study of biological samples. While scanning electron microscopy (SEM) focuses on the external characteristics of specimens [2], transmission electron microscopy (TEM) explores their interior [1].

Biological sample preparation for both SEM and TEM are quite similar and follow three conventional steps: fixation in buffered aldehyde, post-fixation using specific solutions, and dehydration in ethanol or acetone [1]. Subsequent steps differ, and, in the case of SEM, involve transferring the sample to a drying apparatus and mounting it on a stub to sputter coat with a heavy metal [2], whereas for TEM, the sample is infiltrated with an epoxy resin, and an ultramicrotome is used to cut ultrathin sections of the sample that are finally contrasted for examination [1]. Such procedures

Erin E. McClelland (ed.), *Cryptococcus neoformans: Methods and Protocols*, Methods in Molecular Biology, vol. 2775,
https://doi.org/10.1007/978-1-0716-3722-7_10,

have been used to analyze a variety of biological samples, such as eukaryotic fungal cells [3], although adaptations are sometimes necessary for special situations, as in the case of *Cryptococcus neoformans* [3, 4].

Fungal cells, such as *C. neoformans*, are surrounded by a plasma membrane followed by a protective outer coat, the cell wall, that is responsible for protection, cell shape, and intercellular communication. The cell wall is structured in multilayers composed of polysaccharides (PS) and proteins [5–8]. Finally, external to the cell wall and anchored to it is a PS capsule, the main virulence factor of this fungus [9].

The cryptococcal PS capsule is highly hydrophilic, 99% of the total mass of the capsule corresponds to water [10], and the high degree of hydration is a major obstacle in its study. Despite the homogeneous appearance, when visualized using optical microscopy, a heterogeneous spatial organization is observed where the proximal part of the capsule has a higher density in relation to the distal region [11, 12], making it difficult for macromolecules, including antibodies and complement proteins, to penetrate close to the cell wall. These features have raised numerous questions about how the capsule structure is engineered, regulated, and maintained and also pose a challenge for electron microscopy.

This chapter will describe the step-by-step procedures used to prepare *C. neoformans* specimens for SEM and TEM to better preserve their surface and interior ultrastructures.

2 Materials

2.1 Before You Begin

1. Determine if you can use a less hazardous substance.
2. Review manufacturer's safety data sheet and additional chemical information.
3. Be familiar with general environment, health, and safety (EHS) university/institute emergency procedures (*see* **Note 1**).

2.2 TEM Using Microwaves

1. Beakers (a 500 mL glass unit and a 50 mL plastic unit).
2. Graduated cylinders (one 25 mL cylinder and a 50 mL cylinder).
3. Volumetric graduated flasks (two 500 mL flasks).
4. Disposable glass Pasteur pipettes, 5.75″ length, short tip (15 pipettes).
5. Glass bottle (sample vials with rubber-lined screw cap—the number required depends on the sample quantity), for example, Electron Microscopy Sciences (catalog number 64250) or TedPella Clear Wheaton® (catalog number 410).
6. Glass stirring rods 6″ length 5 mm (two rods).

7. Glass petri dishes (two dishes).
8. Laboratory precision balance.
9. Calibrated pH meter.
10. Magnetic stirrer with hot plate and stirring bar.
11. Laboratory centrifuge.
12. Laboratory microwave.
13. Glow discharge (e.g., TedPella PELCO easiGlow™, catalog number: 91000S).
14. An oven adjusted to 60 °C (140 °F) and another adjusted to 70 °C (158 °F).
15. Diamond knives (e.g., DiATOME® Ultra Diamond Knife Wet 35°. Manufacturer, DiATOME US; catalog number, 18-UL).
16. Ultramicrotome (e.g., Leica EM UC7).
17. Phosphate-buffered saline (PBS) [osmolality 280 – 320 mOsm/kg], pH 7.3 ± 0.1.
18. Sodium cacodylate trihydrate[1] ($C_2H_6AsNaO_2.3H_2O$—F.W. 214.02 g/mol).
19. Hydrochloric acid for pH adjustment (HCl—F.W. 36.458 g/mol).
20. Osmium tetroxide crystal[2] (OsO_4—F.W. 254.20 g/mol). Locate and verify that appropriate spill cleanup materials are present. *Neutralizer*: corn oil (because of its high percentage of unsaturated bonds). *Absorbent material*: vermiculite or kitty litter.
21. Aqueous glutaraldehyde electron microscopy (EM) grade 25% ($OCHCH_2CH_2CH_2CHO$—F.W. 100.12 g/mol).
22. Aqueous paraformaldehyde EM grade 16% (methanol-free solution) (CH_2O—F.W. 30.03 g/mol).
23. Acetone (C_3H_6O—F.W. 58.08 g/mol).
24. Low-viscosity kit (Spurr) (e.g., Electron Microscopy Sciences, catalog number: 14300) that contains the following: nonenyl succinic anhydride (NSA) modified, vinyl cyclohexene dioxide (ERL 4221), diglycidyl ether polypropylene glycol (DER 736), dimethylaminoethanol (DMAE), three 10 mL syringes, and one 5 mL syringe.
25. UranyLess, substitute for uranyl acetate.
26. Lead citrate ($Pb(NO_3)_2$—F.W. 331.21 g/mol).

---

[1] Arsenic compound. Highly toxic and hygroscopic.

[2] Osmium tetroxide (OsO_4) is extremely toxic and volatile. Use appropriate personal protective equipment (PPE) and be familiar with general university/institute emergency procedures in the EHS Lab Emergency Response Guide.

27. Thiocarbohydrazide ($H_2NNHCSNHNH_2$—F.W. 106.15 g/mol).
28. Quantitative filter paper circles 11 μm, 10.5 s/100 mL/sq inch flow rate, grade 1, 15 mm diameter (e.g., Whatman catalog number: F1000-01A).
29. Wipes for delicate tasks (e.g., Kimtech Science™ Kimwipes™, catalog number: 34120).
30. Poly(vinyl formal) (e.g., Sigma-Aldrich, catalog number: 09823).
31. Parafilm M® Laboratory Film.
32. Carbon steel blades (e.g., TedPella, catalog number: 121-1).
33. Perfect loop for ultramicrotomy (e.g., TedPella, catalog number: 13064).
34. Square copper grid with 100 mesh (e.g., TedPella StrataTek™, catalog number: 12414-CU).
35. Silicone rubber molds (e.g., PELCO® Double End Mold TedPella PELCO, catalog number: 10590).
36. Sodium hydroxide reagent grade, ≥98%, pellets (anhydrous) (NaOH—F.W. 40.00 g/mol).
37. Grid storage box (e.g., Electron Microscopy Sciences, catalog number: 71140).
38. A sealed glass container for storing osmium tetroxide with an unbreakable secondary container.

2.3 SEM

1. PBS [osmolality 280 – 320 mOsm/kg], pH 7.3 ± 0.1.
2. Aqueous glutaraldehyde EM grade 25% ($OCHCH_2CH_2CH_2CHO$—F.W. 100.12 g/mol).
3. Sodium cacodylate trihydrate ($C_2H_6AsNaO_2.3H_2O$—F.W. 214.02 g/mol).
4. Sucrose ($C_{12}H_{22}O_{11}$—F.W. 342.30 g/mol).
5. Magnesium chloride ($MgCl_2$—F.W. 95.21 g/mol).
6. High-precision tweezers (e.g., Dumont Dumostar TedPella, Inc., catalog number: 537).
7. Coverslip thickness N°.1 (0.13 – 0.17 mm), Ø 12 mm borosilicate glass of hydrolytic class (e.g., Waldemar Knittel Glasbearbeitungs GmbH, catalog number: 1VD100 12 Y1A.01).
8. Poly-L-lysine.
9. Absolute ethanol for analysis EMSURE® (CH_3CH_2OH—F.W. 46.07).
10. Critical point dryer (e.g., Leica EM CPD300).
11. SEM pin stub specimen mounts Ø 12.7 mm × 8 mm pin height (e.g., TedPella, Inc., catalog number: 16111).

12. Conductive adhesives (e.g., PELCO Tabs™ carbon conductive tabs TedPella, Inc., catalog number: 16084-1).
13. Steel gripper tweezers/forceps specially made for handling and transporting grooved SEM pin stub mounts (e.g., PELCO tweezers style pin stub mount gripper, 57° angle, 100 mm long, for 12.7 mm pin stubs TedPella, Inc., catalog number: 1667).
14. Sputter coater (e.g., Q300T Plus from Quantum Design Europe).
15. Desiccated or dry compartment to store SEM specimen stubs.

2.4 Preparation of Solutions

$$\text{Grams} = \text{formula weight}\,\frac{\text{g}}{\text{mol}}\,(\text{F.W.}) \times \text{molar mass constant}\,\frac{\text{g}}{\text{mol}}\,(M_u) \times \text{liter} \quad (1)$$

M_u is 0.999,999,999,65(30) × 10^{-3} kg·mol^{-1}.

One liter is equal to 1000 cubic centimeters (cc), 1000 milliliters (mL).

$$\text{Concentration}_{\text{start}} \times \text{Volume}_{\text{start}} = \text{Concentration}_{\text{final}} \times \text{Volume}_{\text{final}} \quad (2)$$

1. 0.2 M sodium cacodylate buffer, pH 7.2: To prepare 500 mL of 0.2 M sodium cacodylate–HCl buffer solution, add 21,402 g of sodium cacodylate in approximately 350 mL of distilled water (Eq. 1). Adjust the pH to 7.2 by adding 1 N HCl, and bring the volume up to 500 mL (osmolality 700 mOsm/kg). Store between 4 and 8 °C (39.2 and 46.4 °F) (in the refrigerator). Caution: It contains arsenic, which is poisonous and carcinogenic and is absorbed through the skin; wear gloves.
2. Glutaraldehyde fixative: To prepare the desired volume of fixative, mix the 25% stock of EM grade glutaraldehyde solution with the stock solution of 0.2 M sodium cacodylate–HCl buffer, pH 7.2 (Eq. 2). The final concentrations must be 2.5% for glutaraldehyde and 0.1 M for cacodylate buffer.
3. 1% OsO_4 in 0.1 M sodium cacodylate buffer: Carefully add the entire ampoule containing 1 g of OsO_4 into an amber glass bottle with excellent sealing, along with 100 mL of 0.1 M sodium cacodylate buffer. After checking that the vial is hermetically closed, shake the amber glass bottle vigorously until the vial breaks. OsO_4 crystal will dissolve slowly in the buffer. Therefore, it must be prepared at least 1 day in advance. Store the vial containing the solution inside another hermetically sealed and unbreakable secondary container that is not oxidizable. OsO_4 is a strong oxidizer and reacts with easily oxidized organic materials to cause fires or explosions. The solution must be stored at 4 – 8 °C (39.2 – 46.4 °F).

Table 1
Proportion for the preparation of Spurr's resin with medium hardness (by batches of 10 grams, approximately 5 blocks per step), combining very-low-viscosity components for rapid specimen penetration, while also promoting clean cuts and beam stability [13, 14]

Ingredient	Mass in grams (g)
ERL 4221	4.10
Diglycidyl ether of polypropylene glycol (D.E.R. 736)	1.43
Nonenyl succinic anhydride (NSA)	5.90
Dimethylaminoethanol (DMAE)	0.1

Caution: Osmium tetroxide (OsO_4) is extremely toxic and volatile. It is essential that osmium solutions are handled in a fume hood (NOT a biological safety cabinet) and used osmium solutions be properly disposed. If in doubt, always consult the EHS of your university/institution (*see* **Note 2**).

4. Post-fixation solution: 0.1 M sodium cacodylate, 0.2 M sucrose, and 2 mM magnesium chloride in ultrapure water (*see* **Note 3**). This step aims to eliminate excess fixative while effectively stabilizing osmolarity.
5. Dehydration series: To prepare the ethanol or acetone dehydration series, use (Eq. 2), and dilute the desired amount in distilled water to obtain 30, 50, 70, and 95% solutions for ethanol and 30, 50, 70, and 90% solutions for acetone.
6. Embedding media—Spurr's resin: To prepare the resin to a firm consistency (neither hard nor soft), on a precision balance using a 50 mL plastic beaker, add the ingredients of Table 1. Accurate weights should be used for best performance and reproducibility. DMAE should be added last, after gently mixing the other components. The resin must be homogenized carefully, minimizing exposure to atmospheric moisture and bubble formation. If there are bubbles, the resin may be negatively pressurized using a vacuum pump. It is highly recommended that the embedding medium be freshly prepared; however, for convenience, the resin may be stored in airtight aliquots in a freezer *without* the DMAE.

3 Methods

3.1 SEM

1. *Cryptococcus neoformans* cells are grown suspended in liquid culture medium. Thus, collect 1 mL of a solution containing 1×10^6 cells/mL, and add it to a 1.5 mL polypropylene graduated microcentrifuge tube with snap cap.

Table 2
Percent, time, and number of repetitions of increasing ethanol dehydration

Ethanol (V:V)	Minutes	Repetition
30%	5	1
50%	5	1
70%	5	1
95%	10	2
Absolute	10	2

2. Use a centrifuge at 3000 relative centrifugal force (RCF)[3] to wash cells three times with PBS pH 7.3 ± 0.1 to completely remove the culture medium.
3. After the last wash, centrifuge again, and replace the PBS with the fixative solution (2.5% glutaraldehyde in 0.1 M sodium cacodylate buffer). Resuspend the cells and incubate for 1 h at room temperature (25 °C [77 °F]).
4. Centrifuge the cells, and replace the fixative solution with the post-fixation solution. Wash the cells three times with the post-fixation solution, for 10 min intervals (*see* **Note 3**).
5. Pour 200 μL of the solution containing cells onto round glass coverslips (thickness N°.1 (0.13 – 0.17 mm), Ø12 mm borosilicate glass of hydrolytic class 1) previously coated with 0.01% (w/v) poly-L-lysine for 30 min. Incubate at least 20 min to allow cells to adhere to coated coverslips.
6. Next, dehydrate the adhered yeast in a series of freshly made solutions of graded absolute ethanol, as described in Table 2 (*see* **Note 4**).
7. Add the coverslips to a critical-point dryer following the manufacturer's instructions. Wait until the entire process is complete.
8. Remove the dried coverslips from the critical-point dryer, and attach them to specimen pin stubs using thick conductive adhesive PELCO Tabs™ (12 mm outer diameter).
9. Take the specimen stubs to the sputter-coater and follow the manufacturer's instructions. After that, the samples are ready to visualize in a scanning electron microscope.

[3] g Force (RCF) = $(\text{RPM})^2 \times 1.118 \times 10^{-5} \times r$; radius ($r$).

3.2 TEM Using Microwaves

1. *Cryptococcus neoformans* cells are grown suspended in liquid culture medium. Thus, collect 1 mL of a solution containing 1×10^6 cells/mL, and add it to a 1.5 mL polypropylene graduated microcentrifuge tube with a snap cap.
2. Wash three times with PBS, for 10 min intervals. Use a centrifuge at 3000 RCF to separate the cells from the solution during washes.
3. After the last wash, centrifuge again, and replace the PBS with the fixative solution (2.5% glutaraldehyde fixative). Resuspend the cells, and incubate for 1 h at room temperature (25 °C [77 °F]).
4. Centrifuge the cells and replace the fixative with a 0.1 M cacodylate solution, pH 7.2. Wash the cells three times, for 10 min intervals, using this solution.
5. After the last wash, centrifuge again, and replace the cacodylate solution with the post-fixative solution (1% osmium tetroxide). Add the sample to a microwave, and perform a 15 s pulse at a power of 350 watts[4] [W] (*see* **Note 5**).
6. Wash three times with 0.1 M cacodylate pH 7.2 solution, for 10 min intervals. Use a centrifuge at 3000 RCF to separate the cells from the solution during washes.
7. After the last wash, centrifuge again, and replace the cacodylate solution with 1% thiocarbohydrazide. Add the sample to a microwave, and perform a 5 s pulse at a power of 350 [W].
8. Wash three times with 0.1 M cacodylate, pH 7.2 solution, for 10 min intervals. Use a centrifuge at 3000 RCF to separate the cells from the solution during washes.
9. After the last wash, centrifuge again, and replace the cacodylate solution with 1% osmium tetroxide. Add the sample to a microwave, and perform a 15 s pulse at a power of 350 [W].
10. Wash three times with 0.1 M cacodylate, pH 7.2 solution, for 10 min intervals. Use a centrifuge at 3000 RCF to separate the cells from the solution during washes.
11. Subsequently, dehydrate the yeast in a series of freshly made solutions of graded absolute acetone, as described in Table 3. The sample is centrifuged at 3000 RCF between each step to allow for solvent exchanges. After the last step in 100% acetone, the infiltration procedure starts. Centrifuge again, and replace the solution with a new solution made of 5 parts acetone and 1 part Spurr's resin (*see* **Note 6**). Follow the steps described in Table 4 to complete the procedure.
12. Carefully pour the sample into silicone molds, and let dry for 12 h at 70 ± 1 °C (158 ± 2 °F). *See* Fig. 1.

---

[4] The watt [W] is the unit of power in the International System of Units (SI), equal to 1 joule per second 1 W = 1 J/s = 1 N.m/s = 1 kg.m^2.s^{-3}.

Table 3
Percentage, time, power, and temperature of increasing dehydration by acetone

Step	Description	Time	Power [W]	Temp (°C)	Temp (°F)
1	30% acetone	10 s	350	60	140
2	50% acetone	10 s	350	60	140
3	70% acetone	10 s	350	60	140
4	90% acetone	10 s	350	60	140
5	100% acetone	10 s	350	60	140
6	100% acetone	10 s	350	60	140

Table 4
Resin/acetone ratio, time, power, and temperature for low-viscosity Spurr embedding

Step	Description	Time	Power [W]	Temp (°C)	Temp (°F)
7	5 parts acetone: 1 part Spurr	3 min	350	60	140
8	3 parts acetone: 1 part Spurr	3 min	350	60	140
9	2 parts acetone: 1 part Spurr	3 min	350	60	140
10	1 parts acetone: 1 part Spurr	3 min	350	60	140
11	1 part acetone: 2 parts Spurr	3 min	350	60	140
12	1 part acetone: 3 parts Spurr	3 min	350	60	140
13	1 part acetone: 5 parts Spurr	3 min	350	60	140
14	Pure Spurr	12 h	–	75	167

13. After proper polymerization, the resin block containing the sample is removed from the mold, roughly trimmed, and sanded to acquire the conformational shape of a trapezoid (classical shape). It is subsequently trimmed to mold the trapezoid faces in a regular and symmetrical way.
14. Next, the block containing the trapezoid is partitioned using a diamond knife into several ultrathin sections (60 ± 5 nm—gray/silver on the color scale for section thickness) [14] using an ultramicrotome Leica Ultramicrotome (EM UC7), following the manufacturer's instructions. Note that the ultrathin sections are made from the trapezoid region.
15. Carefully collect the ultrathin sections, and place them in EM grids using a "Perfect Loop" or with the grid directly in the liquid.

    The grids are scaffolds for delicate sections of resin-infiltrated material, just as the slide is for light microscopy. Therefore, this grid should provide good support for the

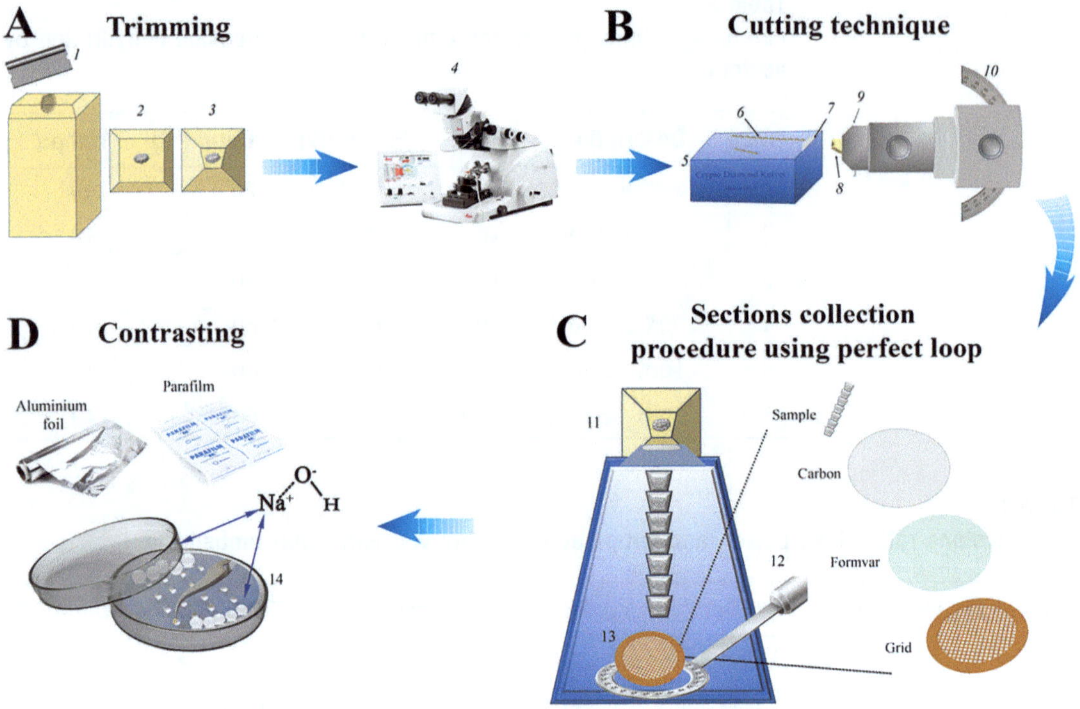

Fig. 1 Illustrative scheme of the main post-polymerization steps:
(**a**) A trimming block is imperative as the shape and the surface of the sample determine the quality and the thickness of sections that may be achieved: (*1*) Blade. (*2*) The shape of the trimmed block should be trapezoidal with a flat top; this shape reduces block compression, smoothes section surfaces, and improves structural preservation, as well as mitigates damage to the knife edge. (*3*) Fountain view of the block trimmed in the form of a trapeze with a smooth top (area that will be perpendicular to the knife). (*4*) Ultramicrotome
(**b**) Ultramicrotome sectioning assembly: (*5*) Knives with boat. (*6*) The collection of the section is collected on the reflecting pool boat. (*7*) Knives (usually diamond, but may be high-quality glass or tungsten carbide). (*8*) Sample. (*9*) Chuck Holder. (*10*) Specimen goniometer.
(**c**) The collection of the slices: (*11*) Front view of the block containing the sample. The collection of the slices may be performed using tweezers holding a grid directly or preferably with a perfect loop (*12*). Then, owing to the high surface tension of water, slices may float into a body of water until the perfect loop is left over a grid that is on a filter paper inside a petri dish lined with parafilm. (*13*) The grid must have a mesh compatible with the cuts and well-defined grid bars. First, the grid is coated with 10 nm Formvar and 20 – 30 nm carbon (this step aims to obtain a scaffold for the sample, thus preventing the sample from dropping between the mesh/ slits).
(**d**) Contrast in TEM is mainly produced by electron scattering from the specimen; heavy elements are most commonly used to increase contrast. Electron-dense dyes may be added during resin infiltration (osmium tetroxide or uranyl acetate en bloc staining) or after ultramicrotomy (denominated post-staining). For *Cryptococcus* spp., the post-staining dyeing method proposed by Reynolds [13] is highly effective. (*14*) On a petri dish entirely covered with aluminum foil and coated with parafilm, float a grid, section side down, on a drop of a uranyl acetate solution for 10 min. After blotting off the stain, the grid is rinsed thoroughly with CO_2-free water (double distilled) to remove any residual unbound stain, followed by 5 min of lead citrate staining, and then is again rinsed with water
To prevent precipitation of the metallic salts during the staining process, especially the precipitation of lead citrate by exposure to CO_2, it is crucial to add NaOH pellets around the petri dish to reduce the presence of atmospheric carbon dioxide

material without affecting the integrity of the sample or interfering with the electron/sample interaction. One way to provide this support is to overlay the grid with Formvar® (polyvinyl formal) with thickness ranging from 5 to 50 nm. Follow the procedures below:

16. Cleaning grids. Place the grids in a beaker containing 50 mL of EM acetone, and sonicate for 5 min. Let the grids air-dry on filter paper. Dry all clean grids onto a petri dish with filter paper.
17. Coating grids. Clean a microscope slide by wiping with wipes for delicate tasks (e.g., Kimtech Science™ Kimwipes™) and absolute ethanol. Gently submerge the clean, dry slide into the vial of Formvar. Quickly raise the slide and hold for approximately 30 s. With a blade, gently cut the perimeter of the slide to facilitate its removal. You can do this on both sides of the slide, but it takes practice to subtract the Formvar film from both sides. Immediately after the previous step, submerge the slide with an incline of 30° (in relation to the water mirror) slowly and at constant speed in a beaker containing distilled water to detach the Formvar film from the slide. If you want to get both sides, the slide must be placed perpendicular to the water mirror; it's tough, but not impossible. Remove the freshly cleaned grids with antistatic tweezers, and place them neatly on the Formvar film, leaving approximately 4 ± 1 mm between the grids. Next, with a piece of parafilm, push one end of the Formvar film into the water, and remove it. The Formvar film will adhere to the parafilm. Next, dry the parafilm and store in a glass petri dish at room temperature. Before using the grids, it is necessary to cut the perimeter with the help of a blade to detach from the rest of the Formvar.
18. Stain with lead citrate and uranyl acetate.

    Lead citrate has high electron density and broad affinity for cell ultrastructure and improves the cell membrane system and lipid contrast [15].

    Uranyl acetate increases the electrode density of the sample at higher density due to the relatively high atomic weight of uranium ($_{92}U$), with 92 protons and 92 electrons, providing excellent image contrast without debris. Uranyl ions bind proteins and lipids with sialic acid carboxyl groups, such as glycoproteins and gangliosides, and nucleic acid phosphate groups of DNA and RNA [14]. Gangliosides are glycolipids and are concentrated on cell surfaces, whereas glycoproteins are abundant in all membranes and part of the glycocalyx [15, 16]. Therefore, glycoproteins provide good contrasting results with membranes, nucleic acids, and nucleic acids containing protein complexes, such as ribosomes [3].
19. Contrasting procedure. Cover the glass petri dish lid with aluminum foil. Coat a petri dish with parafilm. Around the parafilm add NaOH (sodium hydroxide) pellets. Add a drop

of 1% lead citrate (approximately 20 μL) around the NaOH pellets, one drop for each sample (Fig. 1d-14). With the help of antistatic tweezers, add the grid with the face containing the sample so that it is in contact with the drop, and incubate it for 5 – 6 min. In a small beaker (50 mL), add distilled water (ideally distilled water with neutral pH and CO_2-free). Rinse the grid in this water, and dry the edge by gently touching the filter paper (*see* **Note 7**). In another glass petri dish (also covered with aluminum foil, coated with parafilm and NaOH pellets), add a drop of 3% UranyLess (approximately 20 μL) around the NaOH pellets, one drop for each sample, and incubate for 15 min. In another beaker (distinct from step 4), add distilled water. Rinse the grid in this water and dry the edge by gently touching the filter paper. Finally, wait for at least 30 min for the grids to dry, face up on filter paper. After that, the samples are ready to visualize in a transmission electron microscope.

4 Notes

1. All glassware must be exclusive for microscopy to avoid contamination or damage to other sectors (e.g., cell culture).
2. A given volume of osmium tetroxide solution may be disposed of by combining with two to three times the volume of corn oil. Spilled osmium tetroxide solution should be covered with dried milk powder and then transferred to a container for disposal in accordance with local restrictions. If in doubt, always consult the EHS of your institution/department.
3. Sucrose is used to maintain the osmotic flow and the tone of the cell body since the capsule and cell wall mostly comprise carbohydrates. However, magnesium chloride stabilizes the divalent cations of the capsule and keeps its structure stable.
4. Dehydration is the crucial step to preserve the integrity of the fragile polysaccharide fibers, so it must be performed in a meticulous way, where the collapse of the fibers by air-drying or extraction of the polysaccharides by exceeding the time in dehydration must be prevented.
5. The microwave is used to increase the kinetic energy of molecules (fixatives and resins).
6. Spurr's resin low-viscosity (180 cP) embedding mixture is recommended because of its excellent penetration qualities for biological samples, which provide good and rapid infiltration [17].
7. To attenuate lead citrate precipitation from CO_2 exposure, NaOH pellets are added to the petri dish to reduce atmospheric

CO_2. Also, ideally, the rinse water should be CO_2-free (double distilled). One of the most critical processing steps for TEM is contrasting, due to the "unpredictable" risk of precipitation of metallic salts, mainly lead precipitation. Thus, the use of NaOH pellets minimizes the likelihood of precipitation.

Acknowledgments

This work was supported by the Brazilian agencies Conselho Nacional de Desenvolvimento Científico e Tecnológico (CNPq); Coordenação de Aperfeiçoamento de Pessoal de Nível Superior (CAPES)—Finance Code 001; and Fundação Carlos Chagas Filho de Amparo à Pesquisa do Estado do Rio de Janeiro (FAPERJ).

References

1. Bozzola JJ (2007) Conventional specimen preparation techniques for transmission electron microscopy of cultured cells. Methods Mol Biol 369:1–18
2. Bozzola JJ (2014) Conventional specimen preparation techniques for scanning electron microscopy of biological specimens. Methods Mol Biol 1117:133–150
3. Araújo GR d S, Alcantara C d L, Rodrigues N *et al.* (2021) Ultrastructural study of *Cryptococcus neoformans* surface during budding events. Front Microbiol 12:609244
4. Araújo GR, Fontes GN, Leão D *et al.* (2016) *Cryptococcus neoformans* capsular polysaccharides form branched and complex filamentous networks viewed by high-resolution microscopy. J Struct Biol 193:75–82
5. Cantu D, Carl Greve L, Labavitch JM *et al.* (2009) Characterization of the cell wall of the ubiquitous plant pathogen *Botrytis cinerea.* Mycol Res 113:1396–1403
6. Fontaine T, Simenel C, Dubreucq G *et al.* (2000) Molecular organization of the alkali-insoluble fraction of *Aspergillus fumigatus* cell wall. J Biol Chem 275:27594–27607
7. Roncero C (2002) The genetic complexity of chitin synthesis in fungi. Curr Genet 41:367–378
8. Ruiz-Herrera J, Martínez AI, Sentandreu R (2002) Determination of the stability of protein pools from the cell wall of fungi. Res Microbiol 153:373–378
9. Casadevall A, Coelho C, Cordero RJB *et al.* (2019) The capsule of *Cryptococcus neoformans.* Virulence 10:822–831
10. Maxson ME, Cook E, Casadevall A *et al.* (2007) The volume and hydration of the *Cryptococcus neoformans* polysaccharide capsule. Fungal Genet Biol 44:180–186
11. Bryan RA, Zaragoza O, Zhang T *et al.* (2005) Radiological studies reveal radial differences in the architecture of the polysaccharide capsule of *Cryptococcus neoformans.* Eukaryot Cell 4: 465–475
12. Gates MA, Thorkildson P, Kozel TR (2004) Molecular architecture of the *Cryptococcus neoformans* capsule. Mol Microbiol 52:13–24
13. Reynolds ES (1963) The use of lead citrate at high pH as an electron opaque stain in electron microscopy. J Cell Biol 17:208–212
14. Lee PD (1958) A study of section thickness and physical distortion produced during microtomy. J Biophysic Biochem Cytol 4(3): 233–242
15. Santhana Raj L, Nur Dayana M, Norzarila S *et al.* (2021) New preparation method of Reynolds' stain for transmission electron microscope for liver samples. Int J Med Nano Res 8:033
16. Lin YW (2020) Uranyl binding to proteins and structural-functional impacts. Biomolecules 10(3):457
17. Spurr AR (1969) A low-viscosity epoxy resin embedding medium for electron microscopy. J Ultrastruct Res 26(1):31–43

Part V

Interactions with Macrophages

Chapter 11

Assessing Phagocytosis of *Cryptococcus neoformans* Cells in Human Monocytes or the J774 Murine Macrophage Cell Line

Emmanuel Lafont, Aude Sturny-Leclère, Carolina Coelho, Fanny Lanternier, and Alexandre Alanio

Abstract

Monocyte/macrophage cells play a central role in innate immunity against *C. neoformans* and *C. gattii*, species known to cause human disease. *Cryptococcus* is the only fungal genus known to possess such a large extracellular polysaccharide capsule, which impacts interactions of innate cells with the yeast. This interaction results in different fates, such as phagocytosis and intracellular proliferation and, as the interaction progresses, vomocytosis, cell-to-cell transfer, lysis of macrophages, or yeast killing. Differentiating internalized versus external *Cryptococcus* cells is thus essential to evaluate monocyte–macrophage phagocytosis. We describe here a protocol that allows quantification of *Cryptococcus* spp. phagocytosis using quantitative flow cytometry in human monocytes and a murine macrophage cell line (J774).

Key words *Cryptococcus neoformans*, Human monocyte, Murine cell line (J774), Flow cytometry, Fluorescent microscopy, Phagocytosis, Proliferation

1 Introduction

Cryptococcus neoformans is an environmental encapsulated yeast and a major fungal invasive pathogen for patients with cellular immune defects, mostly the ones associated with AIDS [1–3]. Every year, this fungus is responsible for more than 200,000 meningoencephalitis and/or pneumonia cases, with a mortality of approximately 80% [3]. *C. neoformans* is a facultative intracellular fungal organism [4]. Monocytes/macrophages represent the main cellular effector of innate immunity against *Cryptococcus* spp. [5]. Monocytes are also described as vehicles that allow yeast invasion of the central nervous system by crossing the blood–brain barrier through a "Trojan horse" mechanism [6, 7]. The capsule, mainly comprised of glucuronoxylomannan (GXM), galactoxylomannan (GalXM), and mannoproteins [8], is the main virulence factor of *Cryptococcus*

Erin E. McClelland (ed.), *Cryptococcus neoformans: Methods and Protocols*, Methods in Molecular Biology, vol. 2775, https://doi.org/10.1007/978-1-0716-3722-7_11,

neoformans. The capsule as a whole and GXM in particular have antiphagocytic properties with less than 1% of phagocytosis of the yeast by human or murine monocyte/macrophage in the absence of opsonization [9–14]. Three main opsonins are used to increase yeast uptake by macrophages: complement [15], anticapsular antibody [16, 17] (both contained in serum), and surfactant protein D [18]. Even though *C. neoformans* is not an organism that causes major inflammation and cell toxicity, cell damage induced by *C. neoformans* has been observed at different levels (molecular, cellular, tissue, and whole organism) [19]. Cellular damage includes non-lytic or lytic exocytosis of yeast [1, 19].

To assess phagocytosis, it is necessary to differentiate intracellular and extracellular yeast after the interaction. In the literature, determination of *Cryptococcus neoformans* phagocytosis and proliferation indexes by J774 cells [20–23] and by human monocytes [24–26] relies mainly on microcopy, which is a fastidious and limited method for quantification, except if using very specialized and costly automated microscopes [27, 28]. This protocol focuses on the quantification of phagocytosis upon interaction of human monocytes or J774 murine macrophages and *Cryptococcus neoformans* using quantitative flow cytometry. This protocol can be adapted to other phagocytic cells, including amoeba, by replacing the culture media to the one most appropriate to the phagocytic cell in question.

2 Materials

2.1 Media

1. J774 medium: Dulbecco's modified Eagle's medium with glutamine (DMEM-GlutaMAX, Gibco®) supplemented with 10% heat-inactivated fetal calf serum (FCS) and 1% penicillin–streptomycin (murine complete medium).
2. Monocyte medium: Roswell Park Memorial Institute medium 1640 with glutaMAX (Gibco®) supplemented with 10% FCS (human complete medium).
3. Sabouraud agar (SA).
4. Yeast extract–peptone–glucose (YPD) broth: Mix 10 g of yeast extract, with 20 g of peptone, and 20 g of D-glucose in 1 L of distilled water. Autoclave YPD medium for 20 min at 105 °C. Store at 4 °C.

2.2 Strains and Host Cells

1. *Cryptococcus neoformans* reference strain: frozen stock of H99 strain in 40% glycerol at −80 °C.
2. J774A.1 cells (murine macrophage-like cell line, hereafter J774) derived from a reticulum sarcoma [29].
3. Human monocytes isolated from peripheral blood mononuclear cells.

2.3 Yeast Staining and Labelling Reagents (Fig. 1)

1. Phosphate-buffered saline (PBS).
2. Calcofluor White (CFW) dye at 10 μg/mL in PBS.
3. Anticapsular monoclonal antibody (mAb) E1.
4. Anticapsular mAb 18B7.
5. Guinea pig or mouse sera.
6. Hemocytometer.
7. Fluorescein isothiocyanate (FITC)-labeled horse anti-mouse IgG (anti-IgG–FITC) antibody, final concentration 7.5 μg/mL.
8. Allophycocyanin (APC)-labeled goat anti-mouse IgG (anti-IgG–APC) antibody, final concentration 7.5 μg/mL.
9. Positive selection beads kits (e.g., Miltenyi Biotec, ref 130-050-201).

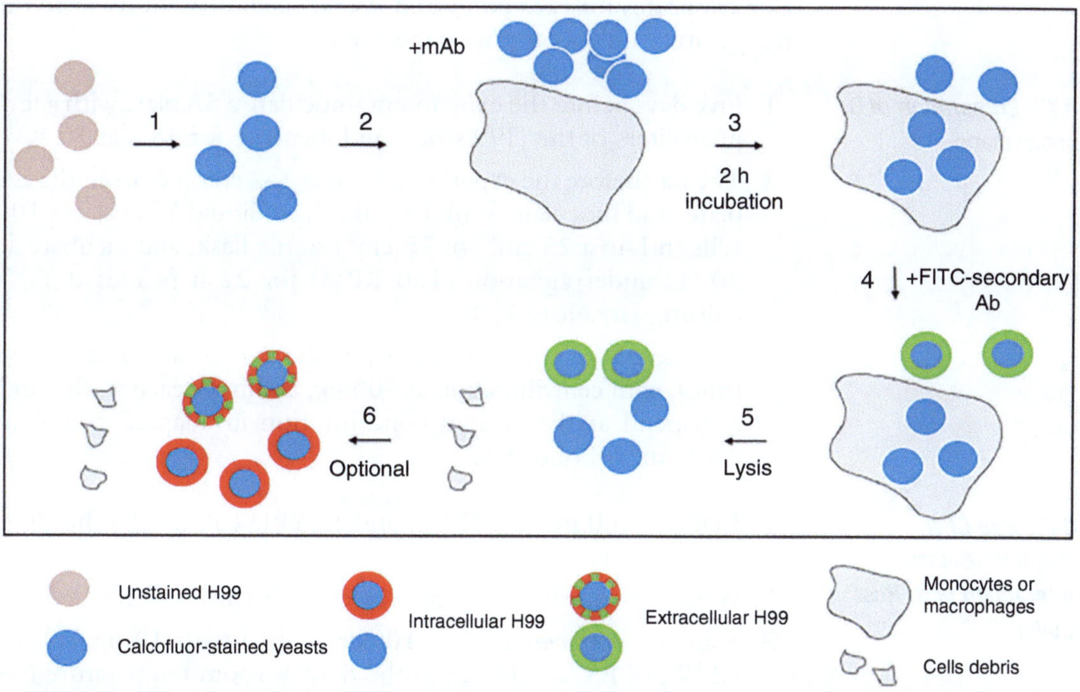

Fig. 1. Schematic representation of *Cryptococcus neoformans* (Cn) labeling steps for flow cytometry analysis of yeast-macrophage interactions
(1) Stain *C. neoformans* with Calcofluor White (CFW) dye. (2) Add 18B7 or E1 monoclonal antibody (mAb) (for yeast opsonization) and CFW-stained *C. neoformans* suspension to the human monocyte (or J774 murine cell) monolayer. (3) After 2 h of incubation, some yeasts are internalized in host cells. (4) Directly stain residual extracellular yeast using a FITC–anti-IgG secondary antibody. Extracellular yeasts harbor high FITC fluorescence, while intracellular yeasts will have negative/low FITC fluorescence. (5) Lyse host cells with water + 0.05% sodium dodecyl sulfate (SDS). (6) After lysis of human monocytes (or J774 murine cells), stain all yeast with APC–anti-IgG (optional). This helps to differentiate yeast from cell debris. Extracellular yeast will be $APC^{high}/FITC^{high}$, intracellular yeast will be $APC^{high}/FITC^{low}$, and cell debris will be $APC^{low}/FITC^{low}$

10. Microbeads conjugated to monoclonal anti-human CD14 antibodies (MACS).
11. MACS columns.
12. MACS magnetic separator.
13. Buffer for positive selection bead kits: PBS, pH 7.2, 0.5% bovine serum albumin, 2 mM EDTA. Keep at 4 °C.
14. AccuCheck Counting Beads.

2.4 Reagents for Phagocytosis Assay

1. 0.05% sodium dodecyl sulfate (SDS) diluted in distilled water.
2. 1.5 mL Eppendorf tube.
3. 1% paraformaldehyde (PFA).

3 Methods

Use triplicates for each condition. Perform all procedures at room temperature unless otherwise specified.

3.1 Preparation of C. neoformans

1. Five days before the experiment, inoculate a SA plate with a few microliters of the H99 stock, and incubate for 48 h at 30 °C.
2. The day before the experiment, select 1–2 colonies from the SA plate, and inoculate 5 mL (or 10 mL) of liquid YPD at 1×10^7 cells/mL in a 25 cm^2 (or 75 cm^2) sterile flask, and incubate at 30 °C under agitation (150 RPM) for 22 h (standard YPD culture) (*see* **Note 1**).
3. On the day of the experiment, wash yeast cells in PBS (three times, with centrifugation at 700× *g*, 5 min for each wash), and resuspend at the desired concentration in adapted complete medium without FCS.

3.2 H99 CFW Staining (Before Interaction with Host Cells)

1. Remove 500 μL of a H99 standard YPD culture (Subheading 3.1, **steps 1**–**3**).
2. Wash pre-cultured yeast cells twice with 1 mL of PBS.
3. Stain *C. neoformans* (1×10^7 cells/mL) with 10 μg/mL of CFW in PBS for 10 min in the dark at room temperature (*see* **Note 2**), and then wash three times with PBS.
4. Count stained yeasts (*see* **Note 3**) and check CFW fluorescence using a UV filter.

3.3 Preparation of Opsonized H99

1. For interaction with J774 cells, add E1 mAb [17] or another opsonic antibody in murine complete medium *without* FCS to H99 at a final concentration of 10 μg/mL (*see* **Note 4**). For interaction with human monocytes, add 18B7 mAb [16] in human complete medium without FCS to H99 at a final

concentration of 5 μg/mL (*see* **Note 4**). For complement opsonization, incubate H99 with 10–20% complement (guinea pig or mouse) for 10 min with end-over-end rotation, and then add the appropriate inoculum to phagocytic cells.

3.4 Preparation of Host J774 Cells

1. Three days before the experiment, thaw one vial of J774A.1 cells (murine macrophage-like cell line, hereafter J774) derived from a reticulum sarcoma [29], previously frozen in 95% FCS and 5% dimethyl sulfoxide in liquid nitrogen, in a 75 cm^2 cell tissue culture flask, and incubate in 5 mL of murine complete medium at 37 °C in the presence of 5% CO_2 (*see* **Note 5**). Cells can be used between 10 and 35 passages.
2. The day before the experiment, detach the cells with a cell scraper and count them.
3. Seed 1×10^5 J774 cells in 24-well plates in murine complete medium (500 μL/well), and incubate at 37 °C in the presence of 5% CO_2. Appropriate scaling-up is calculated by maintaining # cells/area of tissue culture well/dish.

3.5 Preparation of Human Monocytes

1. Use 20 mL heparinized venous blood per patient (to obtain $1–2 \times 10^6$ monocytes). Store blood at room temperature for the first 24 h after collection.
2. Dilute venous blood from one healthy donor with pre-warmed PBS at 37 °C in a 1:1 ratio in a 50 mL tube.
3. Add gently diluted blood onto Ficoll-Hypaque in a 1:1 ratio in a 50 mL tube.
4. Centrifuge at 1000× *g* at 20 °C without brake for 20 min (total time 30 min).
5. Gently remove the peripheral blood mononuclear cell (PBMC) layer.
6. Wash cells three times with 50 mL of PBS using centrifugation steps at 300× *g* at 20 °C without brake for 30 min.
7. Isolate monocytes from PBMCs using positive selection beads kits following the manufacturer's instructions.
8. Count and check viability of the cells, and adjust cell concentration to 1×10^6 cells/mL in human complete medium without FCS at 4 °C (*see* **Note 3**).
9. Add 0.8×10^5 cells in 0.1 mL per well in a 96-well plate in human complete medium without FCS (*see* **Note 6**).
10. Add 10 μL per well of AccuCheck Counting Beads (initial concentration of 1×10^6 microbeads per milliliter) to allow calculation of monocyte/mL at the end of the experiment.

3.6 Host Cell/ Cryptococcus neoformans Interaction Assay with J774 Cells

1. Use a well to remove the cells and quantify their number to adjust the quantity of yeast adapted for the interaction.
2. Remove the murine complete medium from the cells and wash once in warm PBS. Use murine complete medium without 10% FCS during the 2 h phagocytosis process (*see* **Note** 7).
3. Add 250 μL of E1 Mab (2×, final concentration of 100 μg/mL for phagocytosis experiment) (*see* **Note 8**) and 250 μL of opsonized CFW-stained *Cryptococcus neoformans* suspension to the J774 cell monolayer, at a ratio of 5 yeast for 1 macrophage (5:1). Incubate 24-well plates at 37 °C and 5% CO_2 from 15 min to over 1 day, but generally 2 h is a good interval of time to measure maximal phagocytosis. Final volume should be 0.5 mL per well (*see* **Note 9**).
4. After 2 h incubation, remove the medium, including nonadherent extracellular yeast, by gently washing the cell layer twice with PBS.

3.7 Host Cell/ Cryptococcus neoformans Interaction Assay with Human Monocytes

1. Use a well to remove the cells and quantify their number to adjust the quantity of yeast adapted for the interaction.
2. The day of the experiment, add the following to each well, at the same time: 18B7 mAb (*see* **Note 8**), opsonized CFW-stained *Cryptococcus neoformans* suspension (2×10^5 *Cryptococcus neoformans* per well, with a yeast/monocyte ratio of 2.5:1), and human monocytes (0.8×10^5 per well) in human complete medium without FCS. Final volume should be 100 μL per well (*see* **Note 6**). Use human complete medium without 10% FCS during the 2 h phagocytosis process (*see* **Note** 7).
3. Incubate 96-well plates at 37 °C and 5% CO_2 for 2 h.
4. After the 2 h incubation, remove nonadherent extracellular yeast cells, and perform one wash gently with PBS (*see* **Note 10**).

3.8 FITC Labelling Yeast (Before Lysis of Host Cells)

Anti-IgG–FITC is used to stain extracellular yeast after 2 h of phagocytosis and distinguish extracellular ($FITC^{positive}$) from intracellular ($FITC^{negative}$) yeast.

1. At the desired time point after phagocytosis, add anti-IgG–FITC antibody, final concentration 7.5 μg/mL, to the infected cells (*see* **Notes 8** and **11**).
2. Incubate 20 min in the dark at room temperature and then wash cells three times in PBS (Fig. 1).

3.9 APC Labelling Yeast (After Lysis of Host Cells)

Anti-IgG–APC is used to stain all yeasts and distinguish yeasts ($APC^{positive}$) from cell debris ($APC^{negative}$).

1. At the desired time point after phagocytosis, add the following to the infected wells: anticapsular E1 mAb (2×, final concentration of 5 μg/mL for J774) or 18B7 mAb (2×, final concentration of 5 μg/mL for human monocytes) (*see* **Note 8**) and anti-IgG–APC antibody (2×, final concentration of 7.5 μg/mL) (*see* **Note 12**).
2. Incubate 20 min in the dark at room temperature, and then wash cells three times in PBS (Fig. 1).

3.10 Measuring Phagocytosis for Interaction with J774 Cells

1. After the 2 h phagocytosis incubation and the PBS wash step (*see* **step 2** of Subheading 3.2), directly stain residual extracellular yeasts using anti-IgG–FITC (1×, final concentration of 7.5 μg/mL) (*see* **Note 11**) in 500 μL per well for 20 min in the dark at room temperature.
2. Wash three times with 500 μL of PBS per well.
3. Harvest J774 cells with a cell scraper in cold 0.05% SDS diluted in distilled water, and transfer cells into a 1.5 mL Eppendorf tube (*see* **Note 13**).
4. Resuspend cells in cold 0.05% SDS diluted in distilled water in a 1.5 mL tube, and vortex each tube for 10 s.
5. Centrifuge at 1500× *g* for 5 min and remove the supernatant.
6. Resuspend cells in 200 μL of 1% PFA in PBS (*see* **Note 14**).

3.11 Measuring Phagocytosis for Interaction with Human Monocytes

1. Just after washing with PBS (**step 2** of Subheading 3.2), directly stain residual extracellular yeast using FITC–anti-IgG (1×, final concentration of 15 μg/mL) in 100 μL per well for 20 min in the dark at room temperature (*see* **Note 8**).
2. Wash gently three times with 1 mL of PBS per tube (*see* **Notes 6** and **15**).
3. Harvest human monocytes in cold 0.05% SDS diluted in distilled water, by pipetting up and down, and transfer the cells to a 1.5 mL Eppendorf tube.
4. Centrifuge at 1500× *g* for 5 min and remove the supernatant.
5. Add 1 mL of cold 0.05% SDS diluted in distilled water to pellets, and incubate 5 min in the dark at room temperature. Vortex each tube for 10 s and centrifuge at 1500× *g* for 5 min. Remove the supernatant and resuspend cells in 200 μL of 1% PFA in PBS (*see* **Note 14**).

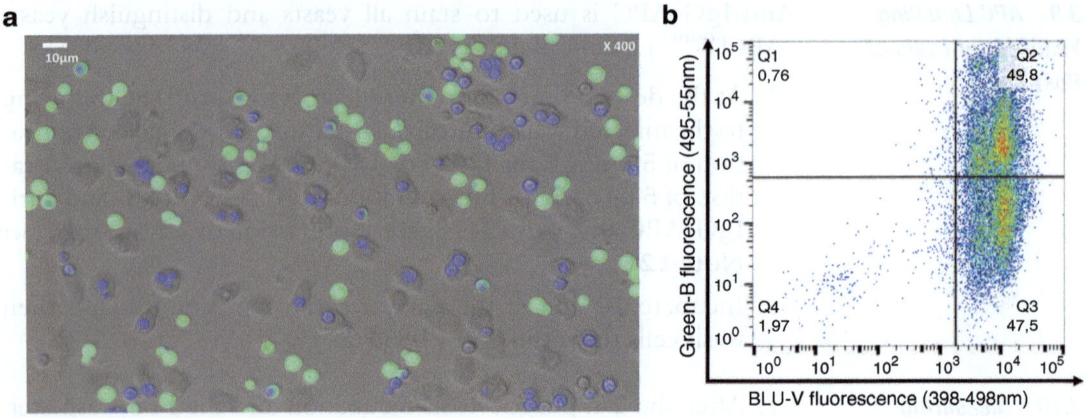

Fig. 2. Picture and flow cytometry dot plot after 2 h of *C. neoformans*/human monocyte interaction. Left panel: Merged picture taken by fluorescent microscopy (bright field + UV + FITC channel) showing non-fluorescent host cells (human monocytes) and internalized (blue) and extracellular yeast (blue and green). Right panel: Calcofluor/FITC flow cytometry dot plot. In the Q2 gate, 49.8% of cells are extracellular yeast ($CFW^{high}/FITC^{high}$), and in the Q3 gate, 47.5% of cells are intracellular H99 ($CFW^{high}/FITC^{low}$). No cell debris are observable in this experiment ($CFW^{low}/FITC^{low}$)

Flow cytometry analyses should allow absolute quantification to determine the exact concentration of yeast. We have successfully used the MACSQuant analyzer cytometer using MACSQuantify software 2.0 (Miltenyi Biotec) and a Guava cytometer with the Guava Incyte software (Merck). Analyses shown in Fig. 2b have been performed using a Guava cytometer, a Guava Incyte™, and the FlowJo 10.1™ software (*see* **Note 16**).

All samples are vortexed and filtered with a cell strainer (50 μm) before analysis and transferred to 96-well plates, 200 μL per well, for flow cytometry analysis. Every sample should be analyzed by flow cytometry at the same time, to reduce inter-experimental variability.

3.12 Flow Cytometry Control Analysis

1. Prepare single-stained yeast controls for flow cytometry analysis (*see* **Note 17**). Controls help to exclude doublets and aggregate cells (*see* **Note 18**). Using single-stained controls, it is possible to assess basic fluorescence levels (FITC, APC, and Calcofluor) and differentiate the following:
   (a) Yeast from cells debris (*see* **Note 19**).
   (b) Extracellular yeast from intracellular yeast (*see* **Note 20**).

With the fluorochrome combination used here, compensation is minimal. Other color combinations are possible. Single-color controls are critical for compensation.

3.13 Flow Cytometry Analysis of Samples from Phagocytosis Experiments

1. To calculate the number of intracellular yeast per condition, gate on the intracellular population $CFW^{positive}/FITC^{negative}$ ($\pm APC^{positive}$).
2. Assess the number of host cells (J774 cells or human monocytes) (*see* **Notes 22** and **23**).
3. Calculate the phagocytosis index (PI) using the following formula (Fig. 2):

   PI = (number of internalized yeast per well × 100)/number of initial host cells per well.

4 Notes

1. Twenty-two hours of growth is needed for *Cryptococcus neoformans* to reach a stationary phase in YPD. At this stage, yeasts are viable and mostly exist as single cells, and very few buds attached to mother cells are visible.
2. CFW dye specifically stains chitin contained in the cell wall. At a final concentration of 10 μg/mL, CFW does not affect *Cryptococcus neoformans* viability, multiplication, and opsonization by E1 mAb or secondary staining with anti-IgG–FITC or anti-IgG–APC.
3. Counts may also be done using an automatic counter or a hemocytometer. Conserve cells at 4 °C to avoid activation of monocytes.
4. *Cryptococcus neoformans* is not phagocytosed well without opsonization (≤1%). E1 and 18B7 mAb are two murine monoclonal anti-glucuronoxylomannan IgG1 antibodies (the major capsular component of the polysaccharide in the *Cryptococcus neoformans* capsule). 18B7 mAb may be used in both murine and human host cell interactions, but E1 is only used in murine interactions. Complement opsonization may also be used.
5. The type of flask used depends on the number of cells desired. It is also possible to use T25 cm^2 flasks (with a final volume of 1 mL).
6. Seed human monocytes just after their isolation by Ficoll density gradient and positive selection. Addition of cytokines is not required. If cytokines are added to polarize or activate the cells, it may introduce additional variability factors between donors. Early interactions between human monocytes and *Cryptococcus neoformans* immediately after isolation prevent unpredictable attachment/detachment of the cells to the bottom of the well. Cell washing may be performed in wells or in tubes, depending on monocyte adherence. Human monocytes are less adherent than J774 cells. Nonetheless, monocyte adherence will increase with time. Use a well with yeasts and monocytes (in a 2.5:1

ratio) to test their adherence after 2 h of incubation. If adherence is good, wash cells directly in wells as done with J774 cells. If adherence is not good, gently harvest human monocytes by gently pipetting up and down, and wash them in tubes.

7. FCS should be heat-inactivated to remove complement components. FCS provides essential nutrients for J774 cells and human monocytes. FCS should be added to J774 cells pre-culture. FCS should not be added during the 2 h phagocytosis interaction between *Cryptococcus neoformans* and J774 cells or human monocytes.
8. It is not necessary to add anticapsular antibody (E1 for J774 interactions; 18B7 for human monocyte interactions) after 2 h of incubation because all yeasts, even intracellular yeast, remain labeled by anticapsular antibody (used as an opsonin). However, if the incubation is protracted, it is necessary to add E1 or 18B7 again to allow anti-IgG–FITC and anti-IgG–APC binding to unopsonized newly produced daughter yeast.
9. For the proliferation assay, a 2.5:1 (*Cryptococcus*/monocyte) ratio is optimal.
10. Human monocytes, after 2 h of incubation, are semi-adherent, but this adherence varies between donors. Washing should be done gently to avoid losing cells. Check with a microscope before and after each wash.
11. Anti-IgG–FITC binds opsonized yeast. This step is introduced to differentiate extracellular yeast (FITC^{high}) from intracellular yeast (FITC^{low}).
12. This step to include APC–anti-IgG staining to help differentiate yeast from cell debris is optional, but is required if differentiation between CFW-positive and CFW-negative events is expected (proliferation assay including generation of CFW^{low} yeast). Indeed, CFW is used to follow yeast multiplication during macrophage interactions where part of the yeast mother cell wall is transmitted to the daughter cell. The daughter cell will have a lower CFW fluorescence than mother cells with generation of CFW-negative yeasts that will only be differentiated from cell debris if this additional APC staining is performed [23, 30].

    Only yeast cells will be APC^{high}. Extracellular yeast cells will have an APC^{high} FITC^{high} fluorescent signal. Cell debris will be APC^{low} FITC^{low}. APC–anti-IgG staining is not affected by previous FITC–anti-IgG staining.
13. In these conditions, 0.05% SDS, in association with centrifugation and vortexing, will lyse monocytes/macrophages. The integrity of the *Cryptococcus neoformans* cell wall will be preserved.

14. Samples fixed in 1% PFA will allow conservation of fluorescence signals until flow cytometry analysis, up to seven days after fixation. All samples and single-color controls should be analyzed in the same run.
15. Use PBS at room temperature or pre-warmed at 37 °C (cold PBS will affect fragile/sensitive cells).
16. Green B, Blue V, and Red R channels (on the flow cytometer) correspond, respectively, to fluorescence signals of FITC (from 495 to 555 nm), CFW (from 398 to 498 nm), and APC (from 646 to 676 nm).
17. Controls help to perform compensations.
18. Doublets consist of yeast aggregates or of budding yeasts. It is important to exclude doublets for final analysis to obtain a homogenous population. Inclusion of doublets will lead to bias.
19. APC staining will help differentiate yeast (APC^{high}) from cells debris (APC^{low}) if this optional step is performed.
20. Extracellular yeasts have CFW^{high}, APC^{high}, and $FITC^{high}$ signals, while intracellular yeasts have CFW^{high}, APC^{high}, and $FITC^{low}$ signals.
21. For adherent J774 cells: Use one well seeded with cells that do not undergo interaction with yeast, incubated for 2 h at 37 °C and 5% CO_2. Remove supernatant, detach the cells by scraping, and count them using a hemocytometer or flow cytometry.
22. For semi-adherent monocytes: Use AccuCheck Counting Beads (initial concentration of 1×10^6 microbeads per milliliter). Add 10 μL (1×10^4 microbeads) per well at the beginning of the interaction. Incubate plates following the described protocol. During flow cytometry analysis, count the remaining microbeads in each well (microbeads are smaller than yeast on the FCS/SSC dot plot). Based on the proportion of the microbeads lost during washing, calculate the number of remaining human monocytes. For example, if there are 5×10^3 microbeads per well remaining at the end of the experiment corresponding to 50% of the initial quantity, the total remaining monocytes per well would be 4×10^4 (50%).

Acknowledgments

The authors thank Marion Benazra for her helpful support and Professor Arturo Casadevall for providing the 18B7 mAb.

References

1. Bliska JB, Casadevall A (2009) Intracellular pathogenic bacteria and fungi – a case of convergent evolution? Nat Rev Microbiol 7(2): 165–171
2. Bitar D, Lortholary O, Le Strat Y et al (2014) Population-based analysis of invasive fungal infections, France, 2001–2010. Emerg Infect Dis 20(7):1149–1155
3. Rajasingham R, Smith RM, Park BJ et al (2017) Global burden of disease of HIV-associated cryptococcal meningitis: an updated analysis. Lancet Infect Dis 17(8): 873–881
4. Feldmesser M, Kress Y, Novikoff P et al (2000) *Cryptococcus neoformans* is a facultative intracellular pathogen in murine pulmonary infection. Infect Immun 68(7):4225–4237
5. Kronstad JW, Attarian R, Cadieux B et al (2011) Expanding fungal pathogenesis: *Cryptococcus* breaks out of the opportunistic box. Nat Rev Microbiol 9(3):193–203
6. Charlier C, Nielsen K, Daou S et al (2009) Evidence of a role for monocytes in dissemination and brain invasion by *Cryptococcus neoformans*. Infect Immun 77(1):120–127
7. García-Rodas R, Zaragoza O (2012) Catch me if you can: Phagocytosis and killing avoidance by *Cryptococcus neoformans*. FEMS Immunol Med Microbiol 64(2):147–161
8. Jesus MD, Nicola AM, Chow S-K et al (2010) Glucuronoxylomannan, galactoxylomannan, and mannoprotein occupy spatially separate and discrete regions in the capsule of *Cryptococcus neoformans*. Virulence 1(6):500–508
9. Levitz SM, DiBenedetto DJ (1989) Paradoxical role of capsule in murine bronchoalveolar macrophage-mediated killing of *Cryptococcus neoformans*. J Immunol 142(2):659–665
10. McQuiston T, Luberto C, Del Poeta M (2011) Role of sphingosine-1-phosphate (S1P) and S1P receptor 2 in the phagocytosis of *Cryptococcus neoformans* by alveolar macrophages. Microbiol Read 157(Pt 5):1416–1427
11. Ben-Abdallah M, Sturny-Leclère A, Avé P et al (2012) Fungal-induced cell cycle impairment, chromosome instability and apoptosis via differential activation of NF-κB. PLoS Pathog 8(3):e1002555
12. Kozel TR, Wilson MA, Pfrommer GS et al (1989) Activation and binding of opsonic fragments of C3 on encapsulated *Cryptococcus neoformans* by using an alternative complement pathway reconstituted from six isolated proteins. Infect Immun 57(7):1922–1927
13. Kozel TR, Pfrommer GS (1986) Activation of the complement system by *Cryptococcus neoformans* leads to binding of iC3b to the yeast. Infect Immun 52(1):1–5
14. Keller RG, Pfrommer GS, Kozel TR (1994) Occurrences, specificities, and functions of ubiquitous antibodies in human serum that are reactive with the *Cryptococcus neoformans* cell wall. Infect Immun 62(1):215–220
15. Zaragoza O, Taborda CP, Casadevall A (2003) The efficacy of complement-mediated phagocytosis of *Cryptococcus neoformans* is dependent on the location of C3 in the polysaccharide capsule and involves both direct and indirect C3-mediated interactions. Eur J Immunol 33(7):1957–1967
16. Casadevall A, Cleare W, Feldmesser M et al (1998) Characterization of a murine monoclonal antibody to *Cryptococcus neoformans* polysaccharide that is a candidate for human therapeutic studies. Antimicrob Agents Chemother 42(6):1437–1446
17. Dromer F, Salamero J, Contrepois A et al (1987) Production, characterization, and antibody specificity of a mouse monoclonal antibody reactive with *Cryptococcus neoformans* capsular polysaccharide. Infect Immun 55(3): 742–748
18. Carreto-Binaghi LE, Aliouat EM, Taylor ML (2016) Surfactant proteins, SP-A and SP-D, in respiratory fungal infections: their role in the inflammatory response. Respir Res 17(1):66
19. Casadevall A, Coelho C, Alanio A (2018) Mechanisms of *Cryptococcus neoformans*-mediated host damage. Front Immunol 9:855
20. Taborda CP, Casadevall A (2002) CR3 (CD11b/CD18) and CR4 (CD11c/CD18) are involved in complement-independent antibody-mediated phagocytosis of *Cryptococcus neoformans*. Immunity 16(6):791–802
21. Coelho C, Souza ACO, Derengowski L d S et al (2015) Macrophage mitochondrial and stress response to ingestion of *Cryptococcus neoformans*. J Immunol 194(5):2345–2357
22. Ikeda-Dantsuji Y, Ohno H, Tanabe K et al (2015) Interferon-γ promotes phagocytosis of *Cryptococcus neoformans* but not *Cryptococcus gattii* by murine macrophages. J Infect Chemother 21(12):831–836
23. Alanio A, Desnos-Ollivier M, Dromer F (2011) Dynamics of *Cryptococcus neoformans*-macrophage interactions reveal that fungal background influences outcome during cryptococcal meningoencephalitis in humans. mBio 2(4):e00158–e00111

24. Levitz SM, Farrell TP (1990) Growth inhibition of *Cryptococcus neoformans* by cultured human monocytes: role of the capsule, opsonins, the culture surface, and cytokines. Infect Immun 58(5):1201–1209
25. Walenkamp AM, Scharringa J, Schramel FM et al (2000) Quantitative analysis of phagocytosis of *Cryptococcus neoformans* by adherent phagocytic cells by fluorescence multi-well plate reader. J Microbiol Methods 40(1): 39–45
26. Alvarez M, Burn T, Luo Y et al (2009) The outcome of *Cryptococcus neoformans* intracellular pathogenesis in human monocytes. BMC Microbiol 9:51
27. Coelho C, Tesfa L, Zhang J et al (2012) Analysis of cell cycle and replication of mouse macrophages after *in vivo* and *in vitro Cryptococcus neoformans* infection using laser scanning cytometry. Infect Immun 80(4): 1467–1478
28. Srikanta D, Yang M, Williams M et al (2011) A sensitive high-throughput assay for evaluating host-pathogen interactions in *Cryptococcus neoformans* infection. PLoS One 6(7):e22773
29. Mukherjee S, Feldmesser M, Casadevall A (1996) J774 murine macrophage-like cell interactions with *Cryptococcus neoformans* in the presence and absence of opsonins. J Infect Dis 173(5):1222–1231
30. Alanio A, Vernel-Pauillac F, Sturny-Leclère A et al (2015) *Cryptococcus neoformans* host adaptation: Toward biological evidence of dormancy. mBio 6(2):e02580–e02514

Chapter 12

Interaction Between Macrophages and *Cryptococcus neoformans*: Distinguishing Phagocytosed Versus External Fungi

Amanda Pereira Rocha, Thaís Silva Alves, Adrielle Veloso Caixeta, Patrícia Albuquerque, and André Moraes Nicola

Abstract

The interaction between macrophages and *Cryptococcus neoformans* is crucial in the pathogenesis of cryptococcosis. These phagocytes are important immune effectors, but also a niche in which facultative intracellular parasites, such as *C. neoformans*, thrive. Consequently, phagocytosis of cryptococcal cells and its outcomes are very frequently studied. One major issue with several of the tests used for this, however, is that macrophage–*C. neoformans* interaction does not always result in phagocytosis, as fungi may be attached to the external surface of the phagocyte. The most used methodologies to study phagocytosis of cryptococcal cells have varying degrees of precision in separating fungi that are truly internalized from those that are outside macrophages. Here we describe two assays to measure phagocytosis that can differentiate internal from external *C. neoformans* cells.

Key words *Cryptococcus neoformans*, Cryptococcosis, Macrophage, Phagocytosis, Fluorescence microscopy, Flow cytometry

1 Introduction

The global incidence of and mortality from cryptococcosis has been slowly decreasing due to advances in the prevention of HIV infection and the increased use of antiretrovirals. Yet HIV-associated cryptococcosis still kills more than a hundred thousand people each year [1], and advances in medicine (e.g., transplants, chemotherapy, and new immunosuppressors) lead to increased numbers of immunocompromised people that are susceptible to cryptococcosis. Understanding the immune response to *C. neoformans* is thus very important, as it could lead to new prophylactic, diagnostic, and therapeutic strategies to improve patient outcomes.

Macrophages are the most important immune effector against *C. neoformans*, so assays that measure ingestion of cryptococcal

Erin E. McClelland (ed.), *Cryptococcus neoformans: Methods and Protocols*, Methods in Molecular Biology, vol. 2775,
https://doi.org/10.1007/978-1-0716-3722-7_12,

cells by macrophages and the phagocytosis outcomes are very common. These assays vary from simple and cheap to complex experiments that require specialized equipment, such as microscopes, fluorimeters, and flow cytometers. A range of tests is necessary because the simplest ones have limitations, among which the most important is the uncertainty if a cryptococcal cell interacting with a macrophage is really internalized or just adhered to the phagocyte's external surface. In this chapter, we will present detailed protocols for two assays that differentiate phagocytosed fungi from those that are simply attached to macrophages.

Perhaps the simplest test for this is a phagocytosis assay, which we have described in detail previously [2]. We present the protocol for this assay again below, with improvements that we have added throughout the last decade. In this assay, macrophages and *C. neoformans* cells interact and are then fixed and stained. Through direct observation or microphotography, the proportion of macrophages with internalized fungi is scored. The second assay is more complex, requiring a flow cytometer. It involves staining the fungi and macrophages with different fluorophores and using Uvitex 2B to evaluate true internalization [3]. This cell wall-binding probe is unable to penetrate intact cell membranes, so it stains extracellular fungi, but not those that are inside live macrophages [4]. In comparison, the microscopy assay is simpler and less expensive, requiring few specialized reagents and mostly equipment that is readily available in microbiology and immunology laboratories. However, it depends on the visual determination of complete fungal internalization, which is subjective. Moreover, the process of analyzing macrophages microscopically is time-consuming and difficult to scale up for multiple different conditions. The flow cytometric assay, in comparison, requires expensive reagents and specialized flow cytometers. Learning to operate a flow cytometer and to analyze the data is also not straightforward. However, this method ascertains with little subjectivity if the fungal cells have indeed been internalized, counts thousands of macrophages in minutes, and is far more amenable to large-scale experiments with multiple conditions.

In addition to these two protocols, which we have used the most, we will briefly present below other alternatives in the literature to aid the readers in their choice of assay:

Assay: Plating extracellular and intracellular *C. neoformans* for CFUs

Rationale: In a tissue culture plate, macrophages are infected with live *C. neoformans*. After the interaction, the supernatant is collected, attached macrophages are lysed, and this lysate is also collected in a different tube. Fungi are quantified by colony-forming units on the supernatant and on macrophage lysate, representing respectively the extracellular and internalized fungi.

Pros: Cheap and easy to perform, requires no complex equipment.

Cons: Overestimates phagocytosis due to fungi that remain attached to macrophages. Cryptococcal killing by macrophages can mislead the results. Results are thus not very reliable.

Assay: Quenching of fluorescein isothiocyanate (FITC) by trypan blue [4]

Rationale: *C. neoformans* cells are initially stained with FITC, which results in covalent labeling of cellular proteins. Labeled fungi are then incubated with macrophages, and after phagocytosis, the cells are treated with trypan blue. This vital stain quenches FITC fluorescence, but does not reach fungi inside live macrophages. Extracellular *C. neoformans* thus loses fluorescence, while internalized fungi retain it.

Pros: Reagents are cheap and easy to obtain. Assay can be adapted to most common fluorescence microscopes and flow cytometers. Little subjectivity and bias in determining if the fungi are truly internalized.

Cons: Must be done with dead *C. neoformans*, as progeny of live fungal cells will not retain the FITC label. FITC is less fluorescent in acidic pH, so the phagolysosomal environment may lead to underestimation of the true rate of phagocytosis.

Assay: Detecting the acidic phagosome using pHrodo Red succinimidyl ester

Rationale: *C. neoformans* cells are labeled with pHrodo Red and then incubated with macrophages. Cells that are internalized end up in the acidic phagolysosome, and the low pH increases pHrodo Red fluorescence. Thus, extracellular *C. neoformans* are not fluorescent, whereas internalized fungi are bright red. pHrodo Red may be substituted for other pH-sensitive probes, with different excitation and emission spectra.

Pros: Assay can be adapted to common fluorescence microscopes and even to flow cytometry, although only specialty cytometers are equipped with the lasers that excite pHrodo Red. Little subjectivity and bias in determining if the fungi are truly internalized.

Cons: Very expensive reagents. Truly internalized fungi might not fluoresce in some cases, as *C. neoformans* is known to neutralize the acidic phagosome [5].

2 Materials

2.1 Phagocytosis Assay: Bright-Field Microscopy

1. J774 cells (*see* **Notes 1** and **2**).
2. J774 cell culture medium: Dulbecco's Modified Eagle Medium (DMEM) with high glucose, L-glutamine, phenol red, sodium pyruvate, and sodium bicarbonate and without HEPES + 10% fetal calf serum + 1% nonessential amino acids.
3. Petri dishes or flasks for tissue culture. We prefer using those with non-treated surfaces for J774 maintenance (*see* **Note 3**).
4. 96-well tissue culture plates. Ensure that the surfaces are vacuum plasma treated (*see* **Note 3**).
5. Opsonin option 1: 18B7 anti-capsule antibody. This can be obtained from Sigma.
6. Opsonin option 2: Complement from fresh (or frozen) serum (*see* **Note 4**).
7. Nonenzymatic cell dissociation solution. This is available from many vendors, such as Sigma, Thermo Fisher, or Corning. Alternatively, you may use trypsin or other cell dissociation solutions.
8. Phosphate-buffered saline (PBS): 137 mM NaCl, 2.7 mM KCl, 10 mM Na_2HPO_4, 1.8 mM KH_2PO_4.
9. Sabouraud dextrose media: 1% peptone, 2% dextrose. Sterilize by autoclaving for 20 min, 121 °C.
10. Hemocytometer (Neubauer chamber).
11. Trypan blue solution, 0.4%, in PBS. This can be purchased or prepared in house, by diluting the dye in PBS. Sterilize by filtration, and make several small aliquots, as the solution is frequently contaminated by mold. It can be stored at room temperature for long periods.
12. Hematology staining kit based on the Romanowsky stain. Also called panoptic or Diff-Quik staining kits.
13. Inverted microscope for bright-field observation. The microscope should have 20× or long working distance 40× objectives. Although not necessary, the experiment will be easier on microscopes equipped with a digital camera.
14. Image analysis software. We use both the commercially available Adobe Photoshop and the freeware ImageJ, which can be downloaded at https://imagej.nih.gov/ij/download.html.
15. Statistical analysis software. We use the commercially available GraphPad Prism, but there are many other paid and free options.

2.2 Flow Cytometry Phagocytosis Assay

1. 24-well and 12-well tissue culture plates. Ensure that the surfaces are vacuum plasma treated (*see* **Note 3**).
2. 10% Uvitex 2B, dissolved in water. The solution will be cloudy and is very stable at −20 °C. Excitation and emission maxima are 350 nm and 435 nm, respectively. For flow cytometry, use the Pacific Blue channel.
3. CellTracker Deep Red dye (*see* **Note 5**). Dissolve in anhydrous dimethyl sulfoxide (DMSO) to 1 mM immediately before using. Excitation and emission maxima are 630 nm and 650 nm, respectively. For flow cytometry, use the allophycocyanin (APC) channel.
4. 5-Chloromethylfluorescein diacetate (CMFDA) (*see* **Note 5**). Dissolve in anhydrous DMSO to 1 mM immediately before using. Excitation and emission maxima are 492 nm and 517 nm, respectively. For flow cytometry, use the FITC channel.
5. 7-Aminoactinomycin D (7-AAD), 1 mg/mL in DMSO. The dissolved dye is stable for long periods at −20 °C. Excitation and emission maxima are 546 nm and 647 nm, respectively. For flow cytometry, use the phycoerythrin (PE)-Cy5 channel.
6. Flow cytometer: The equipment requires at least 405 nm, 488 nm, and 633/640 nm lasers to perform the basic flow cytometry phagocytosis assay.
7. Flow cytometry analysis software. We use FlowJo, but there are many other options available.

3 Methods

For all protocols below, day 0 is the day when macrophages are infected with *C. neoformans*.

3.1 Prepare for the Microscopy Phagocytosis Assay Experiment (Days −3 to −1)

An outline of this assay is presented in Fig. 1.

1. On day −3, split J774 cells into a new Petri dish so that they will be 60–80% confluent on day −1.
2. On day −2, inoculate a single colony from a *C. neoformans* Sabouraud agar plate into 5 mL of Sabouraud broth. Grow for 2 days, shaking at 200 rpm at 37 °C.
3. On day −1, prepare the 96-well plate in which the experiment will be performed. For each experimental condition, prepare at least one well (*see* **Note 6**).
4. Detach J774 cells from the Petri dish and transfer to a sterile tube. If you use dishes with a non-treated surface to routinely passage the cells, simply aspirate, and eject media a few times with a serological pipette to detach the cells.

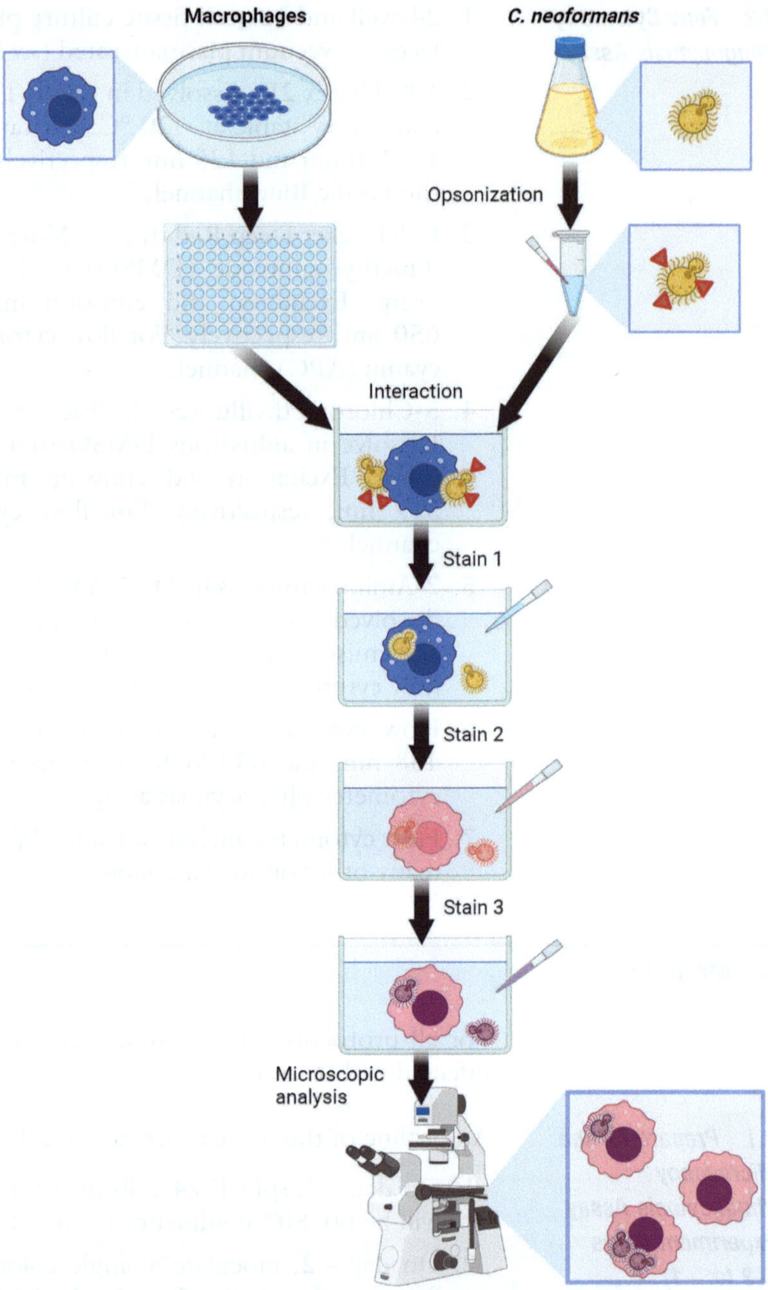

Fig. 1 Outline of the microscopy phagocytosis assay. J774 and *C. neoformans* cells are prepared and co-incubated in 96-well plates. After 2 h, the cells are stained and observed on a microscope. Phagocytosis is determined by visually inspecting the cells

5. In a microfuge tube, mix 10μL of the cell suspension with 10μL of the trypan blue solution. Count the cells within 5 min, as trypan blue toxicity to cells will lead to an underestimation of

cell viability after that. Use a hemocytometer to count the absolute number of cells and the number of live cells, which are colorless. Expect up to 1×10^7 J774 cells with a viability of approximately 95% from a nearly confluent 100 mM Petri dish.

6. Dilute the cell suspension in fresh, warm (37 °C), supplemented DMEM to a concentration of 2×10^5 live cells/mL. You will need 100μL per well.
7. Dispense 100μL of this suspension into each well of the 96-well plate. Incubate the 96-well plate for 24 h at 37 °C and 5% CO_2.

3.2 Infect the J774 Cells with *C. neoformans* (Day 0)

1. After the 24 h incubation, the J774 cells should be firmly attached and have nearly doubled in number. Check the plates on an inverted microscope to confirm that the cells look healthy and have no visible contamination. Healthy J774 monolayers are a mix of round cells and elongated, spindle-shaped cells and may have visible pseudopods. The medium should not contain any suspended particles.
2. Collect 1 mL from the *C. neoformans* culture in Sabouraud broth into a microfuge tube. Centrifuge for 5 min at 1000 × *g* at room temperature.
3. Discard the supernatant and resuspend the pellet in 1 mL of PBS to wash the cells. Briefly vortex to resuspend the cells, and centrifuge again for 5 min at 1000 × *g* at room temperature.
4. Discard the supernatant and repeat the washing procedure a second time. When finished, resuspend again in 1 mL of PBS. Dilute the cells 1:100 by pipetting 10μL into a different microfuge tube containing 990μL of PBS. Count using a hemocytometer, expecting approximately 1×10^8 cells per mL of the original culture.
5. Prepare a tube for opsonization, containing fungi and an opsonin: IgG antibody or complement from fresh serum. For each well in your experiment, prepare 100μL of the suspension containing 8×10^5 cells per mL, which will result in a multiplicity of infection (MOI) of 2. If using antibody to opsonize, also add 18B7 to a final concentration of 10μg/mL. Alternatively, add fresh or frozen serum to a final concentration of 10% (*see* **Note 7**). Incubate at 37 °C for 30 min, preferably with shaking.
6. Add 100μL of the opsonized fungal suspension to each well in the 96-well plate, and return the plate to the CO_2 incubator. Incubate for 2 h.

3.3 Stain the Cells (Day 0)

1. After the 2 h incubation, wash each well three times. To do so, aspirate the contents of each well and immediately add 200μL of pre-warmed (37 °C) PBS (*see* **Note 8**).

2. After the third wash, stain macrophages with the fast Romanowsky staining kit (*see* **Notes 9** and **10**). Start by removing all the PBS from each well and adding 100μL of the first solution, containing a fixative. Incubate for precisely 15 s.
3. Remove this solution from each well, and immediately add 100μL of the second solution, which contains an eosin dye. Incubate for precisely 17 s.
4. Remove this solution from each well, and immediately add 100μL of the third solution containing basophilic dyes. Incubate for precisely 10 s.
5. After incubating with the last dye, remove it immediately, and wash each well with distilled water until the liquid becomes clear. To perform these washes, add 200μL of water to each well, swirl the plate, aspirate the liquid, and discard it. When finished, leave 200μL of water in each well.
6. The plate with fixed, stained cells may be analyzed immediately or stored at 4 °C for several days. Do not let the liquid on the plates dry.

3.4 Analyze the Results

If the plate was stored at 4 °C, let it rest at room temperature until there is no condensation on the bottom of the plate. The internalization of *C. neoformans* cells can then be observed directly on a microscope or quantified from photographs taken from each well. An outline of the experiment is presented in Fig. 1.

The following are important suggestions regarding the analysis:

1. Use a 20× or 40× objective to observe the cells. For the 40×, you might need a long working distance objective due to the plate thickness.
2. Avoid the very edges of the wells, where the plastic distorts the image and the J774 cells tend to be too close together.
3. To differentiate *C. neoformans* cells that are internalized from those that are only attached to the external surface of the macrophage, look at the capsule surrounding the fungal cell. Internalized fungi are surrounded by a clear halo that is not stained by the dyes that color the macrophage cytoplasm, whereas in extracellular fungi the capsule is not visible (*see* Fig. 2).
4. To prevent observing and counting the same macrophage more than once, scan the plate on a zigzag pattern.
5. To avoid artifacts, do not take into consideration macrophages that are not fully visible within the edges of the field (when observing directly on the microscope) or within the edges of the photographs.

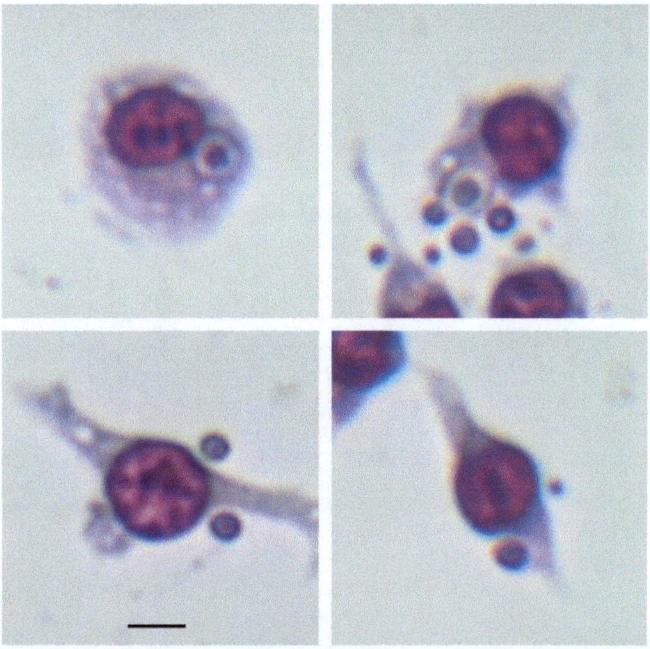

Fig. 2 Examples of microscopy phagocytosis assay photographs. The four panels show cells that were observed in a microscopy phagocytosis assay. The upper left panel shows a macrophage with a fully internalized *C. neoformans* cell; note the unstained halo around the yeast cell, caused by the fungal capsule. The upper right panel shows a macrophage that is in the process of internalizing a *C. neoformans* mother cell with a bud. The two lower panels show macrophages to which the fungi are attached, but not fully internalized

6. Two parameters can usually be counted from these plates. For *percent phagocytosis*, count the number of macrophages with at least one internalized *C. neoformans* cell and the number of macrophages with no internalized yeasts. For the *phagocytic index*, also count the number of fungi inside each of the macrophages that have phagocytosed at least one *C. neoformans* cell.
7. We usually count at least 100 macrophages per well. With this number, a power analysis shows that there is a power of 80% to detect differences in phagocytosis rates of approximately 20%.
8. After counting the cells, analyze the data using statistical analysis software. For percent phagocytosis, use the Fisher's exact test or the χ^2 test, as phagocytosis is a proportion variable. For more complex experiments, use logistic regression for a multivariate analysis. For the phagocytic index, use Poisson regression, as it is a count (discrete) variable.

3.5 Flow Cytometry Phagocytosis Assay

This experiment spans several days and requires many controls, so the principle is presented here and in Fig. 3 to help you organize the experiment.

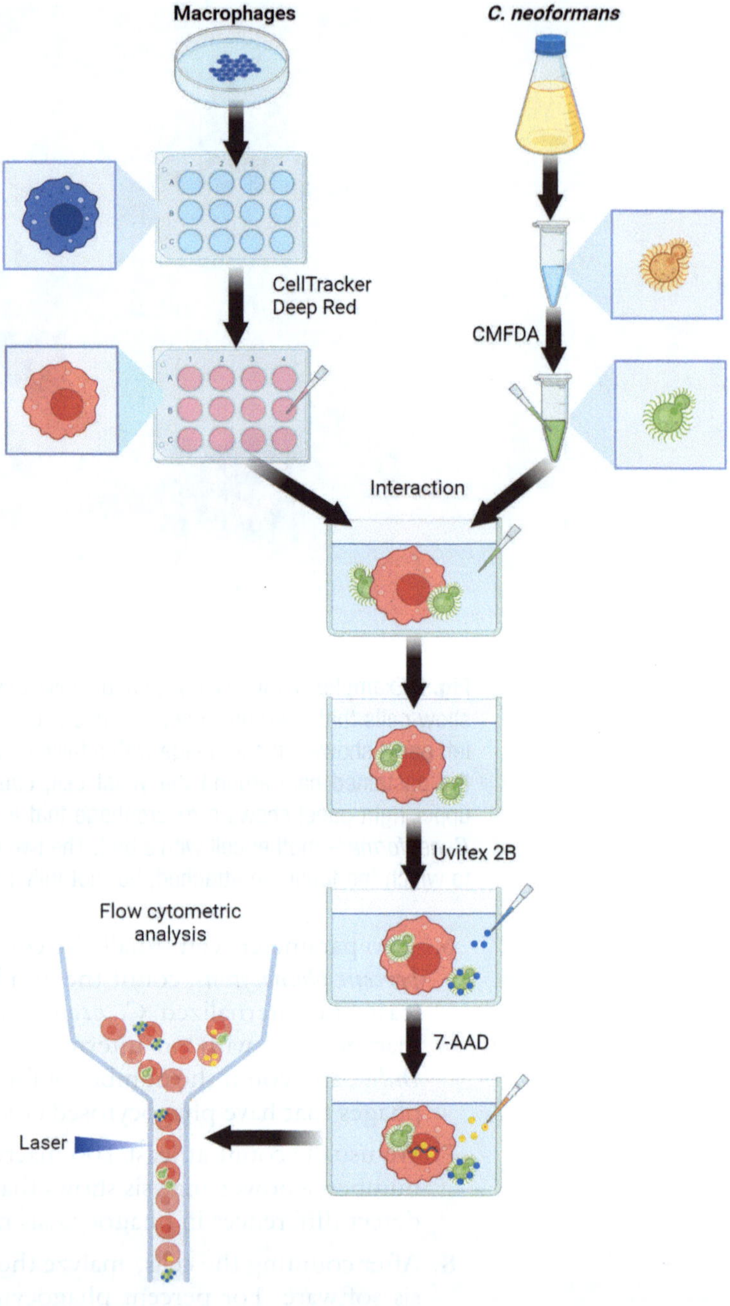

Fig. 3 Outline of the flow cytometry phagocytosis assay. J774 and *C. neoformans* cells are cultured and stained with CellTracker Deep Red and CMFDA, respectively, prior to co-incubation. After 2 h, the infected monolayer is stained with Uvitex 2B and 7-AAD, and the cells are analyzed by flow cytometry

Rationale: J774 cells and *C. neoformans* are stained with different fluorophores. After these cells interact, Uvitex 2B is used to stain fungi that remain extracellular, and 7-AAD is used to stain dead macrophages. The cells are then analyzed using flow

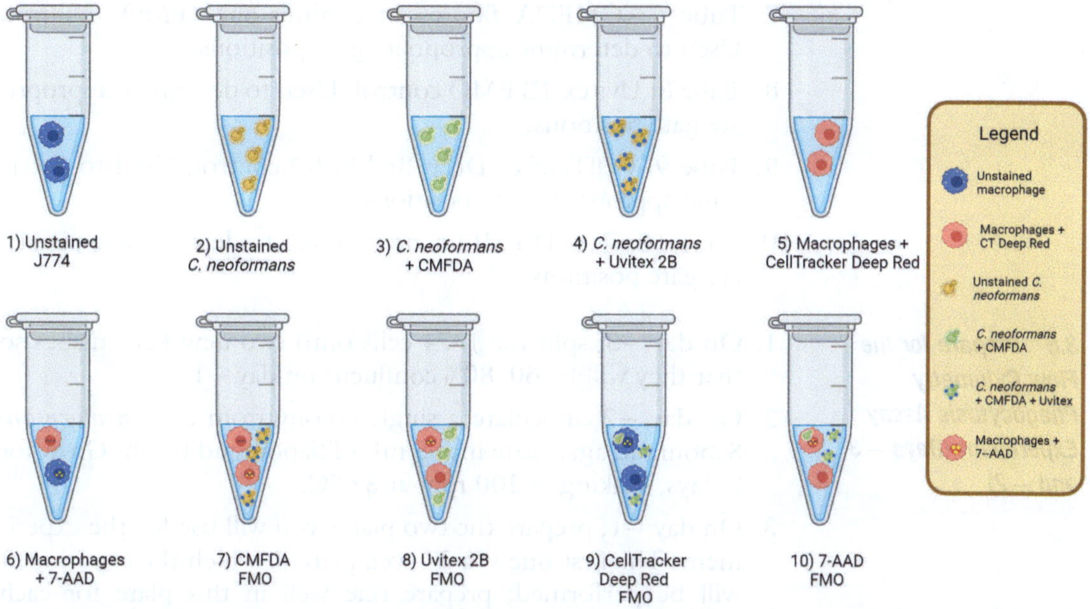

Fig. 4 Flow cytometry phagocytosis assay controls. Schematic representation of the three types of controls used in the experiment: unstained, single-color, and fluorescence-minus-one (FMO)

cytometry to quantify macrophage viability and *C. neoformans* internalization.

Major steps: The first step in the experiment is to grow both J774 cells and *C. neoformans* and prepare them for the staining procedures and interaction. The day of the experiment begins with staining fungi and macrophages, which are then allowed to interact. In parallel, three types of flow cytometry controls are prepared. After the experimental and control samples are collected, they are run on a flow cytometer. The last step is to analyze the resulting data.

The flow cytometry experiment will require ten control tubes, outlined below and in Fig. 4 to also help organize the experiment:

1. Tube 1: Unstained J774 cells.
2. Tube 2: Unstained *C. neoformans* cells.
3. Tube 3: CMFDA-stained *C. neoformans* cells. Single-color control used for compensation.
4. Tube 4: Uvitex 2B-stained *C. neoformans* cells. Single-color control used for compensation.
5. Tube 5: CellTracker Deep Red-stained J774 cells. Single-color control used for compensation.
6. Tube 6: 7-AAD-stained J774 cells. Single-color control used for compensation.

7. Tube 7: CMFDA fluorescence-minus-one (FMO) control. Used to determine appropriate gate positions.
8. Tube 8: Uvitex 2B FMO control. Used to determine appropriate gate positions.
9. Tube 9: CellTracker Deep Red FMO control. Used to determine appropriate gate positions.
10. Tube 10: 7-AAD FMO control. Used to determine appropriate gate positions.

3.6 Prepare for the Flow Cytometry Phagocytosis Assay Experiment (Days −3 and −2)

1. On day −3, split the J774 cells onto two new Petri dishes so that they will be 60–80% confluent on day −1.
2. On day −2, inoculate a single colony from a *C. neoformans* Sabouraud agar plate into 5 mL of Sabouraud broth. Grow for 2 days, shaking at 200 rpm at 37 °C.
3. On day −1, prepare the two plates you will use for the experiment. The first one is a 24-well plate in which the experiment will be performed; prepare one well in this plate for each experimental condition. The second plate is a 12-well plate used for the controls mentioned above; you will need 8 wells with J774 cells in this plate for the controls.
4. To prepare these plates, detach the J774 cells from the Petri dishes, and transfer to a sterile tube. If using dishes with a non-treated surface, simply aspirate and eject media with a serological pipette to detach the cells. In a microfuge tube, mix 10μL of the cell suspension with 10μL of trypan blue solution. Count the cells within 5 min, as the trypan blue toxicity to cells will lead to an underestimation of cell viability after that. Use a hemocytometer to count the absolute number of cells and the number of live cells, which are colorless. Expect approximately 1×10^7 J774 cells with a viability of approximately 95% from a nearly confluent 100 mm Petri dish.
5. Dilute the cell suspension in fresh, warm (37 °C) supplemented DMEM to a concentration of 2×10^5 live cells/mL. You will need 500μL per well for the 24-well plate and 1 mL per well for the 12-well plate. Dispense the appropriate volume into the plates, and incubate them at 37 °C and 5% CO_2 for 24 h.

3.7 Stain the *C. neoformans* Cells (Day 0)

1. Collect 1 mL from the *C. neoformans* culture in Sabouraud broth into a microfuge tube. Centrifuge for 5 min at 1000 × *g* at room temperature.
2. Discard the supernatant and resuspend the pellet in 1 mL of PBS to wash the cells. Briefly vortex to resuspend the cells, and centrifuge again for 5 min at 1000 × *g* at room temperature.
3. Discard the supernatant and repeat the wash procedure a second time. When finished, resuspend the cells in 1 mL of PBS.

Dilute the cells 1:100 by pipetting 10μL into a different microfuge tube containing 990μL of PBS. Count using a hemocytometer, expecting approximately 1 × 10^8 cells per mL of the original culture.

4. Dilute the fungal suspension in 3 mL of pre-warmed (37 °C) PBS to a concentration of 1 × 10^7 cells per mL. Split this suspension in three microfuge tubes. One will be used for CMFDA staining, and the other two will be used for flow cytometry controls.
5. Remove a tube with CMFDA from the freezer, and wait for it to reach room temperature before opening. Once the CMFDA tube has fully thawed, dilute it with anhydrous DMSO to a final concentration of 1 mM. Add 2μL of this concentrated CMFDA stock solution to one of the tubes containing 1 mL of *C. neoformans* suspension. Vortex immediately, and incubate for 30 min at 37 °C, protected from light and preferably with agitation.
6. After this incubation, wash the stained cells twice with PBS by centrifuging and aspirating the supernatant, as described above. When finished, resuspend the CMFDA-stained *C. neoformans* cells in 1 mL of J774 culture medium, and maintain at 37 °C until used to infect J774 cells.

3.8 Stain the J774 Cells (Day 0)

1. After the 24 h incubation, the J774 cells should be firmly attached and have nearly doubled. Check the plates on an inverted microscope to confirm that the cells look healthy and have no visible contamination. Healthy J774 monolayers are a mix of round cells and elongated, spindle-shaped cells and may have visible pseudopods. The medium should not contain any suspended particles.
2. For every well containing J774 cells in the 24-well plate, remove the medium, and wash two times with pre-warmed PBS (*see* **Notes 11** and **12**). After washing, leave the cells in 1 mL of warm PBS per well.
3. Add 1μL of 1 mM CellTracker Deep Red dye into each well of the 24-well plate containing macrophages (*see* **Note 13**). Mix immediately by swirling the plate. Incubate for 10 min at 37 ° C, 5% CO_2 (*see* **Note 14**).
4. After the cells are stained, wash the wells twice with J774 culture medium to remove unreacted CellTracker Deep Red dye. After the washes, keep the cells in 500μL of J774 culture medium in the CO_2 incubator until ready to infect (*see* **Note 15**).

3.9 Infect the J774 Cells with C. neoformans (Day 0)

1. To opsonize CMFDA-stained *C. neoformans* cells, pipette an appropriate amount into a new sterile tube. For each well you test, you will need 500μL of a 1.6 × 10^6 cells per mL

suspension, which will result in a MOI of 2. As described above, you may choose to opsonize the cells with the 18B7 monoclonal antibody at a concentration of 10μg/mL or with fresh or frozen serum at a concentration of 10%. Whatever the opsonin chosen, incubate the CMFDA-stained *C. neoformans* suspension with it for 30 min at 37 °C. Wash the cells twice with PBS, as described in Subheading 3.7, steps **1–3**.

2. Add 500μL of the opsonized cell suspension to each well containing stained J774 cells in the 24-well plate. Incubate for 2 h at 37 °C and 5% CO_2.

3.10 Prepare the Flow Cytometry Controls (Day 0)

1. As mentioned above, the flow cytometry experiment requires several controls (Fig. 4). They are described separately here for clarity, but most of them can be prepared along with the steps above. Alternatively, you can prepare them during the 2 h phagocytosis period when the infection is happening. The first group of controls is the unstained cells, prepared as outlined below:
2. Tube 1: Unstained J774 cells. Detach the J774 cells from one of the wells of the 12-well plate by washing the well twice with warm PBS and then adding 500μL of nonenzymatic cell dissociation solution. After incubating the cells at 37 °C for 5 min, detach the cells by pipetting up and down a few times. Collect the cells into a flow cytometry tube and store on ice.
3. Tube 2: Unstained *C. neoformans* cells. This control is made from one of the three tubes containing 1×10^7 *C. neoformans* cells per mL described above on Subheading 3.7, **step 4**. Transfer 500μL these cells into a flow cytometry tube and maintain on ice.
4. A second set of controls will be used for compensation. These are the four single-color controls:
5. Tube 3: CMFDA single-color control. Pipette into a flow cytometry tube that contains 400μL of PBS and 100μL of the 1×10^7 cells per mL suspension of CMFDA-labeled *C. neoformans* described above on Subheading 3.7, **step 6**. Maintain the tube on ice.
6. Tube 4: Uvitex 2B single-color control. To the third tube containing 1×10^7 *C. neoformans* cells per mL described above on Subheading 3.7, **step 4**, add Uvitex 2B to a final concentration of 0.01%. Incubate for 30 min and wash the cells twice with PBS. Resuspend in 1 mL of PBS, and add 500μL of these stained cells into a flow cytometry tube, which will be kept on ice.
7. Tube 5: CellTracker Deep Red single-color control. In one of the wells from the six-well plate, remove the medium, and wash the cells twice with warm PBS. After the last wash, add 2 mL of

warm PBS. Then add 2μL of the CellTracker Deep Red stock solution in DMSO. Swirl the plate to mix and stain for 10 min. After staining, wash the well twice with warm PBS, detach the cells with 500μL of the nonenzymatic cell dissociation solution, and transfer the cells to a flow cytometry tube on ice.

8. Tube 6: 7-AAD single-color control. Detach cells from 2 wells of the 12-well plate using 500μL per well of the nonenzymatic cell dissociation solution. Transfer the cells from each well into a different microfuge tube. One of the tubes is stored on ice; to the other one, add 500μL of 4% paraformaldehyde, and incubate at 37 °C for 30 min. After fixing the cells, wash the cells twice with PBS to remove the formaldehyde, and resuspend in 500μL of PBS. Mix these dead cells with the other tube of live cells that was on ice. To this tube, add 1μL of the 7-AAD DMSO stock solution. Mix well and transfer 500μL of this suspension to a flow cytometry tube to be kept on ice.
9. The third set of control tubes are the FMO controls. All four are identical to the experimental condition, except that one different fluorophore is omitted in each of them. *See* **Note 16**.
10. Tube 7: CMFDA FMO control. For this control, start by staining J774 cells in one of the wells of the 12-well plate with CellTracker Deep Red, as described above on Subheading 3.10, **step** 7. Infect the J774 cells as described above in Subheading 3.9, but instead of using CMFDA-stained *C. neoformans*, use the unstained *C. neoformans* cells. After the infection, stain the cells with Uvitex 2B, detach the cells, and add 7-AAD.
11. Tube 8: Uvitex 2B FMO control. Stain another well of the 12-well plate with CellTracker Deep Red. Infect these cells with CMFDA-labeled *C. neoformans* (as in Subheading 3.9, **step 2**). After the infection period, do not stain the cells with Uvitex 2B; instead, detach the cells and add 1μL of the 7-AAD DMSO stock solution.
12. Tube 9: CellTracker Deep Red FMO control. This FMO control is the only one that uses a well of the 12-well plate containing unstained J774 cells. Infect these cells with CMFDA-labeled *C. neoformans* (as in Subheading 3.9, **step 2**) and after the infection period stain with Uvitex to a final concentration of 0.01%. Detach the cells and add 1μL of the 7-AAD DMSO stock solution.
13. Tube 10: 7-AAD FMO control. Stain one more well of the 12-well plate with CellTracker Deep Red. Infect these cells with CMFDA-labeled *C. neoformans* (as in Subheading 3.9, **step 2**). After the infection period, stain with Uvitex 2B to a final concentration of 0.01%, and detach the cells, but do not add 7-AAD.

3.11 Stain with Uvitex 2B and Detach the Cells

1. When the 2 h period of infection has passed, remove the supernatant containing most of the *C. neoformans* cells that were not phagocytosed (extracellular *C. neoformans*).
2. Immediately add 500μL of Uvitex 2B dissolved to 0.01% in warm PBS. Incubate for exactly 1 min (*see* **Note 17**) and then remove the Uvitex 2B solution.
3. Carefully wash the wells twice with pre-warmed (37 °C) PBS. After the last wash, remove the PBS, and add 500μL of the nonenzymatic cell dissociation solution to each well in the 24-well plate that contains cells.
4. Incubate the plate for 5 min at 37 °C. Detach the cells by pipetting up and down vigorously, and transfer the contents of each well to a pre-labeled flow cytometry tube.
5. Add 0.5μL of 7-AAD to each tube, and store the cells on ice until they are analyzed using a flow cytometer.

3.12 Analyze the Controls and Experimental Samples on a Flow Cytometer

1. Start the flow cytometer, and perform all adjustments and calibrations per manufacturer instructions and your facility policies. It is helpful to start the equipment and add sample names and create plots during the time macrophages and *C. neoformans* cells are co-incubating.
2. Create the following four dot plots or density plots to help you visualize the experiment:
3. FSC-A (forward scatter area) × SSC-A (side scatter area).
4. FSC-W (forward scatter width) × FSC-H (forward scatter height).
5. CMFDA × CellTracker Deep Red.
6. Uvitex 2B × 7-AAD.
7. Start adjusting photomultiplier tube (PMT) voltages using the unstained J774 control (Tube 1). These cells usually form two distinct populations with similar side scatter but different forward scatter. In our experience, both populations correspond to healthy, functional macrophages. For the fluorescence channels, aim to keep these negative controls with a median fluorescence between the first and second decades of the plot.
8. Quickly check each of the single-color controls. Sometimes the staining can be so bright that you will need to decrease the PMT voltages to avoid events that are above the detection limit, especially in the CellTracker Deep Red channel.
9. Once all channel voltages are set, start collecting data from the unstained and compensation controls. We usually collect 20,000 events per sample. We also usually do offline compensation on FlowJo, but at this time you can use the

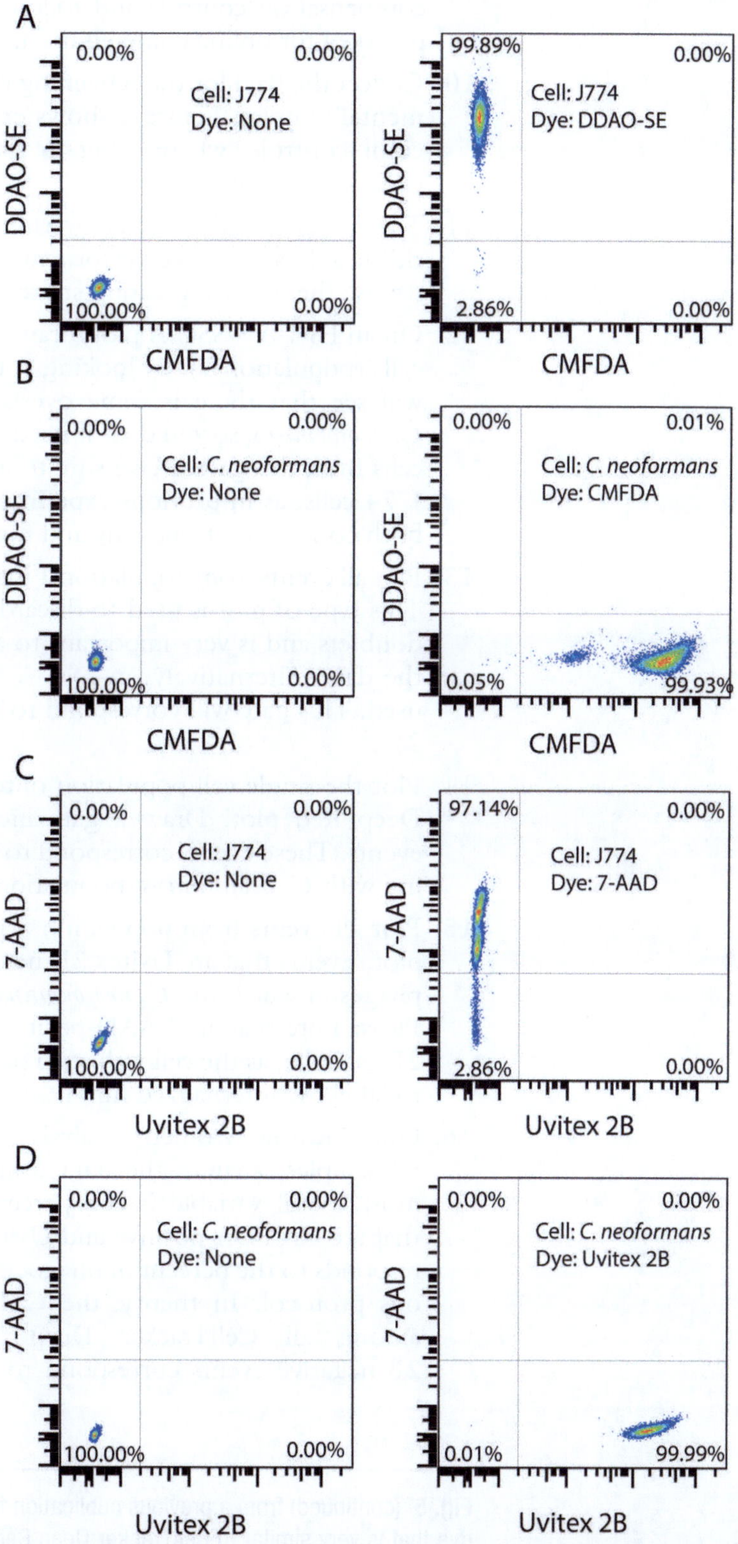

Fig. 5 Example plots for the single-color controls. In panels **a**–**d**, the left column depicts unstained cells, whereas the right column depicts single-color controls

compensation controls and follow your cytometer software's protocol for online compensation. *See* **Note 18**.

10. Collect the data for the remaining control tubes and all experimental samples. Figure 5 shows example plots for the single-color controls, whereas Fig. 6 shows example plots for an experimental sample.
11. To analyze the data, start by running a compensation algorithm unless you have performed online compensation. Next, we use the following gating strategy:
12. On an FSC-A × SSC-A plot, draw a gate that includes all J774 cells (population 1). By looking at the unstained controls, you will see that there is some overlap between J774 cells and *C. neoformans*, so you do not need to try and exclude all fungal cells from this gate. Make sure to include both populations of J774 cells, as in previous experiments we have seen that they both correspond to healthy and functional cells.
13. Plot all events from population 1 into an FSC-W × FSC-H plot. This type of plot is used to discard events that correspond to doublets and is very important to avoid artifacts that can bias the data. Alternatively, an FSC-A × FSC-H plot may also be used. This gate will correspond to the single cells (population 2).
14. Plot the single cell population onto a CMFDA × CellTracker Deep Red plot. Draw a gate including all double-positive events. These events correspond to J774 cells that are interacting with *C. neoformans* (population 3).
15. Plot all events from population 3 into a Uvitex 2B × 7-AAD plot. Events that are Uvitex 2B-negative correspond to macrophages in which the *C. neoformans* cells are fully internalized. Please note that all 7-AAD-positive events will also be Uvitex 2B-positive, as the cell wall stain penetrates dead macrophages to label the internalized fungi.
16. Once the flow cytometry analysis is complete for all experimental samples, analyze the data using statistical software. The most useful variable is the percentage of live macrophages that are CMFDA-positive and Uvitex 2B-negative, which corresponds to the percent of phagocytosis described in the previous protocol. In theory, the CMFDA median fluorescence among all CellTracker Deep Red-positive and Uvitex 2B-negative events correspond to the phagocytic index, but

Fig. 5 (continued) from a previous publication from our group [3]. DDAO-SE is a dye that is very similar to CellTracker Deep Red. The original figure from which this was adapted was published under a Creative Commons Attribution-Noncommercial-ShareAlike 3.0 Unported license

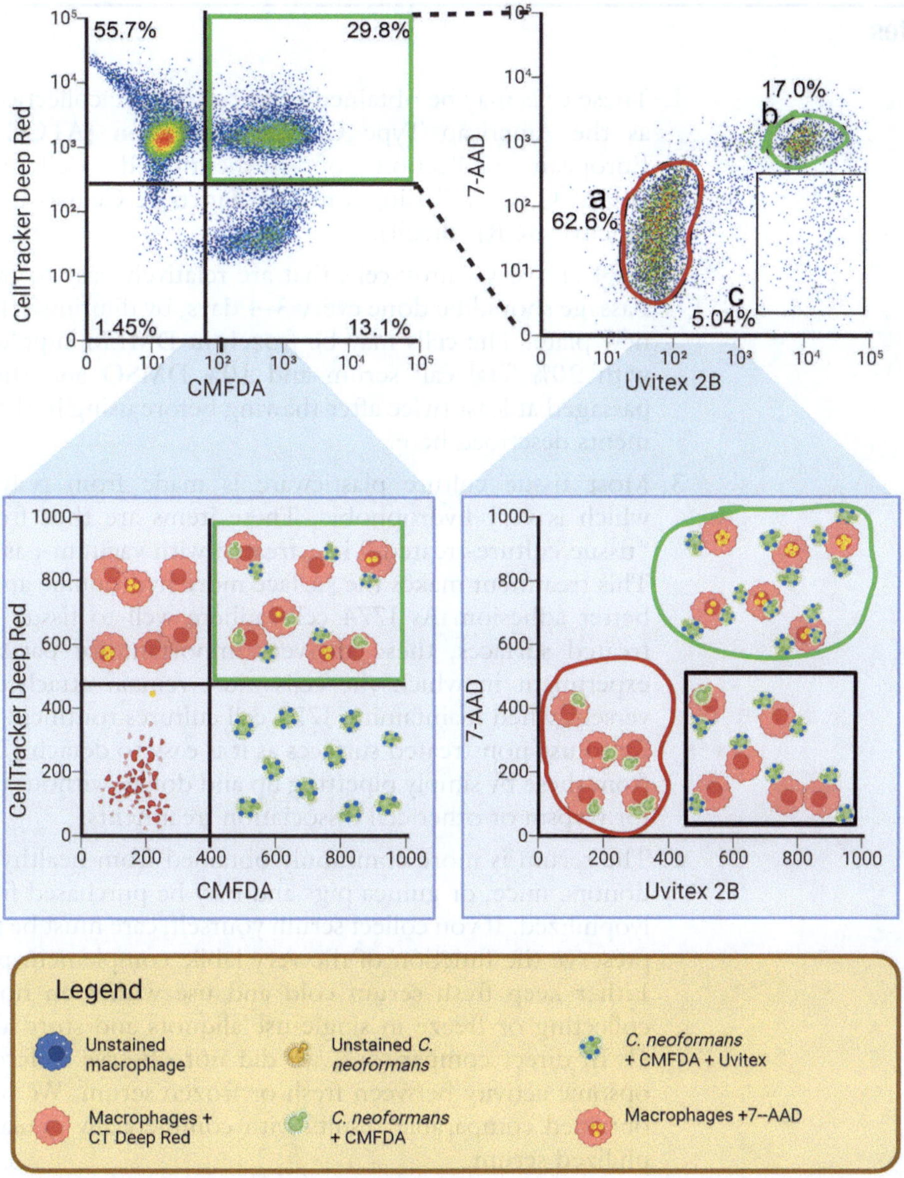

Fig. 6 Example plots of a flow cytometry phagocytosis experiment. The top panels show results from an actual experiment, adapted with permission from a previous publication from our group [3]. The right plot shows only events that were within the green gate shown in the left plot. The two bottom panels represent schematically what each population in the plots above contains. The original figure from which this was adapted was published under a Creative Commons Attribution-Noncommercial-ShareAlike 3.0 Unported license

we have never observed significant differences in this parameter in several experiments.

17. Modifications to this protocol may be used to measure non-lytic exocytosis [3, 6], a process in which macrophages expel phagocytosed *C. neoformans* such that both the fungi and phagocyte remain viable [7, 8]. *See* **Note 19**.

4 Notes

1. These cells may be obtained from cell culture collections, such as the American Type Culture Collection (ATCC, USA), European Collection of Authenticated Cell Cultures (ECACC, United Kingdom), or Banco de Células do Rio de Janeiro (BCRJ, Brazil).
2. J774 are very sturdy cells that are relatively easy to maintain. Passage should be done every 3–4 days, by diluting 1:10 into a new plate. The cells may be frozen in DMEM supplemented with 20% fetal calf serum and 10% DMSO and should be passaged at least twice after thawing before using in the experiments described here.
3. Most tissue culture plasticware is made from polystyrene, which is very hydrophobic. These items are thus frequently "tissue culture-treated," i.e., treated with vacuum-gas plasma. This treatment makes the surface more hydrophilic and allows better adhesion. As J774 cells adhere well to tissue culture-treated surfaces, these are very important for parts of the experiment in which the cells must remain attached. Conversely, when maintaining J774 cell cultures routinely, we prefer to use non-treated surfaces as it is easy to detach J774 cells from these by simply pipetting up and down, without the need for trypsin or other cell dissociation treatments.
4. The serum is more commonly obtained from healthy human donors, mice, or guinea pigs and may be purchased frozen or lyophilized. If you collect serum yourself, care must be taken to preserve the function of the very labile complement proteins. Either keep fresh serum cold and use within an hour after collecting or freeze in single-use aliquots and store at −80 °C. In direct comparisons, we did not observe differences in opsonic activity between fresh or frozen serum. We have also obtained comparable results with commercially available lyophilized serum.
5. CellTracker Deep Red and CMFDA are reactive dyes that result in covalent labeling of proteins with fluorescent dyes. The reactive groups are very sensitive to moisture and degrade quickly in aqueous solutions. Thus, it is better to purchase single-use vial formats. Wait for each vial to reach room temperature before opening, and use only anhydrous DMSO to dissolve these dyes, as regular DMSO contains trace amounts of water.
6. It is very easy to make mistakes when pipetting liquids into 96-well plates. It is highly desirable to thoroughly plan the experiment ahead of time. One way of doing so is to print a plate map, a 12×8 table in which all the conditions in each well

of the plate are clearly written. Another helpful tool is a card with well numbers that can be put beneath the plate, so you can more easily identify each well. These cards are available commercially or may be printed.

7. Instead of opsonizing *C. neoformans* cells in a separate tube and then adding the opsonized fungi into the plate with J774 cells, you may add both opsonin and fungi directly to the 96-well plate. We have not observed significant and reproducible differences between prior or simultaneous opsonization.
8. Add PBS slowly, ejecting liquid with the pipette tip touching the wall. Pipetting that is too vigorous will detach too many J774 cells. However, you need to wash the J774 cells well enough to remove most extracellular *C. neoformans.* Optimally washing the wells to remove extracellular *C. neoformans* takes some practice.
9. Staining the cells is a crucial step in the experiment. The most sensitive parameters are the staining times, and these will probably vary depending on the precise composition of the dye solutions used by each manufacturer. The times we describe in the protocol were optimized in our laboratory using a specific kit from a single manufacturer, so different kits might require different staining times. Note that these times are not the same as those that were specified on the manufacturer's instruction sheet, as the manufacturer optimized times to stain cells on a slide instead of in a 96-well plate.
10. As the staining times are very short and you need to be precise, it is better to use a multichannel pipette and stain one row (or column) at a time.
11. It is important at this point to use pre-warmed (37 °C) PBS to avoid detaching the J774 cells. Cold PBS might also activate stress signaling, interfering with the experiment.
12. The CellTracker Deep Red dye is a fluorophore that is chemically reactive with cellular proteins. However, it will react with any protein, which is why it is important to thoroughly wash the J774 cells to remove proteins (fetal calf serum) from the medium.
13. As noted above (*see* **Note 5**), the reactive group quickly loses reactivity in aqueous solutions. This is why we have better results when pipetting the DMSO-dissolved stock dye solution directly into the wells in which the cells are resuspended in warm PBS, instead of dissolving the dye first in PBS and then adding this solution to the wells.
14. The time and dye concentration are very important when staining J774 cells with CellTracker Deep Red. The reactions between reactive dyes and proteins may generate toxic

substances, and long incubations in media without nutrients may harm the cells. We have in the past observed decreased cell viability with longer times or higher dye concentrations, so be cautious if you need to optimize the parameters to obtain better staining.

15. J774 cells labeled with this dye will remain fluorescent for a long time, as when the cells divide the dye labeled mother cell proteins are split between the daughter cells. We have successfully used this dye to track cells over 24 h after staining.
16. The FMO controls are not absolutely necessary for the experiment. They do, however, allow an accurate determination of the gate positions when analyzing the data. They are the only controls that involve infecting J774 cells with *C. neoformans*, so they take as long as the experimental samples to prepare. We have organized the preparation of these and the other controls separate from the preparation of the experimental samples in this protocol for the sake of clarity, given its complexity. But in practice, once enough experience with the assay has been accrued, you may prepare the controls along with the experimental samples to save time.
17. Timing the Uvitex 2B staining step precisely is crucial. If the time the cells are in Uvitex 2B is prolonged, small amounts of the dye may start penetrating macrophages and staining intracellular fungi, which defeats the purpose of using it. Be especially careful when testing multiple experimental conditions in parallel, as you might not be able to finish adding the Uvitex 2B solution in the last wells by the time the 1 min incubation of the first wells is done.
18. Compensation in this experiment may be tricky. A first suggestion is to not try to do it manually, but instead use compensation algorithms available in several different flow cytometry analysis programs. Another suggestion is to always run compensation controls for each experiment you perform, as slight differences in staining may change the compensation matrix.
19. Using this protocol to measure non-lytic exocytosis requires a flow cytometer with sorting capabilities. Moreover, given that live *C. neoformans*, a biosafety level 2 pathogen, is being sorted, a number of aerosol containment tools and biosafety equipment are necessary. In terms of the protocol itself, the infection must be done in larger vessels, e.g., in an entire Petri dish. Events corresponding to live macrophages with internalized *C. neoformans* (CMFDA-positive, CellTracker Deep Red-positive, Uvitex 2B-negative, and 7-AAD-negative) are sorted and incubated for 24 h in J774 culture medium at 37 °C and 5% CO_2. After this period, the cells are detached, 7-AAD is added, and the sample is analyzed again. Live

macrophages that are now without *C. neoformans* (CMFDA-negative, CellTracker Deep Red-positive, and 7-AAD-negative) are the ones that underwent non-lytic exocytosis. This technique is very powerful because it permits measuring non-lytic exocytosis in thousands of J774 cells (in comparison with dozens in a usual time-lapse microscopy experiment) and because it is more amenable to high-throughput analyses. However, this experiment is much more complex and expensive, as only very few centers have the appropriate sorters in a safe biocontainment environment.

Acknowledgments

This work was funded by FAP-DF (Fundação de Apoio à Pesquisa do Distrito Federal) awards 0193.001048/2015-0193.001561/2017 and the CNPq (Conselho Nacional de Desenvolvimento Científico e Tecnológico) grant 437484/2018-1. We would also like to thank CNPq and CAPES (Coordenação de Aperfeiçoamento de Pessoal de Nível Superior) for graduate student scholarships, Finatec (Fundação de Empreendimentos Científicos e Tecnológicos) for administrative support, and Jéssika Paula Silva Vieira and Mariângela Souza de Oliveira for technical support.

References

1. Rajasingham R, Govender NP, Jordan A et al (2022) The global burden of HIV-associated cryptococcal infection in adults in 2020: a modelling analysis. Lancet Infect Dis 22:1748–1755. https://doi.org/10.1016/S1473-3099(22)00499-6
2. Nicola AM, Casadevall A (2012) *In vitro* measurement of phagocytosis and killing of *Cryptococcus neoformans* by macrophages. In: Ashman RB (ed) Leucocytes, vol 189–197. Methods in Molecular Biology, vol 844. Humana Press, p 9. https://doi.org/10.1007/978-1-61779-527-5_14
3. Nicola AM, Robertson EJ, Albuquerque P et al (2011) Nonlytic exocytosis of *Cryptococcus neoformans* from macrophages occurs *in vivo* and is influenced by phagosomal pH. MBio 2: e00167-11. https://doi.org/10.1128/mBio.00167-11
4. Chaka W, Scharringa J, Verheul AF et al (1995) Quantitative analysis of phagocytosis and killing of *Cryptococcus neoformans* by human peripheral blood mononuclear cells by flow cytometry. Clin Diagn Lab Immunol 2:753–759. https://doi.org/10.1128/cdli.2.6.753-759.1995
5. Tucker SC, Casadevall A (2002) Replication of *Cryptococcus neoformans* in macrophages is accompanied by phagosomal permeabilization and accumulation of vesicles containing polysaccharide in the cytoplasm. Proc Natl Acad Sci U S A 99:3165–3170. https://doi.org/10.1073/pnas.052702799
6. Frazao SO, Sousa HR, Silva LGD et al (2020) Laccase affects the rate of *Cryptococcus neoformans* nonlytic exocytosis from macrophages. MBio 11:e02085-20. https://doi.org/10.1128/mBio.02085-20
7. Alvarez M, Casadevall A (2006) Phagosome extrusion and host-cell survival after *Cryptococcus neoformans* phagocytosis by macrophages. Curr Biol 16:2161–2165. https://doi.org/10.1016/j.cub.2006.09.061
8. Ma H, Croudace JE, Lammas DA et al (2006) Expulsion of live pathogenic yeast by macrophages. Curr Biol 16:2156–2160. https://doi.org/10.1016/j.cub.2006.09.032

Chapter 13

Detection and Quantification of *Cryptococcus* Uptake by Phagocytic Cells Using Imaging Flow Cytometry

Althea Campuzano, Chiung-Yu Hung, and Floyd L. Wormley Jr

Abstract

Cryptococcus neoformans, the predominant etiological agent of cryptococcosis, is an encapsulated fungal pathogen found ubiquitously in the environment that causes pneumonia and life-threatening infections of the central nervous system. Following inhalation of yeasts or desiccated basidiospores into the lung alveoli, resident pulmonary phagocytic cells aid in the identification and eradication of *Cryptococcus* yeast through their arsenal of pattern recognition receptors (PRRs). PRRs recognize conserved pathogen-associated molecular patterns (PAMPs), such as branched mannans, β-glucans, and chitins that are the major components of the fungal cell wall. However, the key receptors/ligand interactions required for cryptococcal recognition and eventual fungal clearance have yet to be elucidated. Here we present an imaging flow cytometer (IFC) method that offers a novel quantitative cellular imaging and population statistics tool to accurately measure phagocytosis of fungal cells. It has the capacity to measure two distinct steps of phagocytosis: association/attachment and internalization in a high-throughput and quantitative manner that is difficult to achieve with other technologies. Results from these IFC studies allow for the potential to identify PRRs required for recognition, uptake, and subsequent activation of cytokine production, as well as other effector cell responses required for fungal clearance.

Key words Imaging flow cytometer, Phagocytosis, Innate immunity, Macrophages, *Cryptococcus*

1 Introduction

1.1 *Cryptococcosis and the Host Response*

Cryptococcus neoformans, the predominant causative agent of cryptococcosis, is an opportunistic fungal pathogen that impacts immunocompetent and immunosuppressed individuals. *C. neoformans* largely causes disease in patients with suppressed T cell-mediated immunity (i.e., AIDS patients, solid organ transplant recipients, and individuals on various immunosuppressive therapies) [1–3]. Cryptococcal meningitis affects approximately 180,000 individuals each year and is responsible for 15% of AIDS-related deaths worldwide [4]. Exposure typically occurs by the inhalation of desiccated spores and yeasts that are found ubiquitously in the environment into the lung alveoli. The first line of defense

Erin E. McClelland (ed.), *Cryptococcus neoformans: Methods and Protocols*, Methods in Molecular Biology, vol. 2775,
https://doi.org/10.1007/978-1-0716-3722-7_13,

following inhalation of *C. neoformans* into the lungs includes innate effector cells, namely, pulmonary macrophages, dendritic cells, and other types of phagocytic cells that survey the pulmonary environment and eliminate potential pathogens [5]. Phagocytosis is a process that may be roughly divided into two stages: association/attachment and internalization. Phagocytic cells detect highly conserved pathogen-associated molecular patterns (PAMPs) via pattern recognition receptors (PRRs). One of the outcomes following PAMP/PPR interactions include the initiation of phagocytosis and activation of effector cellular responses that result in eradication of invading microbes. Using imaging flow cytometry, we can phenotype, visualize, and capture images of effector cells as they associate with and engulf *Cryptococcus* yeast. Additionally, this technology allows for the relatively rapid quantification of the efficiency of association and/or phagocytosis of *C. neoformans* by phagocytic cells that do not contain or have defects in specific host PRRs.

1.2 Advantages of Imaging Flow Cytometry Compared to Alternative Assay Techniques

Quantification of phagocytosis has been assayed using various techniques, including the use of live-cell video, confocal microscopy, and flow cytometry [6–8]. Live-cell video is able to capture phagocytosis in real time. However, live capture is very time-consuming, and keeping cells viable for extended periods may be a limiting factor. Microscopic examination offers an ultrahigh resolution that accurately discriminates between internalization and association of microorganisms by phagocytic cells. However, it lacks objectively quantitative features or high-throughput capacity. Alternatively, flow cytometric analysis of phagocytosis is fast and sensitive, but it lacks spatial resolution to distinguish between associated, internalized, and non-phagocytic pathogens that appear on the same interrogation point of phagocytic cells. Also, images of the interaction cannot be acquired. Nevertheless, we have observed that imaging flow cytometer (IFC) analysis is fast, sensitive, and offers the spatial resolution to accurately and objectively quantify phagocytosis.

IFC combines the speed, sensitivity, and phenotyping abilities of flow cytometers with the detailed imaging and functional insights of microscopes. The technology has recently been applied to many image-based assays, including molecule co-localization, cell cycle analysis, molecule translocation, apoptosis, cell morphology changes, and cell–cell interactions among other analyses [9]. IFC works by converting fluorescence on labeled cells into photons of light that are collected by pixels present in the charge-coupled device (CCD) cameras, converting photons into quantitative digits that represent the features of the analyzed cell/particle. Each image recorded by the IFC is a grayscale two-dimensional image of the cell, allowing for each pixel to have an X and Y coordinate and intensity values corresponding to the amount of light captured in each defined area. A mask that represents the defined area of the analyzed subject is then applied to determine the digital features (e.g., size = 100 μm^2, centroid point X = 50, intensity = 15,000). IFC simultaneously detects digital features for cells and particles ranging from 2 to 100 μm in diameter and up to 12 channels, including 2 bright-field, 1 side-scatter, and 9 fluorochrome-labeled images. Each channel captures up to 85 features, including size, shape, location, texture, signal strength, and combination features using Boolean logic among other analyses. With a 60× objective lens, IFC achieves digital image quality of 0.33 μm per pixel, 2.5 μm depth of field, and 0.9 numerical aperture resolution (~0.3 μm). This imaging and resolution capacity is necessary to examine the spatial relationship of microbial and/or host cells during phagocytosis.

Many IFC systems are equipped with 96-well format autosamplers and/or robotic liquid handling systems that offer the capacity for high-throughput analysis. Ideally, acquisition rates reach approximately 1000 cells per second using a low volume velocity and high-sensitivity program to achieve high resolution when samples are prepared at a concentration of 10^6 cells in 50 μL of liquid solution. Several commercial and free software programs (e.g., IDEAS® software) are available to integrate image data, plots, and statistics with various degrees of automation. The raw data output may be automatically corrected for fluorescent light spillover using a proper compensation setup for each experiment. Batch data analysis automatically generates a statistical report and is available using a consistent analysis template and gating strategy to ensure reliable data comparison. With these unique properties, IFC is a powerful tool to quantify phagocytosis of microbial pathogens and particles. Thus, this technology is ideal for measuring and imaging *C. neoformans* association/internalization by phagocytic cells with normal or defective expression of various PRRs.

2 Materials

2.1 Inoculate Mice with Cryptococcus

1. mCherry-producing mutant, KN99mCH (serotype A, KN99 Mat *α*) [*see* **Note 1**], or fluorescently labelled yeast (see Subheading 3.1, **step 4**).
2. Alexa Fluor 568 (for nonfluorescent *C. neoformans* strains).
3. Yeast extract peptone dextrose (YPD) agar: 1% yeast extract, 2% peptone, 2% dextrose, and 2% Bacto agar.
4. YPD broth: 1% yeast extract, 2% peptone, 2% dextrose.
5. Male and/or female mice (4–6-week-old PPR-deficient and wild-type mouse strains) [*see* **Note 2**].
6. Isoflurane and rodent isoflurane anesthesia device.

7. Roswell Park Memorial Institute (RPMI) complete media: RPMI 1640 supplemented with 10% heat-inactivated fetal bovine serum, 2 mM L-glutamine, 100 U penicillin/mL, 100 μg of streptomycin/mL, and 50 mM 2-mercaptoethanol.

2.2 Preparation of Single-Cell Suspension of Pulmonary Leukocytes

1. Collagenase type IV enzyme.
2. Digestion buffer: RPMI 1640 with phenol red containing 1 mg/mL collagenase type IV.
3. Whirl-Pak (4 oz.).
4. 37 °C shaking incubator.
5. Stomacher (with the capacity for 30 s intervals).
6. Sterile nylon cell strainers (70 μm and 40 μm).
7. Sterile syringe (3 mL).
8. Sterile, endotoxin-free pipettes (10 mL).
9. 1× Hank's balanced salt solution (HBSS) with magnesium and calcium chloride.
10. Red blood cell lysis or NH_4Cl buffer: 0.859% NH_4Cl, 0.1% $KHCO_3$, 0.0372% Na_2EDTA [pH 7.4].

2.3 Enrichment of Pulmonary Macrophages and Dendritic Cells

1. Magnetic cell separator.
2. LS-positive selection columns for cell purification.
3. Magnetic bead buffer: 1× phosphate-buffered saline (PBS), pH 7.2 containing 0.5% bovine serum albumin (BSA), and 2 mM EDTA. Keep cold at 2–8 °C.
4. Fc block (anti-mouse CD16/CD32 antibody; clone: 2.4G2).
5. Anti-CD3e MicroBead mouse kit.
6. Biotinylated anti-F4/80 (clone: BM8) and anti-biotin MicroBead UltraPure kit.
7. Anti-CD11c MicroBead UltraPure, mouse kit.

2.4 Label Murine Macrophages and Dendritic Cell Populations with Fluorescent-Conjugated Antibodies

1. CD45 (clone: 30-F11).
2. CD11b (clone: M1/70).
3. CD11c (clone: N418).
4. F4/80 (clone: BM8).
5. CD64 (clone: 10.1).
6. CD24 (clone: M1/69).
7. Zombie Fixable Viability dye (1:100).
8. Fluorescence-activated cell sorting (FACS) buffer: 1× PBS with 2% BSA.
9. 2% paraformaldehyde.
10. Sterile Eppendorf tubes (1.7 μL).

2.5 Imaging Flow Cytometry and Data Analysis

1. Amnis® ImageStream^{X}-ISX-MKII imaging cytometer equipped with five lasers (405 nm, 488 nm 561 nm, 642 nm, and 785 nm) and INSPIRE® acquisition software.
2. SpeedBeads.
3. Sheath solution: 1× PBS without magnesium and calcium.
4. Sterilizer: freshly made 10% bleach solution.
5. Cleanser: Coulter Clenz®.
6. Debubbler: 70% isopropanol.
7. IDEAS® analysis software.

3 Methods

Perform all experiments in sterile conditions. Centrifugation is conducted using a cold centrifuge at 4 °C at 1,200 × *g* for 3 min unless otherwise specified.

3.1 Mouse Inoculation of Cryptococcus

1. Culture mCherry cryptococcal strain, KN99mCH [10] from 15% glycerol stocks stored at −80 °C and maintained on YPD agar plates. Alternatively, nonfluorescent cryptococcal yeasts from strains of interest may also be used and maintained in YPD agar plates.
2. Inoculate 10 mL of YPD broth with a loop full of *Cryptococcus* taken from agar plates. Incubate the culture for 16–18 h at 30 ° C with shaking (35 RPM).
3. Prepare an inoculum of *Cryptococcus* yeasts by pelleting the culture at 4 °C.
4. Wash yeast cells with 10 mL of 1× sterile PBS three times. For nonfluorescent cryptococcal strains, incubate with 2 μg/mL of Alexa Fluor 568 at room temperature (25 °C) in the dark for 2 h, according to manufacturer's instructions. The organisms are then washed three times with 10 mL of sterile PBS, counted, and resuspended in 10 mL of sterile PBS. Quantify viable yeasts using trypan blue dye exclusion on a hemocytometer. Briefly, add 10 μL of resuspended yeast cells, 50 μL of 0.4% trypan blue solution, and 940 μL of 1× PBS into a 1.5 mL Eppendorf tube. The yeast sample is diluted 100×. Add 10 μL of the diluted and stained cells to each side of the hemocytometer coverslip. Count the viable, unstained cell number (N) within the 5 × 5 double grid on the hemocytometer. Viable cell density of the original sample is calculated by the following formula: N × 10^4 × 100 (dilution factor). Exclude counting trypan blue-stained cells.

5. Yeast may be cultured with phagocytes in vitro or used to inoculate mice. To inoculate mice, anesthetize mice with 2% isoflurane using a rodent anesthesia device.
6. Infect mice by the intranasal route with 1×10^4 to 1×10^5 colony-forming units (CFUs) of *C. neoformans* strain KN99mCH in 50 μL of sterile PBS. Verify the inoculum used for the intranasal challenge by plating tenfold dilutions of the inoculum on yeast-extract peptone dextrose (YPD) agar supplemented with chloramphenicol (100 μg/mL) to confirm the quantity of inoculum used. Plates are then incubated at 30 °C for 48 h to count colony-forming units [11].
7. House infected mice in an ABSL2 laboratory using standard husbandry. Monitor mice daily.

3.2 Prepare Single-Cell Suspension of Pulmonary Leukocytes

1. If the goal is to use cells for in vitro assays, euthanize naïve mice, or, if using infected mice, euthanize on predetermined days post-challenge by CO_2 inhalation followed by cervical dislocation to ensure stress-free death. Excise the lungs, and transfer to a 50 mL conical tube containing 10 mL of digestion buffer on ice.
2. Mince the lungs into ~0.2 cm^3 pieces using a sterile blade so that collagenase properly dissociates the pulmonary cells [*see* **Note 3**].
3. Transfer the lungs and the digestion buffer into a Whirl-Pak bag, and ensure all small tissue segments are successfully transferred.
4. Digest lung tissue in a shaking incubator for 30 min at 37 °C with intermittent (30 s) stomaching of the bag every 10 min. It takes 30 min to ensure complete dissociation of pulmonary cells. However, times greater than 30 min do not result in a greater yield of cells.
5. Filter the dissociated cells through a sterile 70 μm nylon filter into a sterile 50 mL conical tube to obtain a single-cell suspension. Tissues may be gently pressed into the filter using the plunger of a sterile 3 mL syringe to enhance release of cells.
6. Wash the filter once with 5 mL of sterile 1× HBSS.
7. Filter cell suspension through a sterile 40 μm nylon filter into a sterile 50 mL conical tube to obtain a single-cell suspension.
8. Wash the filter once with 5 mL of sterile 1× HBSS and centrifuge cell suspension. Carefully remove the supernatant, leaving behind the cell population.
9. Lyse erythrocytes by suspending the cells in 3 mL of NH_4Cl buffer for 3 min on ice, followed by adding a twofold volume excess of cold sterile 1× PBS. Erythrocytes are auto-fluorescent and skew the flow analysis. Centrifuge cells and carefully remove supernatant.

3.3 Enrich Specific Populations of Pulmonary Macrophages and Dendritic Cells

1. Suspend the cell pellet in a total of 200 μL of magnetic bead buffer to enrich pulmonary phagocytic cells.
2. Count pulmonary leukocytes using trypan blue dye exclusion on a hemocytometer, as described in Subheading 3.1, to ensure that accurate amounts of antibodies and reagents are used during the experiment. Add a 1:500 dilution of Fc block to each cell suspension. Gently vortex the pellet, and let rest for 1 min at room temperature (25 °C) to prevent nonspecific binding of PRRs by the Fc portion of antibodies.
3. Deplete T cells using an adequate amount of biotin-conjugated anti-CD3e antibody provided in the MicroBead kit, according to the manufacturer's recommendations [*see* **Note 4**]. Incubate cells for 10 min at 4 °C, and then wash once with magnetic bead buffer, as directed by the manufacturer's instructions.
4. Resuspend the cell pellet in magnetic bead buffer, and add an appropriate amount of anti-biotin antibodies that are included in the CD3e MicroBead kit. Incubate the sample for 15 min at 4 °C. Wash cells with magnetic bead buffer, as recommended in the manufacturer's instructions, and centrifuge tubes as described in Subheading 3 to obtain the cell pellet. Carefully remove the supernatant, being careful to avoid disturbing the loose cell pellet.
5. Place separation column in magnetic cell separator, and prime the magnetic LS separation column by adding 3 mL of magnetic bead buffer. Collect the flow-through in a sterile 15 mL tube.
6. Suspend the cell pellet in 500 μL of magnetic bead buffer. Add the cell suspension to the separation column, wash the sample tube with an additional 500 μL of magnetic bead buffer, and add to the separation column to increase cell yield.
7. Wash the column three times with 3 mL of magnetic bead buffer. Centrifuge the flow-through that contains the desired phagocytes without $CD3^+$ cells. Discard column containing $CD3^+$ cells.
8. Suspend the flow-through pellet in 100 μL of magnetic bead buffer to enrich pulmonary $F4/80^+$ macrophages. Add a 1:20 dilution of biotinylated F4/80 antibody (clone: MB8), and incubate for 10 min at 4 °C. Following incubation, wash the cell pellet with 1–2 mL of FACS buffer, and centrifuge cells, as described on Subheading 3 [*see* **Note 7**].
9. Add anti-biotin microbeads that have the capacity of binding to 10^9 cells, and incubate for 15 min at 4 °C, according to the manufacturer's recommendations. Wash the sample with magnetic bead buffer once. Centrifuge cells as described in

Subheading 3, and carefully remove the supernatant without disturbing the loose cell pellet.

10. Place a new LS separation column in the magnetic cell separator, and prime with 3 mL of magnetic bead buffer.
11. Wash the cell pellet in 500 μL of magnetic bead buffer and add it to the separation column. Repeat the wash step once more with 500 μL of magnetic bead buffer to ensure complete recovery and transfer of cells into the LS columns. Wash the column three times with 3 mL of magnetic bead buffer. Centrifuge the flow-through as described in Subheading 3, and set it aside for enriching dendritic cells (3.3.13).
12. Remove the LS column from the magnetic stand, and flush the cells out using 5 mL of magnetic bead buffer. Centrifuge the cells as stated in Subheading 3; the flushed cells will contain enriched F4/80$^+$ macrophages [12]. Resuspend the cell pellet containing F4/80$^+$ macrophage cells, and count macrophages using trypan blue dye exclusion on a hemocytometer, as described in Subheading 3.1, and suspend in an appropriate amount of 1× PBS buffer. A small aliquot of macrophages (~5 × 10^5 cells) is used for fluorescent labeling and flow cytometric analysis of purity and viability.
13. Add 400 μL of magnetic bead buffer to the cell pellet from 3.3.11 that is used to enrich dendritic cells using positive selection by labeling cells with CD11c MicroBeads, according to the manufacturer's recommendations.
14. Based on the manufacturer's recommendation, add 100 μL of anti-CD11c antibody per 10^8 cells, and incubate the cells for 15 min at 4 °C, followed by washing with 2 mL of magnetic bead buffer.
15. Prime the magnetic LS separation column by washing the column with 3 mL of magnetic bead buffer. Pass the cell sample through the column, and collect the flow-through in a sterile 15 mL tube. Wash the column three times with 3 mL of magnetic bead buffer.
16. Remove the column from the magnet stand, and add 5 mL of magnetic bead buffer to flush out the bound CD11c$^+$ dendritic cells [11] [*see* **Note 5**]. Centrifuge cells as described in Subheading 3, and count CD11c$^+$ dendritic cells as described in Subheading 3.1. A small aliquot of dendritic cells (~5 × 10^5 cells) is used for fluorescent labeling and flow cytometric analysis of purity and viability.

3.4 In Vitro Phagocytosis Assessment of Cryptococcus

1. Isolate pulmonary phagocytic cells from naïve mice, as described in Subheading 3.3.
2. Pre-opsonize 1–2 × 10^8 cells of the mCherry-expressing strain of *Cryptococcus*, KN99mCH, or labelled *Cryptococcus* in 1 mL of 1× PBS, with 1 μg/mL of anti-GXM antibody (clone: 3C2, IgG1, a kind gift from Thomas Kozel, University of Nevada, Reno NV) for 30 min [13] [*see* **Note 6**].
3. Incubate 100 μL of *Cryptococcus* with 100 μL of pulmonary phagocytes (1 × 10^6 phagocytic cells) in a 5:1 or 10:1 ratio with a final volume of 200 μL in a 96-well plate [*see* **Note 6**].
4. Incubate phagocytes and *Cryptococcus* at various time points (i.e., 30 min, 2, 6, and 24 h). Following these incubations, proceed with labeling step (*see* Subheading 3.5).

3.5 Label Cells with Fluorochrome-Conjugated Antibody for Phenotyping

1. Prepare cells or beads for compensation that are processed along with the experimental samples. Label aliquots of sample for each fluorochrome-conjugated antibody used in this experiment.
2. Add 10^6 enriched phagocytes and compensation cells in 100 μL 1× PBS per well of a 96-well U-bottom plate. Incubate with a 1:500 dilution of Fc block for 1 min at room temperature (25 °C). Add a 1:100 dilution of Zombie viability dye, according to the manufacturer's recommendations [*see* **Note 7**] [14]. Incubate the cells at room temperature (25 °C) in the dark for an additional 15–30 min.
3. Wash wells once with 100 uL of FACS buffer.
4. Add an appropriate amount of titrated antibody cocktail containing accurate amounts of each fluorochrome-conjugated antibodies, according to the manufacturer's recommendation (Table 1). Incubate the plate for 30 min at 4 °C in the dark [*see* **Note 8**].

Table 1
Phenotypes of pulmonary myeloid cells

Marker	Macrophages	Dendritic cells	CD24 cells	Monocytes
CD45	+	+	+	+
CD11b	+	+	+	+
CD11c	±	Intermediate	±	±
F4/80	+	–	±	±
CD24	–	±	+	–
CD64	+	–	+	±

5. Wash each well three times with FACS buffer and fix with 30 μL of fixation buffer. Transfer each sample into a 1.7 μL Eppendorf tube for single-tube configuration, or maintain in the 96-well round-bottom plate for autosampler configuration. Samples are now ready for imaging flow cytometric analysis.

3.6 Data Acquisition Using an Imaging Flow Cytometer

1. Turn on the ImageStream^X MKII system, and top off sheath solution, sterilizer, cleanser, debubbler, and Milli-Q water bottles on the fluidic deck, and install a tube of SpeedBeads that are run together with all samples for background calibration.
2. Run startup program using INSPIRE® software before sample acquisition to ensure the system is functional [*see* **Note 9**].
3. Set up an acquisition program by selecting the required brightfield, side-scatter, and fluorochrome signal channels, magnification lens, flow speed, and data storage domain [*see* **Note 10**].
4. In the work area of INSPIRE software, draw a "histogram" plot for each used channel to monitor optimal data acquisition. Right click on the X-axis to select the feature of fluorochrome intensity (Intensity_MC_Channel name) for each histogram plot.
5. Adjust the laser power for each channel to optimally separate the negative and positive signals within the manufacturer-specified ranges.
6. Cell images are also visible on the image gallery window. Adjust brightness and contrast of each channel to optimally display cell images. Once the optimal laser powers for each channel are determined, the acquisition program is saved as a template for future runs.
7. Run the Compensation Wizard program using cells or beads that are labeled with one of the fluorochromes used in the experiment to establish a compensation matrix [*see* **Note 11**].
8. Use the same established acquisition program to run the entire set of experimental samples that are intended for comparison. Collect enough numbers of events for statistical analysis of the cell population of interest [*see* **Note 12**].
9. Data are automatically saved as raw image files with a (.rif) file extension [*see* **Note 13**].
10. Clean the cytometer and run the shutdown program specified by the manufacturer.

3.7 Phagocytosis Analysis Using IDEAS® Software

1. Create a new compensation matrix that may be fine-tuned to optimally correct the spectral overlap of all fluorochromes that are used in the multicolor assay [*see* **Note 14**]. Follow the "Create new matrix" instructions under the drop-down menu

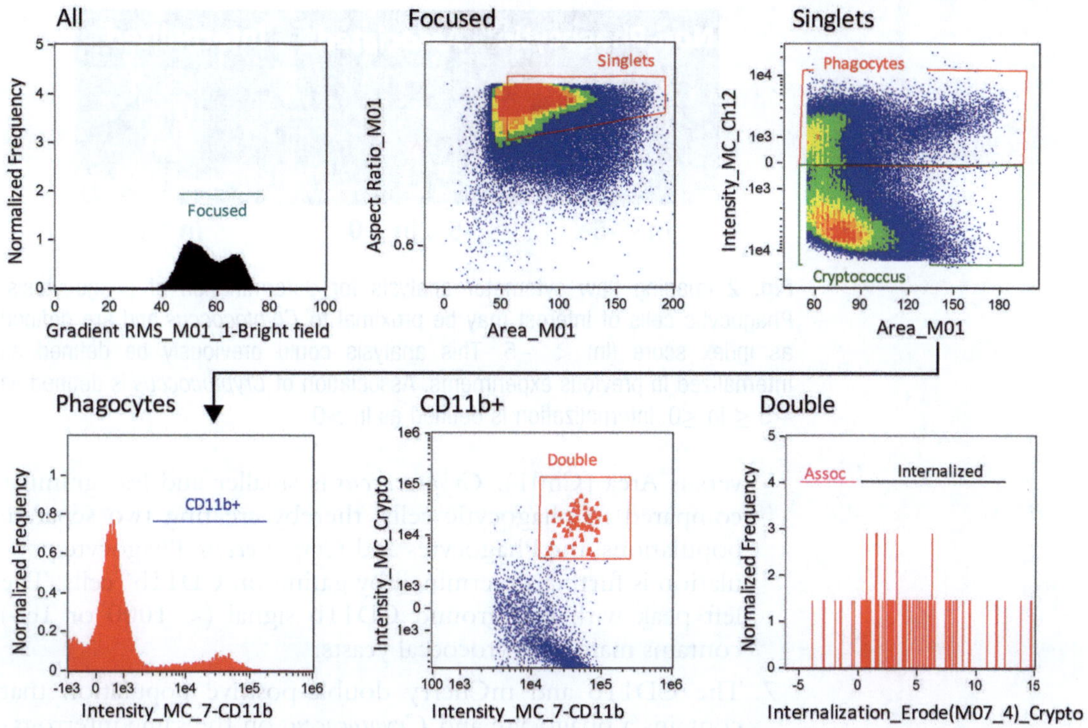

Fig. 1 Gating strategy assessment for internalization/association of *Cryptococcus*

of "Compensation tool." Save the matrix using a name with (.ctm) file extension for easy reference.

2. Create a gating strategy and analysis template for a raw image file (.rif) that contains sufficient target cells, applying the newly created compensation matrix (.ctm) and following the "Internalization" wizard tool. Select Guided Analysis > Wizards > Select Internalization >. Click "OK" [15, 16].
3. The "Internalization" wizard tool contains intuitive step-by-step instructions displayed on the monitor for users to follow. The tool is revised to the following six steps to analyze phagocytosis of *Cryptococcus* yeasts, as shown in Fig. 1.
4. Select focused cells that have a gradient RMS value of 46 or above. The quality of bright-field images may be examined on the image gallery. You may choose a cutoff value that shows the satisfactory quality of cell images by examining the bright field images.
5. Select singlets that have an aspect ratio (minor axe/major axe) above 0.6 for phagocytes and *Cryptococcus*, since they tend to be round in morphology (*see* **Note 15**).
6. Phagocytes and *Cryptococcus* may be further separated by gating on a dot plot displaying Intensity of side scatter (Ch12)

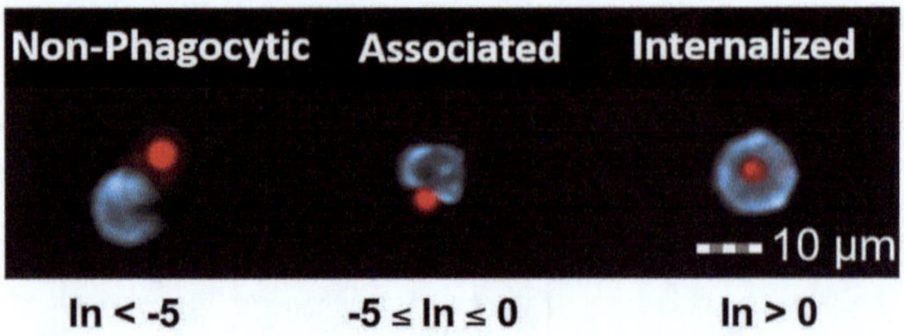

Fig. 2 Imaging flow cytometer analysis for determination of phagocytosis. Phagocytic cells of interest may be proximal to *Cryptococcus* and are defined as index score (In) < -5. This analysis could previously be defined as internalized in previous experiments. Association of *Cryptococcus* is defined as $-5 \leq$ In ≤ 0. Internalization is defined as In >0

versus Area (Ch01). *Cryptococcus* is smaller and less granular compared to phagocytic cells, thereby creating two separate populations, the Phagocytes and *Cryptococcus.* Phagocyte population is further determined by gating on CD11b$^+$ cells. The left peak with background CD11b signal ($<$ 1000 or 1e3) contains mainly cryptococcal yeasts.

7. The CD11b and mCherry double-positive population that contains a phagocyte and *Cryptococcus* on the same interrogation view field is gated on a dot plot displaying Intensity of CD11b versus Intensity of mCherry.
8. Finally, assign CD11b and *Cryptococcus* signal channels for cell images and Internalization objects, respectively, to complete the wizard analysis. The Internalization feature is defined as the ratio of the Intensity of mCherry inside the phagocytes to the Intensity of the entire view field. The ratio is mapped to a log scale to increase the dynamic range to values between ($-\infty$, ∞). Internalized cells typically have positive Internalization index scores (In), while cells with little internalization have negative scores. We define the non-phagocytic, associated, and internalized cryptococcal yeast using In < -5, $-5 \leq$ In ≤ 0, and In >0, respectively, as shown in Fig. 2.
9. Following internalization analysis, phenotypes of phagocytes may be assessed using cell markers listed in Table 1.
10. Cell images of each populations may be viewed in the image gallery and further examined for the accuracy of internalization analysis. The images may also be exported as tiff and png file formats, as shown in Fig. 3.

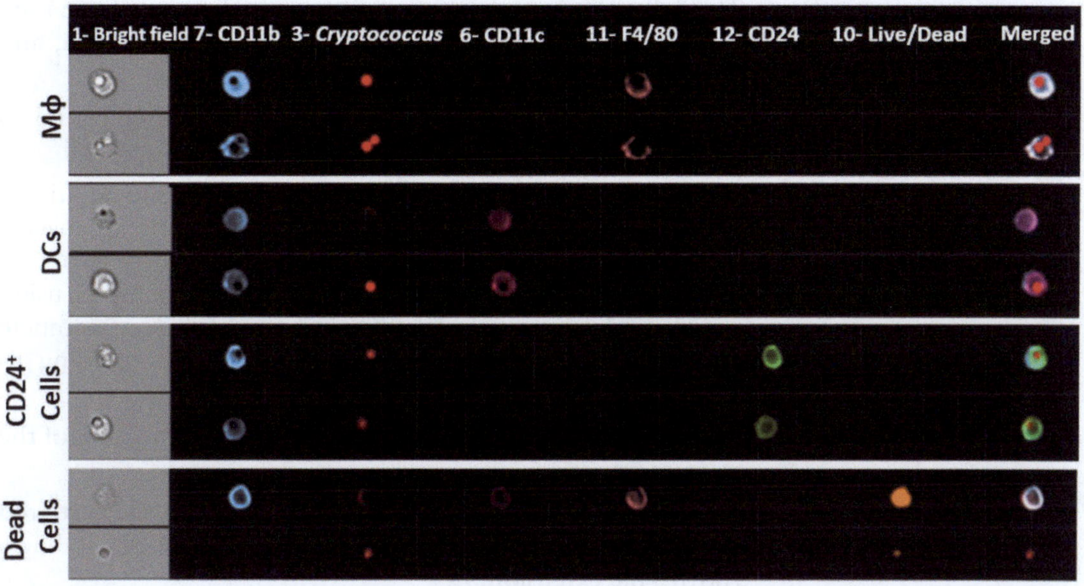

Fig. 3 Imaging flow cytometer analysis of pulmonary leukocyte populations. Imaging flow cytometry may identify specific leukocyte and cryptococcal populations. Phagocytized *Cryptococci* may be identified in macrophages, dendritic cells, $CD24^+$ cells, and monocytes. Viability may also be assessed to ensure exclusion of autofluorescence. CD24 may also be present on eosinophils, neutrophils, plasmacytoid dendritic cells (pDCs), and dendritic cells (DCs)

4 Notes

1. KN99mCH was a kind gift from Dr. Jenny Lodge (Washington University, St. Louis, MO).
2. All animal experiments should be conducted following NIH guidelines and in accordance with the Institutional Animal Care and Use Committee of your university. Animals are allowed one week of acclimation period in the facility before conducting experiments.
3. Be careful to not excise the thymus while removing lung tissue, as cells present in the thymus have yet to be differentiated and may skew the flow cytometry analysis. Maintain the digestion buffer on ice.
4. Antibody amount is based on total cell count.
5. The protocol for enrichment of macrophages and dendritic cells may also be applied to other tissue types, including bone marrow-derived cells, as described in [11]. We have used bone marrow-derived macrophage-to-*Cryptococcus* yeast ratios of 5: 1 or 10:1 to determine internalization or association at various time points.

6. The 5:1 or 10:1 yeast-to-innate immune cell ratio may be used in a variety of cells including dendritic cells, microglia, and eosinophils.
7. Titrate all antibodies used in this experiment prior to use to ensure optimal labeling.
8. CD64 (clone 10.1) and CD24 (clone M1/69) are additional markers recommended for accurate phenotyping of macrophages and dendritic cells, respectively. Use of these two markers may be necessary since macrophages are purified using positive selection with F4/80 antibody that may compete with the fluorochrome-conjugated antibody for binding to F4/80.
9. Avoid bubbles in the sample. This will prevent clogging of the instrument.
10. These data are obtained without bright-field and site-scattered signals, and they are automatically saved with name including a (noBF) file extension.
11. The instrument will automatically flush its fluidic system and run the ASSIST calibration routine to test a comprehensive set of parameters using SpeedBeads. These tests ensure that all internal systems including illumination, filter alignment, data collection, fluidic, and electronic connections are functioning within factory specifications. The routine takes approximately 1 h to run and requires no attendant.
12. Magnification selection is based on the required resolution, sizes of particles, and types of assays. ImageStream MKII is equipped with three types of objective lenses (i.e., 20×, 40×, and 60×) that have resolution capacity of 0.5 NA (~0.55 μm), 0.75 NA (~0.37 μm), and 0.9 NA (~0.32 μm), respectively.
13. Collect data for the target cell population without SpeedBead information to minimize the sizes of image data files. Gate on the cell population of interest using a dot plot displaying Aspect Ratio of_M01 (Bright-filed) and Area_M01. The area of beads is under 50. Specific positive populations for each channel may be gated and selected for by creating a new compensation matrix using the IDEAS® compensation tool. Thus, the newly created matrix may be superior to the one created using INSPIRE® software.
14. Data file organization includes raw image files (.rif), compensated image files (.cif), and data analyzed files (.daf).
15. We use a 561 laser to excite mCherry with maximal excitation and emission wavelength at 587 nm and 610 nm, respectively. The KN99mCH strain expresses a high amount of mCherry that causes the fluorescent signal to mainly fall in Channel 4 (610/30 nm band filter) and also spill over to Channel

3 (577/35 nm band filter) and Chanel 5 (702/85 nm band filter). Avoid using dim fluorochromes with emission signals that fall into Channel 3 (e.g., PE and Cy3) and Channel 5 (e.g., PE-Cy5) for cell markers in the same assays.

References

1. Park BJ, Wannemuehler KA, Marston BJ et al (2009) Estimation of the current global burden of cryptococcal meningitis among persons living with HIV/AIDS. AIDS 23(4):525–530. https://doi.org/10.1097/QAD.0b013e328322ffac
2. Singh N, Dromer F, Perfect JR et al (2008) Cryptococcosis in solid organ transplant recipients: current state of the science. Clin Infect Dis 47(10):1321–1327. https://doi.org/10.1086/592690
3. Bratton EW, El Husseini N, Chastain CA et al (2012) Comparison and temporal trends of three groups with cryptococcosis: HIV-infected, solid organ transplant, and HIV-negative/non-transplant. PLoS One 7(8):e43582. https://doi.org/10.1371/journal.pone.0043582
4. Rajasingham R, Smith RM, Park BJ et al (2017) Global burden of disease of HIV-associated cryptococcal meningitis: an updated analysis. Lancet Infect Dis 17(8): 873–881. https://doi.org/10.1016/S1473-3099(17)30243-8
5. Yang C, Huang Y, Zhou Y et al (2022) *Cryptococcus* escapes host immunity: what do we know? Front Cell Infect Microbiol 12: 1041036. https://doi.org/10.3389/fcimb.2022.1041036
6. Lewis LE, Bain JM, Okai B et al (2013) Live-cell video microscopy of fungal pathogen phagocytosis. J Vis Exp 71:50196. https://doi.org/10.3791/50196
7. Geunes-Boyer S, Oliver TN, Janbon G et al (2009) Surfactant protein D increases phagocytosis of hypocapsular *Cryptococcus neoformans* by murine macrophages and enhances fungal survival. Infect Immun 77(7): 2783–2794. https://doi.org/10.1128/IAI.00088-09
8. Bos H, de Souza W (2000) Phagocytosis of yeast: a method for concurrent quantification of binding and internalization using differential interference contrast microscopy. J Immunol Methods 238(1–2):29–43. https://doi.org/10.1016/s0022-1759(00)00132-0
9. Barteneva NS, Fasler-Kan E, Vorobjev IA (2012) Imaging flow cytometry: coping with heterogeneity in biological systems. J Histochem Cytochem 60(10):723–733. https://doi.org/10.1369/0022155412453052
10. Upadhya R, Lam WC, Maybruck BT et al (2017) A fluorogenic *C. Neoformans* reporter strain with a robust expression of m-cherry expressed from a safe haven site in the genome. Fungal Genet Biol 108:13–25. https://doi.org/10.1016/j.fgb.2017.08.008
11. Campuzano A, Castro-Lopez N, Wozniak KL et al (2017) Dectin-3 is not required for protection against *Cryptococcus neoformans* infection. PLoS One 12(1):e0169347. https://doi.org/10.1371/journal.pone.0169347
12. Leopold Wager CM, Hole CR, Wozniak KL et al (2014) STAT1 signaling is essential for protection against *Cryptococcus neoformans* infection in mice. J Immunol 193(8): 4060–4071. https://doi.org/10.4049/jimmunol.1400318
13. Kelly RM, Chen J, Yauch LE et al (2005) Opsonic requirements for dendritic cell-mediated responses to *Cryptococcus neoformans*. Infect Immun 73(1):592–598. https://doi.org/10.1128/IAI.73.1.592-598.2005
14. Gouwy M, Ruytinx P, Radice E et al (2016) CXCL4 and CXCL4L1 differentially affect monocyte survival and dendritic cell differentiation and phagocytosis. PLoS One 11(11): e0166006. https://doi.org/10.1371/journal.pone.0166006
15. Jenner D, Ducker C, Clark G et al (2016) Using multispectral imaging flow cytometry to assess an *in vitro* intracellular *Burkholderia thailandensis* infection model. Cytometry A 89(4):328–337. https://doi.org/10.1002/cyto.a.22809
16. Phanse Y, Ramer-Tait AE, Friend SL et al (2012) Analyzing cellular internalization of nanoparticles and bacteria by multi-spectral imaging flow cytometry. J Vis Exp. 64:e3884. https://doi.org/10.3791/3884

Chapter 14

Two Methods of Measuring *Cryptococcus neoformans* Fungal Burden in Macrophages

Emily Smith-Peavler, Linda M. Sircy, David E. Nelson, and Erin E. McClelland

Abstract

The ability of *C. neoformans* to survive and replicate within host phagocytes enables it to evade the immune system and allows for persistence of the infection. As such, measuring fungal burden of *C. neoformans* strains—and indeed how drug treatments can influence fungal burden—provides important information about *C. neoformans* pathogenesis. In this chapter, we describe two methods that may be used to appraise fungal burden: a standard end-point colony-formation assay for calculating the average number of yeast per host cell and a fluorescence microscopy-based method that may be used to measure changes in fungal burden in individual living macrophages in real time.

Key words *Cryptococcus neoformans*, Fungal burden, Live cell imaging, Phagocytosis assays, Intracellular fungal pathogen, Macrophage cell lines

1 Introduction

Cryptococcus neoformans is a ubiquitous fungal pathogen that is a leading cause of mortality among the immunocompromised with an estimated 181,000 deaths per year worldwide [1]. In the immunocompromised, *C. neoformans* causes life-threatening acute pulmonary infections, which may progress into fungal meningoencephalitis as fungal spores use phagocytic cells—specifically macrophages—as vehicles to cross the blood–brain barrier to invade the central nervous system [2–5]. Additionally, *C. neoformans* subverts the host immune response and remains within host macrophages in a latent state for many years [2], which is a challenging complication as long-term chronic infections may occur if an existing latent infection revives within an individual.

Emily Smith-Peavler and Linda M. Sircy contributed equally to this work.

Erin E. McClelland (ed.), *Cryptococcus neoformans: Methods and Protocols*, Methods in Molecular Biology, vol. 2775,
https://doi.org/10.1007/978-1-0716-3722-7_14,

Therefore, fungal burden experiments are worthwhile in determining the intracellular replication rate of strains of *C. neoformans.*

Within this chapter, we discuss two ways of conducting in vitro fungal burden assays. The first technique covers the standard method of plating and counting colonies of intracellular *C. neoformans* from lysed macrophages. Potential uses of this technique include studying the impact of genes involved in virulence and observing the effects caused by treatment exposures on *C. neoformans.* This technique has been used to compare differences in colony-forming units (CFU) between different types of tissue, such as brain, lung, and spleen, as well as between strains of *C. neoformans* in in vitro macrophage cell lines [6–8].

The second technique utilizes live cell imaging of macrophages containing intracellular *C. neoformans* and allows the experimenter to obtain time-resolved information about the fungal burden of individual live macrophages, as well as the rate of *C. neoformans* proliferation within the intracellular environment and the extrusion of yeast from host cells. Previous studies have demonstrated success with this technique using a variety of fluorescent stains, fungi constitutively expressing fluorescent proteins, and reporter macrophage cell lines [8–10].

2 Materials

2.1 *Cryptococcus neoformans* Strain Preparation

1. Prepare yeast extract peptone dextrose (YPD) broth medium (or other rich medium for culturing yeast) with deionized water and autoclave to sterilize. Store at room temperature.

2.2 Macrophage Preparation

1. Plastic cell culture dishes or flasks.
2. 50 mL conical tubes.
3. RAW 264.7 (American Type Culture Collection (ATCC) TIB-71) macrophage growth media: Supplement Dulbecco's Modification of Eagle's Medium (DMEM) containing HEPES with 10% fetal bovine serum (FBS), 100 U/mL penicillin, 100 μg/mL streptomycin, 2 mM L-glutamine, and 50 μg/mL gentamicin. Store at 4 °C.
4. J774A.1 (ATCC TIB-67) macrophage growth media: Supplement DMEM containing 4.5 g/L glucose with 10% FBS, 100 U/mL penicillin, 100 μg/mL streptomycin, 2 mM L-glutamine, 10% 1× NCTC-109, and 1× nonessential amino acids. Store at 4 °C.
5. Phosphate-buffered saline (PBS): 137 mM sodium chloride, 2.7 mM potassium chloride, 1.5 mM potassium phosphate, 8.5 mM disodium hydrogen phosphate in deionized water. Autoclave to sterilize and store at room temperature.

6. Lipopolysaccharide (LPS) isolated from *Escherichia coli* O127: B8 is diluted in water distilled in a Milli-Q water purification system (referred to as Milli-Q water) at a stock concentration of 10 mg/mL. The stock LPS is then diluted in Milli-Q water for a working stock concentration of 100 μg/mL. Store at −20 °C.
7. Mouse interferon gamma (IFN-γ) (carrier-free). Store at −80 °C in 10 μL aliquots at a stock concentration of 2.1×10^5 U/mL. However, specific activity of stock IFN-γ is dependent upon source and product lot.

2.3 Opsonization of C. neoformans

1. 18B7 (anti-GXM antibody): 1.98 mg/mL 18B7 mAb (a kind gift of Dr. Arturo Casadevall). Store at 4 °C.
2. Opsonization blocking buffer: Supplement PBS with 0.5% goat serum and 1% bovine serum albumin. Store at 4 °C.

2.4 Calculating Fungal Burden

1. Sodium dodecyl sulfate (SDS) solution: 0.5% $SDS_{(w/v)}$ in deionized water. Filter sterilize before use. Store at room temperature.
2. PBS, as described above (Subheading 2.2, **step 5**).
3. Prepare YPD agar medium with deionized water and autoclave to sterilize. Store agar plates at 4 °C.

2.5 Live Cell Imaging

1. CellTracker™ Red CMTPX Dye (Molecular Probes). Product is a lyophilized powder in 50 μg aliquots. Store at −20 °C. CellTracker dye should be prepared fresh on the day of staining. Lyophilized product is warmed to room temperature before addition of 72.86 μL of dimethyl sulfoxide (DMSO) for a stock concentration of 1 mM. 1 mM stock is diluted to 10 μM in pre-warmed (37 °C) macrophage growth media for staining.

3 Methods

All cell waste should be disposed of in a beaker containing 10% bleach for 24 h to ensure death of yeast and mammalian cells. All yeast and cell culture preparation should take place in a BSL-II certified biosafety hood. All steps are at room temperature unless otherwise noted. PBS and macrophage medium (either J774A.1 or RAW 264.7 depending on the cell line used) should be pre-warmed by placing at 37 °C for 30 min prior to use.

3.1 Cryptococcus neoformans Strain Preparation

1. Pipette 50 mL of YPD into a sterile 125 mL flask (*see* **Note 1**). Add inoculum of desired strain from either a frozen stock or a freshly streaked agar plate.
2. Incubate at 37 °C (*see* **Note 2**) for 36 h with shaking (150 RPM) until culture is in log phase.

3.2 Maintenance of Macrophages in Culture

1. Macrophage cell lines (both J774A.1 and RAW 264.7) are stored at −80 °C or liquid nitrogen vapor in 1 mL aliquots of 90% growth media and 10% DMSO in 2 mL cryovials. For use, stored cell lines are thawed at room temperature, and 9 mL of growth media is added dropwise to the cells. The cell suspension is then centrifuged at 300 × *g* for 5 min at room temperature to pellet the cells and the supernatant decanted, thereby removing the DMSO. The cells are immediately resuspended in growth media and plated in plastic cell culture dishes or flasks. Macrophage cells are passaged when they are 80% confluent (*see* **Note 3**) and are kept for no longer than 35 passages or until their phenotype deteriorates (*see* **Note 4**).
2. To dissociate cells from the surface of the culture vessels to either passage them or plate them for experiments, aspirate the medium, and wash the cells with 5 mL of warm PBS. Add 10 mL of fresh, warm macrophage medium, and scrape cells into the medium using a cell lifter (*see* **Note 5**), gently mixing the cell suspension to break apart clusters of cells. Place cells into a 50 mL conical tube.
3. For standard maintenance, the cells may be split 1:20, adding 0.5 mL of the cell suspension to 9.5 mL of fresh, warm macrophage medium and applying to a new culture dish or flask (*see* **Note 6**). These should be lightly shaken in a figure-8 pattern to ensure cells uniformly populate the entire surface of the dish and incubated at 37 °C and 5% CO_2. If the cells are to be used for a standard or a live cell fungal burden experiment, proceed to Subheading 3.3 or 3.6, respectively.

3.3 Macrophage Well Seeding for Fungal Burden Plating Protocol

1. Determine the concentration of live macrophages in the cell suspension (prepared in Subheading 3.2, **step 2**) using a hemocytometer by diluting cells 1:2.5 in trypan blue (*see* **Note** 7), and dilute to 5×10^5 cells/mL in growth medium. Prepare a sufficient volume of diluted cells so that 200 μL may be added to the required number of wells in a 96-well plate. For each condition to be tested, four wells of cells are needed for quadruplicate measurements. Make sure to include a set of macrophage-only control wells.
2. IFN-γ is used to classically activate the macrophages. Add IFN-γ to a final concentration of 100 U/mL to your diluted cell suspension (*see* **Note 8**).
3. Pipette 200 μL of the IFN-γ-treated cell suspension into the required number of wells in a 96-well plate.
4. Centrifuge cells for 1 min at 300 × *g* to ensure macrophages are on the bottom of the wells for adherence.
5. Incubate the seeded cells at 37 °C and 5% CO_2 overnight.

3.4 Preparation of Opsonized Cryptococcus neoformans Strains and Macrophages for Fungal Burden Plating Protocol

1. Pour prepared *C. neoformans* cultures into 50 mL conical tubes, and centrifuge for 5 min at 872 × *g* to pellet yeast. Decant supernatant.
2. Wash yeast three times with 10 mL of PBS. Centrifuge 5 min at 872 × *g*.
3. Resuspend the yeast in 5 mL of PBS, dilute such that 50–200 yeast are counted for each outer quadrant on a hemocytometer, and then calculate the number of cells/mL (average of four outer quadrants × dilution factor × 10^4/mL).
4. For each culture, prepare 100 μL of yeast suspension at 3.6 × 10^6 cells/mL in macrophage medium for each well that is to be infected. This will produce a multiplicity of infection (MOI) of 2:1 (yeast/macrophages). To opsonize the yeast, add 10 μg/mL 18B7 mAb (or another *C. neoformans* or capsule-specific mAb) to the suspension. Incubate the yeast for 30 min at 37 °C, shaking.
5. After opsonization of *C. neoformans*, centrifuge yeast for 5 min at 872 × *g* to pellet yeast. Decant supernatant to remove opsonization blocking buffer.
6. Wash the yeast three times with 1 mL of PBS. Centrifuge for 5 min at 872 × *g* after each wash to pellet the yeast.
7. Resuspend the final yeast pellet in 1 mL of PBS, dilute such that 50–200 yeast are counted for each outer quadrant on a hemocytometer, and then calculate the number of cells/mL.
8. Prepare 100 μL of yeast suspension at 3.6 × 10^6 cells/mL in macrophage medium for each well that is to be infected. At this stage, 1 μg/mL LPS and 200 U/mL IFN-γ (*see* **Note 8**) should be added to promote phagocytosis when the yeast is eventually added to the macrophages (*see* **Note 9**).
9. Infect macrophages with opsonized yeast. To do so, aspirate the medium from each macrophage-containing well in the 96-well plate, making sure to not disrupt the macrophage monolayer in the wells. To the macrophage-only control wells, add 200 μL of warm macrophage media. For all treatment wells, add 100 μL of warm macrophage media and 100 μL of the appropriate opsonized *C. neoformans* strain containing LPS and IFN-γ. Centrifuge the plate at 300 × *g* for 1 min to bring the yeast into proximity with the macrophages, thereby enhancing the efficiency of phagocytosis. Incubate the plate for 1 h at 37 °C and 5% CO_2 (*see* **Note 10**).
10. After incubation, aspirate the medium, and carefully wash the wells twice with 300 μL of warm PBS, pipetting slowly five times per wash (*see* **Note 11**).

11. Add 200 μL of either warm macrophage medium or warm macrophage medium containing treatment (if testing for impact of a compound on fungal burden). Incubate for 18 or 24 h at 37 °C and 5% CO_2 (*see* **Note 12**).

3.5 Plating Cryptococcus neoformans *and Calculating Fungal Burden*

1. After the incubation period has elapsed, add 100 μL of 0.5% SDS directly to each well to lyse the macrophages, releasing the intracellular yeast. Let sit for 5 min. Verify that the macrophages have lysed by viewing the wells using a standard inverted light microscope (Fig. 1).
2. Transfer the resultant yeast suspension from each well into separate, appropriately labeled 1.5 mL microcentrifuge tubes.
3. Wash each well twice with 100 μL of PBS, and pipette liquid into corresponding microcentrifuge tubes to gather remaining yeast cells.
4. Take 100 μL of the sample, and make 1:10 serial dilutions in 1.5 mL tubes containing 900 μL of sterile PBS. Change tips after each dilution and vortex tubes between dilutions.
5. Plate 100 μL of the original tube and the tubes containing diluted yeast suspensions onto separate YPD plates, and incubate for 2 days at 37 °C.
6. At the end of the incubation period, count and average the number of colonies on each set of plates, and multiply by the dilution factor to determine fungal burden (CFU/mL, Fig. 2).

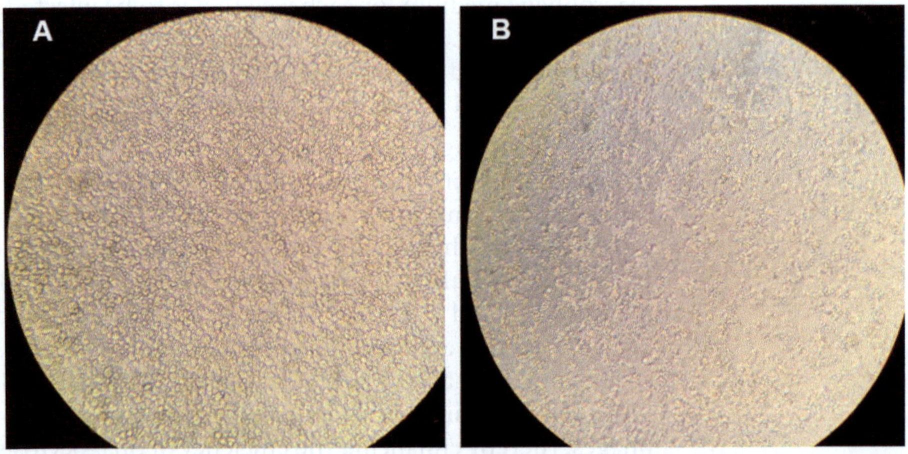

Fig. 1 Demonstration of J774A.1 macrophage lysing using 0.5% SDS after 18–24 h incubation with *C. neoformans*. *C. neoformans*-infected J774A.1 cells imaged (**a**) before and (**b**) after addition of 100 μL 0.5% SDS for 5 min show effective lysing of J774A.1 cells and intact *C. neoformans*. Imaged using an inverted light microscope at 100× magnification

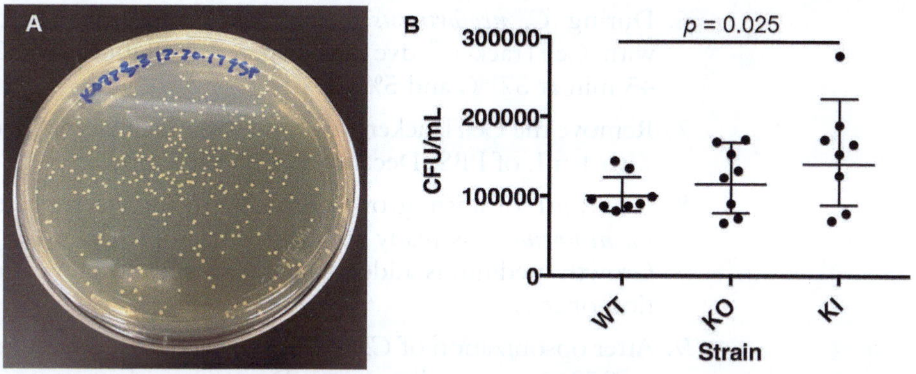

Fig. 2 Results from standard fungal burden plating method. (**a**) YPD plate with *C. neoformans* colony growth after 48 h. Colonies are counted and CFU/mL is calculated to determine fungal burden. (**b**) Graph of results from a fungal burden experiment using the plating method to test for a difference in the fungal burden of three different *C. neoformans* strains: wild-type (WT), knock-out (KO) of gene of interest, and knock-in (KI) of gene of interest. The central horizontal line represents the geometric mean. Error is represented as the geometric SD

3.6 Seeding Macrophages for Live Cell Imaging

1. Determine the concentration of live macrophages in the cell suspension (prepared in Subheading 3.2, **step 2**) using a hemocytometer by diluting cells 1:2.5 in trypan blue (*see* **Note 7**), and dilute to 1×10^5 cells/mL in growth medium. Apply 2 mL of the diluted cells to a 35 mm glass bottom culture dish. The glass constituting the base of the dish should be #1.5 cover glass. Incubate the cells at 37 °C and 5% CO_2 overnight.
2. Classically activate the macrophages by incubating them with 1 μg/mL LPS or 100 U/mL IFN-γ for 24 h at 37 °C and 5% CO_2 (*see* **Note 8**).

3.7 Preparation of Opsonized Cryptococcus neoformans Strains and Macrophages for Live Cell Imaging

1. For the live cell imaging protocol, a *C. neoformans* strain constitutively expressing a fluorescent protein (such as GFP) is required. In this protocol, H99S-GFP, a kind gift from Robin May (University of Birmingham), is used.
2. Pour prepared *C. neoformans* culture into a 50 mL conical, and centrifuge for 5 min at 872 × *g* to pellet yeast. Decant supernatant.
3. Wash the yeast three times with 10 mL of PBS. Centrifuge for 5 min at 872 × *g* after each wash to pellet the yeast.
4. Resuspend the final yeast pellet in 5 mL of PBS, dilute such that 50–200 yeast are counted for each outer quadrant on a hemocytometer, and then calculate the number of cells/mL.
5. Opsonize 1×10^7 cells/mL *C. neoformans* in opsonization blocking buffer supplemented with 10 μg/mL 18B7 mAb for 30 min at 37 °C (*see* **Note 9**).

6. During *C. neoformans* opsonization, incubate macrophages with CellTracker™ dye (as prepared in Subheading 2.5) for 45 min at 37 °C and 5% CO_2.
7. Remove the CellTracker™ dye and wash the macrophages twice with 1 mL of PBS. Decant the PBS each time.
8. Add 2 mL of warm growth medium to the macrophages until *C. neoformans* is ready to add to macrophages for infection. Growth medium is added to macrophages to prevent desiccation of cells.
9. After opsonization of *C. neoformans*, centrifuge yeast for 5 min at 872 × *g* to pellet yeast. Decant supernatant to remove opsonization blocking buffer.
10. Wash the yeast three times with 1 mL of PBS. Centrifuge for 5 min at 872 × *g* after each wash to pellet the yeast.
11. Resuspend the final yeast pellet in 1 mL of PBS, dilute such that 50–200 yeast are counted for each outer quadrant on a hemocytometer, and then calculate the number of cells/mL.
12. Infect CellTracker™ dye-stained macrophages with *C. neoformans* at a MOI of 2:1 (yeast/macrophages) (*see* **Note 13**) by adding opsonized yeast to the macrophages together with LPS at a final concentration of 1 μg/mL (*see* **Note 8**). Incubate the cells at 37 °C and 5% CO_2, allowing 1–2 h for phagocytosis (*see* **Note 10**).
13. After incubation, aspirate the growth medium to remove extracellular *C. neoformans*, and wash wells twice with 1 mL of warm PBS each time before replacing with 2 mL of fresh growth medium.

3.8 Live Cell Imaging of Macrophages to Visualize Fungal Burden of Intracellular C. neoformans

1. CellTracker™ dye-stained macrophages containing phagocytosed H99S-GFP (Fig. 3) may be imaged using a widefield or confocal fluorescence microscope equipped with appropriate filter sets and cell incubation system.
2. For in vitro fungal burden experiments, image acquisition is recommended to be set at 2 h intervals, to be sure to image the 2–3 h doubling time of wild-type *C. neoformans*, for 24 h. However, image acquisition intervals and total imaging times may be modified to meet the demands of the experiment.

4 Notes

1. If time is a limiting factor, cultures may be grown in 10 or 25 mL of YPD for 24 h, as long as the cultures are in log phase when used.

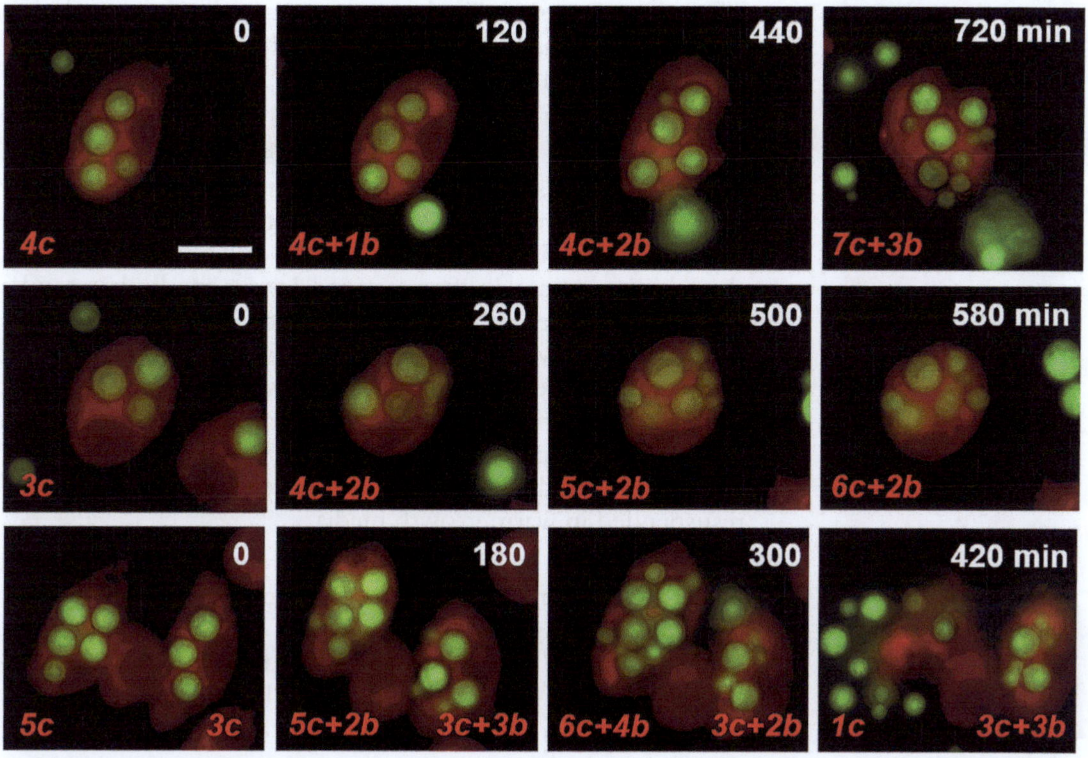

Fig. 3 Fluorescent microscopy images of RAW 264.7 macrophages stained with CellTracker™ Red CMTPX (*red*) dye and intracellular *C. neoformans* strain H99S constitutively expressing GFP (*green*) (H99S-GFP). RAW 264.7 macrophages with intracellular H99S-GFP were imaged at the indicated times with a Nikon Eclipse Ti-E widefield microscope using a CFI Plan Fluor 40× oil immersion (numerical aperture 1.30) objective lens in an environmental chamber (InVivo Scientific). The microscope was controlled with Nikon Elements Software and equipped with a CoolSNAP MYO camera (Photometrics) and the Nikon Ti-E Perfect Focus System to correct focus drift during imaging. GFP of *C. neoformans* was excited through a 465–495 nm excitation filter, and emitted light was detected through a 515–555 nm barrier filter reflected from a 505 nm dichroic mirror. CellTracker™ Red CMTPX fluorescent dye was excited through a 540–580 nm excitation filter, and emitted light was detected through a 600–660 nm barrier filter reflected from a 595 nm dichroic mirror. Post-acquisition images were analyzed by counting intracellular H99S-GFP yeast cells (abbreviated "c") and budding yeast cells (abbreviated "b"). The image acquired at 420 min shows extrusion of H99S-GFP from the RAW 264.7 cell, with one intracellular H99S-GFP yeast cell remaining. Scale bar represents 20 μm

2. We standardly grow our *C. neoformans* strains at 37 °C, but yeast can also be grown at 30 °C, depending on the strain.
3. For our experiments, we look for a confluency of around 80%–90% when passing or splitting cultures. Dishes should be monitored every day for change in media color and confluency under a microscope. As a general rule of thumb, media should be red or pink in color; orange or dark media could indicate cell death, unhealthiness of the cells, or contamination.
4. Macrophage cells are kept no longer than 35 passages, as they begin to exhibit uncharacteristic growth patterns that could

affect data. Occasionally, the macrophage cells begin to deteriorate before passage 35, so the macrophage phenotype should be checked often.

5. There are several methods to removing cells from the bottom of a flask or dish depending on the cells used, the difficulty of getting your cells to dissociate, or the type of plastic used for culture. There are some cases where just the mechanical action of pipetting media will lift cells from the dish or flask. However, the addition of a cell lifter (as used in our methods) may be used. Alternatives to the physical methods described above include using trypsin or versene to dissociate cells. However, if trypsin or versene is used, before passaging or using macrophages for experiments, macrophages must be centrifuged at $300 \times g$ for 3 min and the supernatant containing trypsin or versene removed. The pellet of macrophages may then be resuspended in warm growth media.
6. Incubation time to confluency of your cells may be adjusted based on your timeline. If you are passaging and need confluent cells the following day, a 1:2 dilution of cells to media could be done in place of the 1:20 dilution listed above. Times to confluency will also vary based on the cell line used. The number given is based on a 3 day to 80% confluency model for J774A.1 macrophage cells.
7. Trypan blue is used to differentiate between live and dead cells. Live cells with intact cell membranes should appear clear on a blue background; dead cells without an intact membrane will appear dark blue due to staining. Typically, cell counts should be somewhere around 50–200 cells per 16 grid of the hemocytometer, so the 1:2.5 dilution may be adjusted as necessary.
8. Within this chapter, we demonstrate multiple successful methods to classically activate macrophages, using LPS (Subheadings 3.6, **step 2** and 3.7, **step 12**), IFN-γ (Subheading 3.3, **step 2**), or a combination of the two (Subheading 3.4, **step 8**). These methods are interchangeable and may be implemented based on experimental needs or reagent availability.
9. *C. neoformans* must be opsonized with 18B7 for at least 30 min. Some strains may require longer opsonization times. Extending minimum opsonization times for strains will not negatively impact experimental outcomes.
10. While 1 h is listed as the incubation time, incubation times longer than this have not had an impact on results. Additionally, incubation time may be extended to increase phagocytosis of yeast by macrophages.
11. For this step, we have found it most effective to turn the well plate after the first PBS wash, so we are not removing cells from only one corner during the wash. When washing, the pipette

should be angled against the side of the well instead of directly into the well bottom to minimize lifting of macrophages. After this step, some extracellular *Cryptococcus neoformans* may remain, but the goal is to remove as much as possible.

12. Typically, macrophages are incubated for 18 h post-infection, but this period may be extended up to 24 h depending on the specific needs of the experiment (e.g., if testing the effect of a specific treatment).
13. At least a 2:1 MOI (yeast/macrophages) is used for an end goal of at least 50% of macrophages containing intracellular yeast. If a higher percentage of phagocytosis is needed, the MOI should be increased.

Acknowledgments

This work was supported by funds from the National Institutes of Health (NIAID 1R15AI178461-01) to D.E.N. and E.E.M.

References

1. Rajasingham R, Smith RM, Park BJ et al (2017) Global burden of disease of HIV-associated cryptococcal meningitis: an updated analysis. Lancet Infect Dis 17:873–881
2. Vecchiarelli A, Pericolini E, Gabrielli E et al (2013) Elucidating the immunological function of the *Cryptococcus neoformans* capsule. Future Microbiol 8(9):1107–1116
3. Ianiri G, Idnurm A (2015) Essential gene discovery in the basidiomycete *Cryptococcus neoformans* for antifungal drug target prioritization. MBio 6(2):e02334–e02314
4. Eisenman HC, Casadevall A, McClelland EE (2007) New insights on the pathogenesis of invasive *Cryptococcus neoformans* infection. Curr Infect Dis Rep 9(6):457–464
5. Coelho C, Bocca AL, Casadevall A (2014) The intracellular life of *Cryptococcus neoformans*. Annu Rev Pathol 9:219–238
6. McClelland EE, Ramagopal UA, Rivera J et al (2016) A small protein associated with fungal energy metabolism affects the virulence of *Cryptococcus neoformans* in mammals. PLoS Pathog 12(9):e1005849
7. Mukherjee S, Lee SC, Casadevall A (1995) Antibodies to *Cryptococcus neoformans* glucuronoxylomannan enhance antifungal activity of murine macrophages. Infect Immun 63(2):573–579
8. Nicola AM, Albuquerque P, Martinez LR et al (2012) Macrophage autophagy in immunity to *Cryptococcus neoformans* and *Candida albicans*. Infect Immun 80(9):3065–3076
9. Hayes JB, Sircy LM, Heusinkveld LE et al (2016) Modulation of macrophage inflammatory nuclear factor κB (NF-κB) signaling by intracellular *Cryptococcus neoformans*. J Biol Chem 291(30):15614–15627
10. Alvarez M, Burn T, Luo Y et al (2009) The outcome of *Cryptococcus neoformans* intracellular pathogenesis in human monocytes. BMC Microbiol 9(51):1–9

Part VI

Virulence Factors

Chapter 15

Quantification of *C. neoformans* Capsule Diameter

Tiffany E. Guess, Hoyin Lai, David E. Nelson, and Erin E. McClelland

Abstract

The polysaccharide capsule of *Cryptococcus neoformans* is the primary virulence factor and one of the most commonly studied aspects of this pathogenic yeast. Capsule size varies widely between strains, has the ability to grow rapidly when introduced to stressful or low-nutrient conditions, and has been positively correlated with strain virulence. For these reasons, the size of the capsule is of great interest to *C. neoformans* researchers. Inducing the growth of the *C. neoformans* capsule is used during phenotypic testing to help understand the effects of different treatments on the yeast or size differences between strains. Here, we describe one of the standard methods of capsule induction and detail two accepted methods of staining: (i) India ink, a negative stain, used in conjunction with conventional light microscopy and (ii) co-staining with fluorescent dyes of both the cell wall and capsule followed by confocal microscopy. Finally, we outline how to measure capsule diameter manually and offer a protocol for automated diameter measurement of India ink-stained samples using computational image analysis.

Key words *Cryptococcus neoformans*, Capsule, Virulence, India ink, Fluorescent stain, Image analysis, Yeast

1 Introduction

For decades, capsule has been a major focus of research for both mycologists and clinicians interested in *C. neoformans* and cryptococcosis due to its role as a major virulence factor for the pathogen. Measuring differences in capsule size between strains and under different growth conditions by microscopy provides important information about *C. neoformans* and its responses to various stimuli (i.e., different environmental conditions or potential drug treatments) [1, 2]. Here, we outline a method to induce the growth of capsule in *C. neoformans*, detail two different capsule staining protocols, and give instructions for measurement of the capsule diameter.

To illustrate this, we use two clinically isolated strains, B18 and B52, and H99S, a standard lab strain of known capsule size [3, 4]. The first protocol outlined in this chapter is one of several

Erin E. McClelland (ed.), *Cryptococcus neoformans: Methods and Protocols*, Methods in Molecular Biology, vol. 2775,
https://doi.org/10.1007/978-1-0716-3722-7_15,

standard methods for capsule induction in *C. neoformans* and has been used successfully in yeast labs for many years [5]. Results from both methods of staining are consistent. However, each method has its own inherent advantages. India ink is the most commonly used by researchers because of its affordability, ease of use, and stability. Because India ink is a simple, negative stain, it may be used in conjunction with standard light microscopes, which are more readily available than fluorescent microscopes. While expensive and more time-consuming, fluorescent dyes may be advantageous for their flexibility and, as we show, their sensitivity when measuring strains with smaller capsule (2.9 vs. 4 μm) (*see* **Note 1**) [6]. They are particularly useful for the measurement of capsule size and cell burden in experiments involving *C. neoformans* that have been phagocytosed by immune cells and for the histological examination of *C. neoformans* capsule in tissue samples (a technique not outlined in this chapter) [7].

To quantify capsule size, researchers use what should be a simple measurement, the linear diameter of the capsule. Digital microscopes are used to capture images of multiple *C. neoformans* cells (generally hundreds) stained with either India ink or fluorescent dyes. The size of each cell body and surrounding capsule is measured. The data is compiled, and the average diameter of the capsule is calculated by subtracting the cell body diameter from the whole cell diameter (cell body + capsule). Historically, these measurements have been done manually. AxioVision (Zeiss), NIS-Elements, and Fiji/ImageJ are a few of the software programs regularly used for this purpose [8]. However, the simple nature of *C. neoformans* cell body and capsule—appearing as a circle within a circle—in 2D images makes the measurement of their sizes relatively straightforward to automate using off-the-shelf image segmentation software, such as Aivia (DRVision), or through the use of custom MatLab scripts based on existing algorithms. For example the Circle Hough Transformation (CHT) algorithm has been successfully used to detect the *C. neoformans* capsule and cell body situated within it, enabling automated measurement (not detailed in this chapter) [9]. This chapter includes instructions for manual measurement of *C. neoformans* capsule diameter along with a protocol for automated measurement.

2 Materials

All solutions should be prepared using deionized water, sterilized, and stored at room temperature (unless indicated otherwise).

2.1 Capsule Induction

1. *C. neoformans* cells.
2. Yeast extract–peptone–dextrose (YPD) broth: Dissolve 65 g in 1 L H_2O.
3. Shaking incubator.
4. Phosphate-buffered saline (PBS): This is made in the lab using a standard recipe (137 mM NaCl, 2.7 mM KCl, 10 mM Na_2HPO_4O, 2 mM KH_2PO_4O) [10].
5. Centrifuge.
6. Hemocytometer.
7. Dulbecco's minimal essential media (DMEM): high-glucose formula with L-glutamine (without sodium pyruvate). Store at 4 °C.
8. 6-well plates.
9. CO_2 incubator set to 37 °C.

2.2 Staining

1. Superfrost microscope slides.
2. 13 mm glass coverslips (#1.5 thickness).
3. Clear nail polish or other nontoxic sealant.
4. Microcentrifuge.
5. India ink stain.
6. PBS with 1% bovine serum albumin (BSA): PBS is made using the recipe listed above, and BSA is added to the total volume to create a 1% solution.
7. Calcofluor White (CFW) stain.
8. mAb 18B7 conjugated to AlexaFluor488 (*see* **Note 2**).

2.3 Image Acquisition/ Microscopy

1. Light microscope with oil immersion lens (*see* **Note 3**).
2. Light microscope camera.
3. Microscope immersion oil.
4. Confocal or fluorescent microscope with immersion oil objective (*see* **Note 4**).
5. Confocal microscope camera.

2.4 Manual Measurement of Capsule

1. Cell measurement software, such as AxioVision Version 4.8 (Carl Zeiss).
2. Spreadsheet software, such as Excel (Microsoft Office).

2.5 Automated Measurement of Capsule

1. Automated image segmentation software, such as Aivia (DRVision Technologies).
2. Image-editing software, such as Photoshop (Adobe).

3 Methods

Perform all procedures at room temperature unless otherwise specified. All yeast preparation should take place in a BSL-II certified A2 Biosafety cabinet. All cell waste should be disposed of in a beaker containing 10 % bleach for 24 h to ensure death of yeast.

3.1 Capsule Induction

1. Culture *C. neoformans* in YPD broth (pH 6.5 +/−0.2), and incubate at 37 °C with shaking (150 RPM) until in log phase (approximately 18–36 h) (Fig. 1).
2. Centrifuge cells at 870 × *g* and wash three times with PBS.
3. Use a hemocytometer (or an automated cell counter) to count the cells, and resuspend at 2×10^5 cells/2 mL in DMEM.
4. To induce capsule, incubate at 37 °C and 5% CO_2 for 18 h. The combination of the absence of nutrients in the media and presence of CO_2 stimulates *C. neoformans* capsule growth (*see* **Note 5**).
5. Post-incubation, harvest the cells by centrifuging (as above) and removing all but 5–10 μL of supernatant (*see* **Note 6**). At the same time, harvest a corresponding uninduced cell sample for each strain to be measured. These will be used as negative controls (*see* **Note 7**).
6. Stain with either India ink or fluorescent dyes (Subheading 3.2).

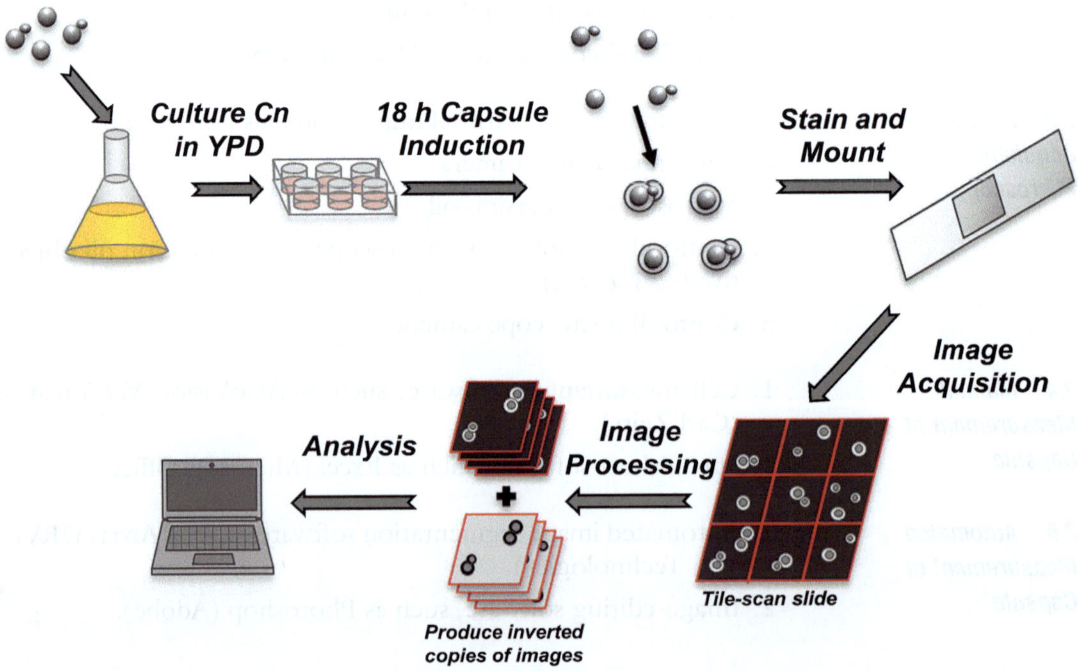

Fig. 1 Schematic of *C. neoformans* capsule induction workflow

3.2 Staining with India ink (Fig. 2)

1. Pipette equal volumes (4 μL) of cell suspension and India ink onto a glass microscope slide (*see* **Note 8**). Gently mix and avoid foaming (*see* **Note 9**).
2. Apply a glass coverslip, and seal edges with a nontoxic sealant (clear nail polish) to prevent the sample from drying out (*see* **Note 10**).

3.3 Staining with Fluorescent Dyes (Fig. 3)

1. Resuspend cell pellet in 50 μL of PBS with 1% BSA.
2. Incubate the culture with an equal volume of CFW (1 μg/mL) and capsule mAb 18B7 conjugated to AlexaFluor488 (AF488) (10 μg/mL) for 30 min at 37 °C (*see* **Notes 11** and **12**).
3. Centrifuge cells at 870 × *g* post-incubation and aspirate supernatant.
4. Resuspend cell pellet in 10 μL of PBS with 1% BSA.
5. Pipette 8 μL of cell suspension onto a glass microscope slide.
6. Apply a 13 mm glass coverslip (#1.5 thickness), and seal edges with a nontoxic sealant (clear nail polish) to prevent the sample from drying out.

3.4 Imaging Cells Stained with India Ink Via Microscopy (see Note 13)

1. Acquire images using a standard light microscope (*see* **Note 14**).
2. Image a minimum of 100 cells per condition (*see* **Note 15**).

3.5 Imaging Cells Stained with Fluorescent Dyes Via Microscopy (Fig. 4)

1. Acquire images using a confocal (recommended) or widefield epifluorescence microscope. If using a confocal microscope, CFW and AF488 may be excited using a 405 nm and 488 nm laser, respectively. If using a widefield epifluorescence microscope, CFW fluorescence is excited through a 340–380 nm excitation filter, and emitted light should be detected through a 435–485 nm barrier filter reflected from a 400 nm dichroic mirror. AF488 fluorescence may be excited through a 465–495 nm excitation filter, and emitted light may be detected through a 515–555 nm barrier filter reflected from a 505 nm dichroic mirror.
2. Image a minimum of 100 cells per condition (*see* **Note 16**).
3. Create maximum intensity projections of each z-stack image series prior to analysis (*see* **Note 17**).

3.6 Manual Measurement of Capsule

1. For cells stained with India ink or fluorescent stains, use a cell measurement software to manually draw a straight line through the widest point of the whole cell (total diameter). AxioVision was used in this protocol.
2. Next, draw a straight line through the cell body (cell body diameter). Repeat this process for all cells.

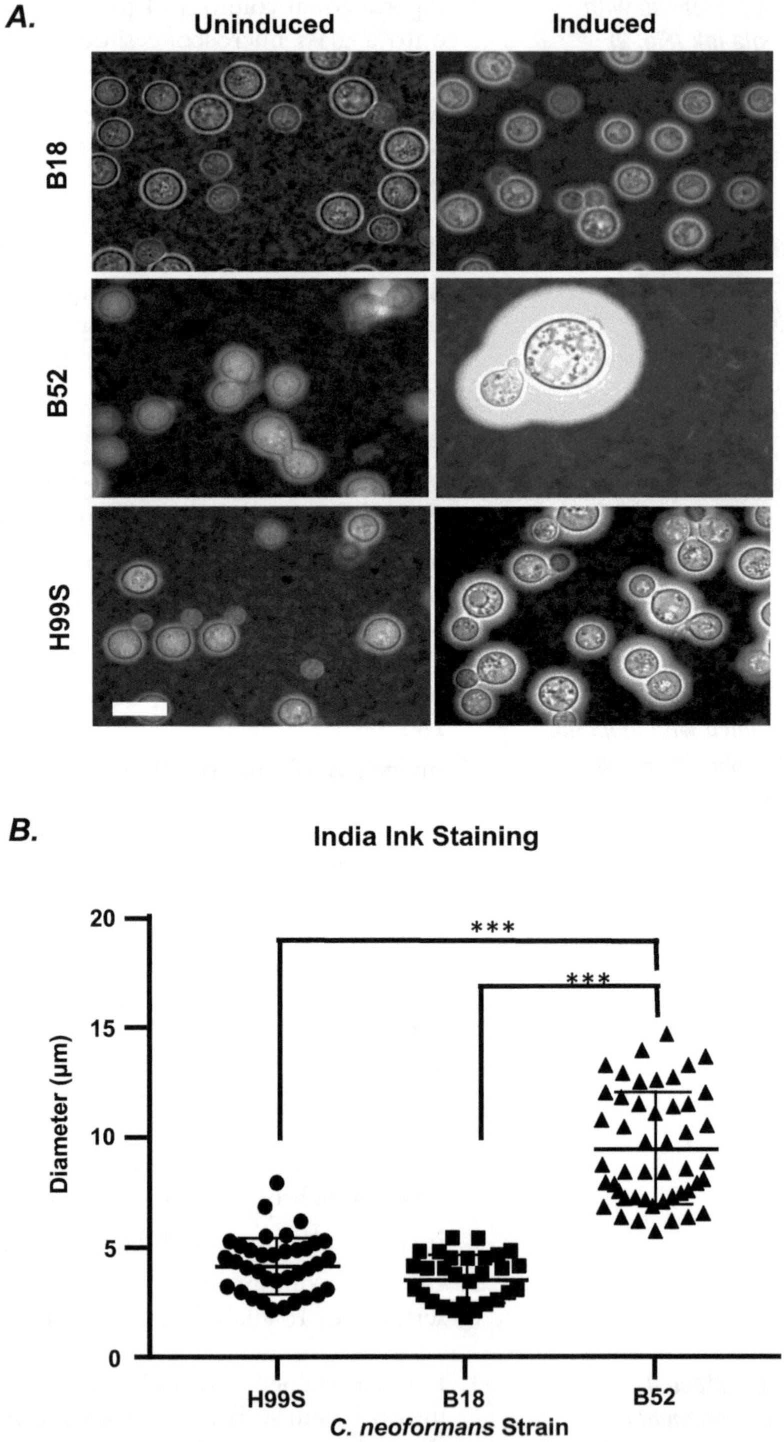

Fig. 2 India ink staining to measure *C. neoformans* capsule. Note: Scale bars represent 10 μM. Scale bars are slightly different in length for the India ink and confocal images—this is correct and is due to slight differences in the magnification (100× vs. 63× with 1.8 digital magnification). (**a**) Images of two clinical strains, B18 and B52, and the common lab strain H99S, all pre- and post-

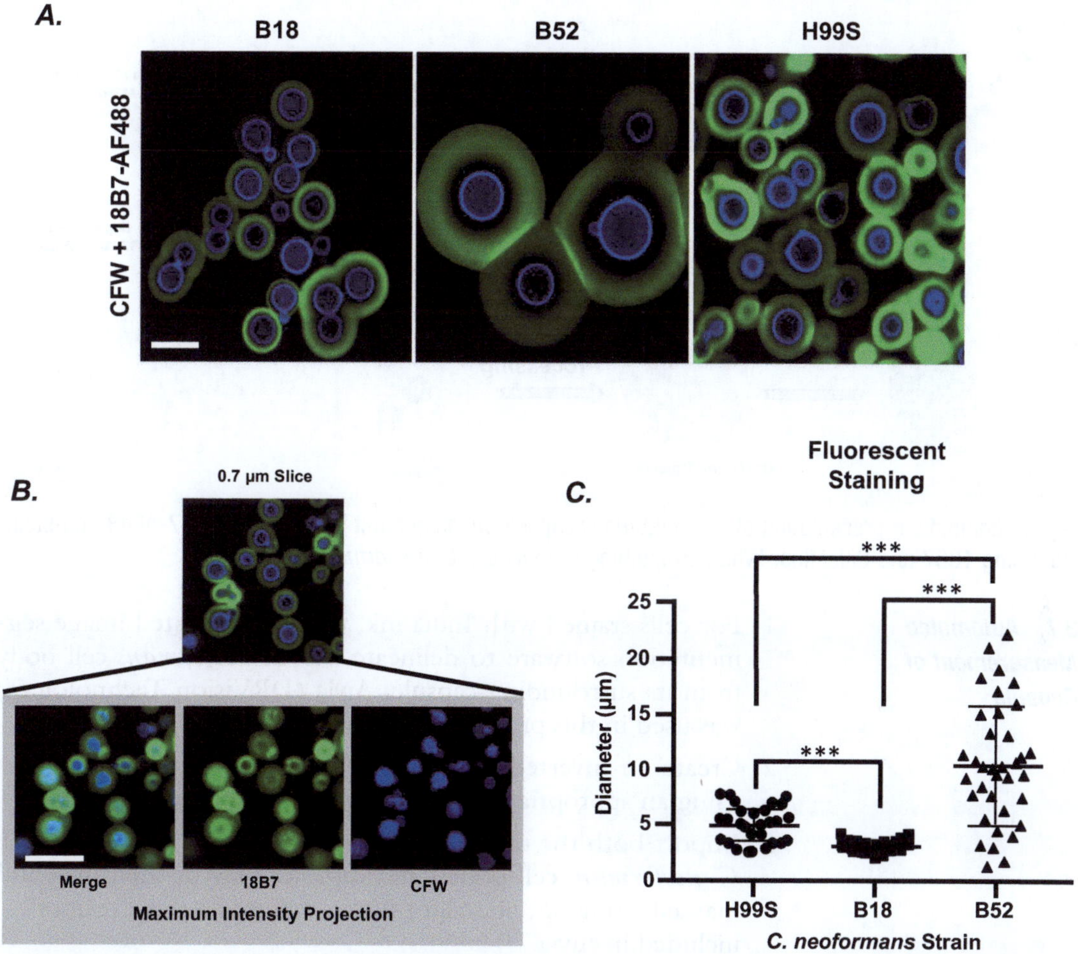

Fig. 3 Fluorescent staining to measure *C. neoformans* capsule. (**a**) Images of two clinical strains, B18 and B52, and the common lab strain H99S, post-capsule induction. Note: Scale bar represents 10 µm. (**b**) Maximum intensity projections of fluorescently labeled *C. neoformans* cell. (**c**) Graph of a representative experiment showing manually measured capsule diameter of all three strains stained with Calcofluor White (CFW) and 18B7-AF488. Statistical significance is indicated as follows: *, $p < 0.05$; **, $p < 0.01$; and ***, $p < 0.001$

3. Export data to a spreadsheet software.
4. Calculate the average capsule diameter using the equation: ([total diameter]-[cell body diameter]/2).

Fig. 2 (continued) capsule induction. (**b**) Graph of a representative experiment showing capsule diameter of all three strains stained with India ink and measured manually. Statistical significance is indicated as follows: *, $p < 0.05$; **, $p < 0.01$; and ***, $p < 0.001$

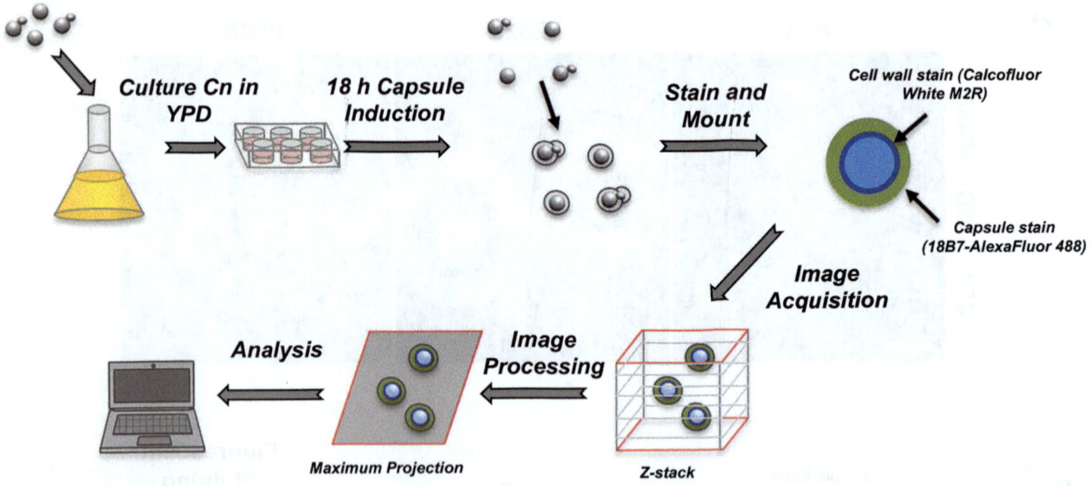

Fig. 4 Schematic representation of *C. neoformans* capsule measurement by CFW and 18B7-AF488 co-staining. Using 18B7 and Calcofluor White co-staining to measure *C. neoformans* capsule

3.7 Automated Measurement of Capsule

1. For cells stained with India ink, use an automated image segmentation software to delineate the *C. neoformans* cell body from its surrounding capsule. Aivia (DRVision Technologies) was used in this protocol (*see* **Note 18**) (Fig. 5).
2. Create an inverted copy of all India ink-stained cell images using an appropriate image-editing software (*see* **Note 19**).
3. Import both the original and inverted images into Aivia. The *C. neoformans* cell bodies and capsules may be identified and masked using specific algorithms—referred to as "recipes"—included in Aivia.
4. Use the Colony Analyzer (Fluorescence) recipe on the inverted image to detect and mask the cell body (*see* **Note 20**).
5. Next, use the Cell Proliferation recipe on the original image to detect and mask the capsule (*see* **Note 21**).
6. Using the recipe, Capsule Partition, create a new mask onto the original image to partition adjacent capsules from one another (*see* **Notes 22** and **23**).
7. Export data to the spreadsheet software of your choice for further analysis.

4 Notes

1. Both staining methods show that the clinical strain, B52, had the largest capsule diameter of the three strains tested. However, results from fluorescent measurements were able to discern a difference between the two strains with smaller capsule size where India ink measurements did not.

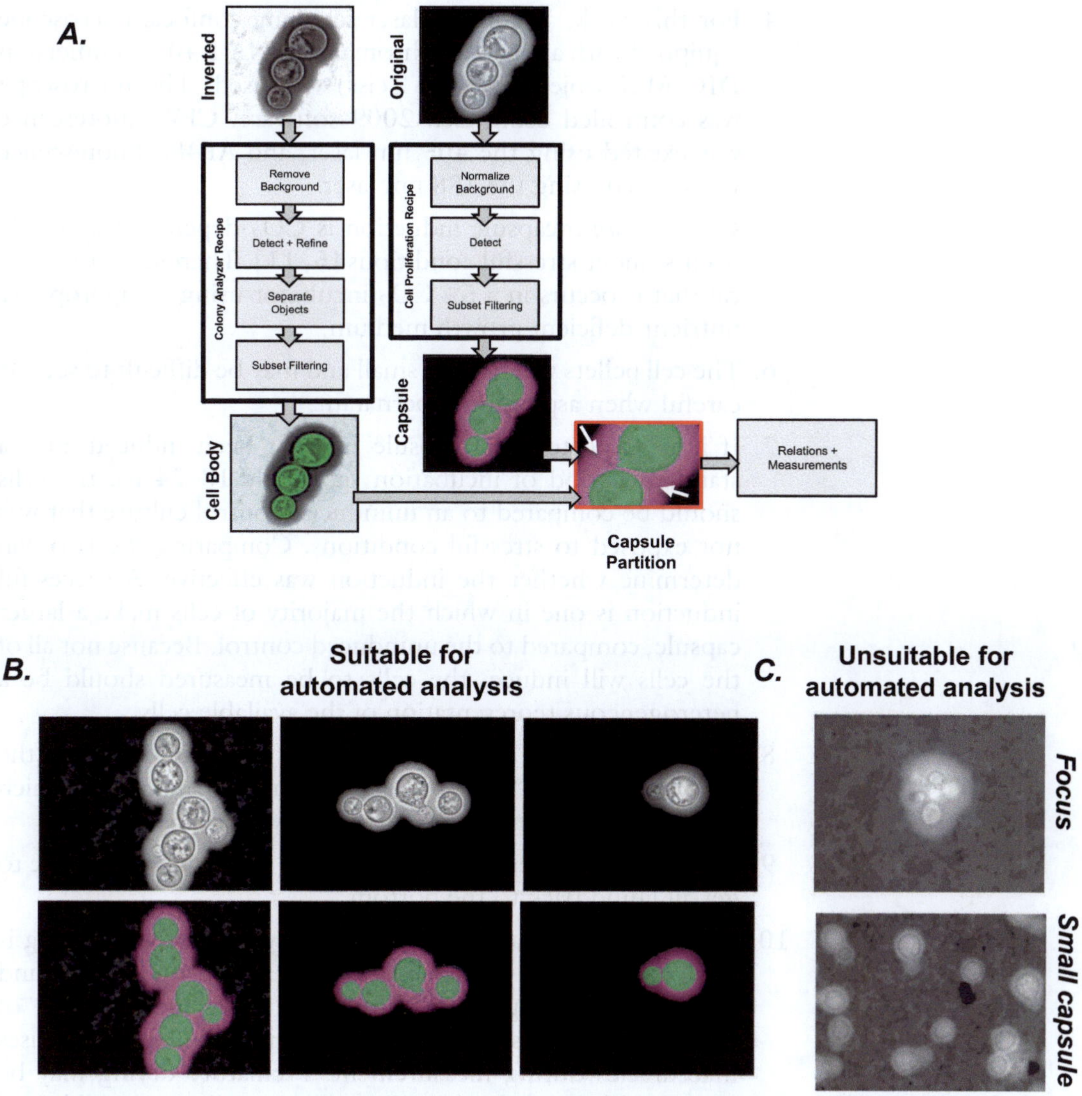

Fig. 5 Computational analysis of *C. neoformans* capsule diameter from images of India ink-stained cells. (**a**) Schematic representation of image segmentation and *C. neoformans* capsule measurement using an automated measurement software. Examples of both (**b**) ideal and (**c**) nonideal images of *C. neoformans*. Nonideal images cannot be successfully measured in the software

2. The monoclonal antibody, 18B7, discussed in this chapter was a kind gift from Dr. Arturo Casadevall (Johns Hopkins University). It was then conjugated to AF488 using a kit from Molecular Probes. Other antibodies may be used, but it should be noted that their epitopes may bind in other places and possibly yield different results.

3. For this work, a Zeiss Axio A1 light microscope equipped with a Plan-NEOFLUAR 100× oil immersion NA 1.30 objective and a Zeiss Axiocam ErC5s camera was used.

4. For this work, a LSM700 laser scanning confocal microscope equipped with a Plan-Apochromat 63× NA 1.40 oil immersion DIC M27 objective (Carl Zeiss) was used. The microscope was controlled using Zen 2009 software. CFW fluorescence was excited using the 405 nm laser, and AF488 fluorescence was excited using the 488 nm laser.
5. *C. neoformans* capsule induction is CO_2-dependent and only occurs under stressful conditions [5, 11]. Therefore, it is critical that it occurs in a 5% CO_2 incubator using an appropriate nutrient-deficient growth medium.
6. The cell pellets will be very small and may be difficult to see. Be careful when aspirating supernatant.
7. If it is suspected that capsule has not been induced after a standard period of incubation (generally 18–24 h), the cells should be compared to an uninduced control culture that was not exposed to stressful conditions. Comparing the two will determine whether the induction was effective. A successful induction is one in which the majority of cells make a larger capsule, compared to the uninduced control. Because not all of the cells will induce, the cells to be measured should be a heterogeneous representation of the available cells.
8. If not used in the correct proportions (India ink/buffer), the stain may create a "feathery" background when imaged, which may complicate automated image analysis.
9. If bubbles do appear, close the lid, and shake down the tube to get all liquid back to the bottom.
10. It is imperative that the sample not dry out before imaging is complete. Drying of the slides changes both the background staining (if using India ink) and the morphology of *C. neoformans* cells, and subsequently its capsule, which causes inaccuracies during measurement. Premature drying may be prevented by completely sealing the coverslip to the slide and imaging cells within 3–4 h.
11. If staining with fluorescent dyes, such as CFW or fluorescently tagged anti-GXM antibodies, is problematic (e.g., insufficient staining), this may generally be overcome by optimizing the staining protocol (increasing/decreasing the concentration and/or incubation time with the dye).
12. The fluorescent antibody, 18B7-AF488, used to stain the capsule depicted in the figures of this chapter was created by conjugating AF488 to mAb 18B7 (a kind gift from Dr. Arturo Casadevall (Johns Hopkins University)) using a kit from Molecular Probes. Other antibodies that bind *C. neoformans* capsule may be used, but it is possible that

their epitopes bind different locations on the capsule and could yield other results.

13. The importance for consistent image acquisition cannot be overstated, especially if automated measurement is the goal. Focus, contrast, and background fluorescence, as well as the bit-depth and image compression of the files, are all important considerations. We have found that 16-bit images with no compression work best. When possible, large clusters of cells that do not sit in a single focal plane should be avoided, although some budding and touching cells are acceptable and may be unavoidable (this rule should be followed for both manual and automated measuring). This problem may be circumvented to a certain degree when using confocal imaging of fluorescently stained cells. The use of z-stacks to produce maximum intensity projections may enable accurate measurement of cells in different focal planes from a single image. Additionally, images should be acquired quickly after preparing the sample. This is particularly important for India ink-stained samples because when they start to dry, the contrast between the background and foreground elements (the cells) may diminish, and undesirable background features may be enhanced, which leads to reduced automated segmentation accuracy.
14. When imaging *C. neoformans* for capsule measurement, it is important to focus on the dark ring at the boundary between the cell wall and capsule. This allows the software to correctly delineate the cell and capsule.
15. The number of cells per field will vary, and this is acceptable. It is important, however, that overlapping cells be kept at a minimum as these are difficult to measure (both manually and with automated software). For these experiments, images were acquired as 16-bit tiffs with exposure times optimized to maximize image contrast while minimizing blur caused by small movements of the cells.
16. If using a fluorescence microscope, acquiring z-stack image series of the cells will help to ensure that the maximum capsule diameter for each cell within a given field is obtained. For the images shown in this chapter, the confocal pinhole was set to 1.0 AU, and an overlapping z-series of 10–20 slices covering 0.7 μm each were acquired at the selected positions.
17. Using maximum intensity projections generated from z-stack images of fluorescently stained cells produces a 2D representation of the captured 3D stack with the maximum cell and capsule diameter apparent for most, if not all, of the cells in the field [12]. This allows researchers to ensure that accurate capsule diameter measurements are made even when the

imaged cells are in different focal planes. Not accounting for these differences could confound the ability of investigators to detect small changes in capsule size or small differences between strains.

18. For successful automated measurement in Aivia, images should be acquired as properly focused, high-contrast 16-bit tiff images.
19. In a black-and-white image (i.e., crypto cell stained with India ink), copy images are inverted changing all black to white and vice versa.
20. In the inverted image, the dark ring defining the boundary of the *C. neoformans* cell becomes brighter than the surrounding capsule. The Colony Analyzer (Fluorescence) recipe in Aivia is used to segment the *C. neoformans* cell bodies from their capsules.
21. By adjusting the parameters of the Cell Proliferation recipe, it is capable of efficiently detecting *C. neoformans* capsule as a foreground object. The algorithm takes the original image as input to generate a second mask output for capsule region detection.
22. The Capsule Partition recipe generates cell and capsule diameter, defined as the average of the longest and shortest axes crossing the center of the cell and capsule, and area measurements from detection masks. The output is located under the spreadsheet tab in Aivia.
23. The recipes' parameters may be adjusted to optimize detection of individual images or optimized for batch detection of multiple images. Additional measurements may be calculated using the recipe for comprehensive characterization of the cell and capsule's morphology.

Acknowledgments

This work was supported by the Biology Department and the Molecular Biosciences program of Middle Tennessee State University.

References

1. McFadden D, Zaragoza O, Casadevall A (2006) The capsular dynamics of *Cryptococcus neoformans*. Trends Microbiol 14(11): 497–505. https://doi.org/10.1016/j.tim.2006.09.003
2. Zaragoza O, Casadevall A (2004) Experimental modulation of capsule size in *Cryptococcus neoformans*. Biol Proced Online 6:10–15. https://doi.org/10.1251/bpo68
3. McClelland EE, Perrine WT, Potts WK et al (2005) Relationship of virulence factor expression to evolved virulence in mouse-passaged *Cryptococcus neoformans* lines. Infect Immun

73(10):7047–7050. https://doi.org/10.1128/IAI.73.10.7047-7050.2005

4. Bisson GP, Nthobatsong R, Thakur R et al (2008) The use of HAART is associated with decreased risk of death during initial treatment of cryptococcal meningitis in adults in Botswana. J Acquir Immune Defic Syndr 49(2): 227–229. https://doi.org/10.1097/QAI.0b013e318183181e
5. Gates MA, Thorkildson P, Kozel TR (2004) Molecular architecture of the *Cryptococcus neoformans* capsule. Mol Microbiol 52(1):13–24. https://doi.org/10.1111/j.1365-2958.2003.03957.x
6. Guess T, Lai H, Smith S et al (2018) Size matters: measurement of capsule diameter in *Cryptococcus neoformans*. J Vis Exp 132: 57171. https://doi.org/10.3791/57171
7. van Teeffelen S, Shaevitz JW, Gitai Z (2012) Image analysis in fluorescence microscopy: bacterial dynamics as a case study. BioEssays 34(5): 427–436. https://doi.org/10.1002/bies.201100148
8. Schindelin J, Arganda-Carreras I, Frise E et al (2012) Fiji: an open-source platform for biological-image analysis. Nat Methods 9(7): 676–682. https://doi.org/10.1038/nmeth.2019
9. Dragotakes Q, Casadevall A (2018) Automated measurement of cryptococcal species polysaccharide capsule and cell body. J Vis Exp 131:56957. https://doi.org/10.3791/56957
10. Roskams J, Rodgers L, Mellick AS (2002) Lab ref: a handbook of recipes, reagents, and other reference tools for use at the bench. Cold Spring Harbor Laboratory Press, Cold Spring Harbor
11. Granger DL, Perfect JR, Durack DT (1985) Virulence of *Cryptococcus neoformans*. Regulation of capsule synthesis by carbon dioxide. J Clin Invest 76(2):508–516. https://doi.org/10.1172/JCI112000
12. Kwon O, Kang ST, Kim SH et al (2015) Maximum intensity projection using bidirectional compositing with block skipping. J Xray Sci Technol 23(1):33–44. https://doi.org/10.3233/XST-140468

Chapter 16

Methods of Cryptococcal Polysaccharide Analysis Using ELISA

Maggie P. Wear, Scott A. McConnell, Seth D. Greengo, Livia Liporagi Lopes, and Arturo Casadevall

Abstract

One of the standard assays for the fungal pathogen *Cryptococcus neoformans* is the glucuronoxylomannan (GXM) ELISA. This assay utilizes monoclonal antibodies targeted against the critical virulence factor, the polysaccharide (PS) capsule. GXM ELISA is one of the most used assays in the field used for diagnosis of cryptococcal infection, quantification of PS content, and determination of binding specificity for antibodies. Here we present three variations of the GXM ELISA used by our group—indirect, capture, and competition ELISAs. We have also provided some history, perspective, and notes on these methods, which we hope will help the reader choose, and implement, the best assay for their research.

While it has long been referred to as the GXM ELISA, we also suggest a name update to better reflect our updated understanding of the polysaccharide antigens targeted by this assay. The Cryptococcal PS ELISA is a more accurate description of this set of methodologies and the antigens they measure. Finally, we discuss the limitations of this assay and put forth future plans for expanding the antigens assayed by ELISA.

Key words *Cryptococcus*, Polysaccharide, Antigen, Enzyme-linked immunosorbent assay, ELISA, Glucuronoxylomannan, GXM, Exopolysaccharide, EPS, Capsule

1 Introduction

1.1 Background and History of Cryptococcal ELISAs

Enzyme-linked immunosorbent assays (ELISAs) have been used for the identification and quantification of antigens since their advent in the 1970s. For the fungal pathogen *Cryptococcus neoformans*, the ELISA has become a mainstay of laboratory research. As the only known encapsulated fungal species, *Cryptococcus* produces two polysaccharide antigens: glucuronoxylomannan (GXM) and glucuronoxylomannogalactan (GXMgal—previously galactomannan or GalXM). When antibodies were first produced against cryptococcal species, two distinct methods were used. Immunizing with capsular material [1], which we now know contains not just capsular polysaccharide (CPS), but also mannoproteins, other proteins,

Erin E. McClelland (ed.), *Cryptococcus neoformans: Methods and Protocols*, Methods in Molecular Biology, vol. 2775, https://doi.org/10.1007/978-1-0716-3722-7_16,

and lipids [2, 3], yielded monoclonal antibodies (mAbs) targeting cryptococcal polysaccharide GXMgal. The earlier and still more successful immunizations utilize the easier to isolate exopolysaccharide (EPS) that is shed by the yeast into the culture media [4]. While techniques for EPS isolation vary from repeated alcohol precipitation to detergent extraction with cetyltrimethyl ammonium bromide (CTAB) and physical isolation, all antibodies resulting from immunization with EPS target the more dominant polymer, GXM [4–7].

The 1990s seminal work by Cherniak and colleagues propelled our understanding of cryptococcal EPS forward [8]. By examining the EPS of 106 cryptococcal strains using NMR and an artificial neural network, they were able to characterize the repeat units of GXM. Consisting of an α-1,3-trimannose with a β-1,2 linked glucuronic acid and varied β-1,2 and β-1,4 linked xylose branches, the GXM repeat varies in size from 4–8 saccharide residues. Varying only by density of xylose substitutions, GXM motifs M1–M6 were expressed in both cryptococcal species—*neoformans* and *gattii*—and across serotypes. While nonspecific, there is a trend connecting cryptococcal serotype to motif expression. Although most strains (78/106) express more than one GXM motif, some strains of serotype D only express the M1 motif and some serotype A strains are 100% M2. Fewer in number, strains of serotypes B and C express only motif M3 or M4, respectively. No strains have been identified that express exclusively the M5 or M6 GXM motifs. The impact of GXM motif on strain virulence and CPS and EPS antigenicity has not been investigated. However, anti-GXM mAbs have a better binding affinity for some GXM motifs over others [9, 10]. Additionally, some anti-GXM mAbs are protective against cryptococcal infections in mice [6, 11–14]. Further studies will be necessary to understand the interplay between GXM motif, polysaccharide size, structure, and flexibility in determining antigenicity, as well as how anti-GXM antibodies bind and if antibody-mediated catalysis plays a role in protective capacity.

Our group has produced and characterized many anti-GXM mAbs [9, 10, 15–21]. While there are always improvements to be made in the future, the method and variations detailed here are used regularly in our laboratory. The origin of our most commonly used anti-GXM mAb set for sandwich ELISAs—IgG1 18B7 and IgM 2D10—is immunization with EPS-tetanus toxoid conjugate from the GXM M2 motif expressing serotype A strain ATCC-24064 [7, 22]. Any pair of anti-GXM mAbs of differing isotype will work for this assay; however, we note that 18B7 has become widely used and is available from Sigma (MABF2069).

2 Materials and Supplies

2.1 Solutions

1. Phosphate Buffered Saline (PBS): 137 mM NaCl, 2.7 mM KCl, 1.5 mM KH_2PO_4, 8.5 mM Na_2HPO_4.
2. Tris-buffered Saline + Tween-20 (TBST): 10 mM Tris–HCl, 150 mM NaCl, 1 mM sodium azide, pH to 7.2.
3. Blocking solution: 1% w/v bovine serum albumin (BSA), 0.1 M sodium azide. Dissolve in TBST and filter sterilize.
4. Substrate buffer: 0.001 M $MgCl_2$ * $6H_2O$, 0.05 M Na_2CO_3, pH to 9.8 (*see* **Note 1**).
5. p-Nitrophenyl Phosphate (pNPP) Stop Solution: 3 N NaOH.

2.2 Reagents/Supplies

1. Yeast peptone dextrose (YPD).
2. Reagent reservoirs.
3. GXM exopolysaccharide (EPS).
4. GXM EPS Standard (*see* Subheading 3.1 for method).
5. pNPP.
6. GXM-specific monoclonal antibody (two) for capture/competition ELISA, both GXM-specific, different isotypes.
7. Alkaline phosphatase-conjugated secondary antibody. Must have specificity to the Fc of the primary antibody (indirect or competition) or second detection antibody (capture ELISA).

2.3 Instruments

1. Multichannel pipet (12 channels).
2. Single pipet (P2, P10, P1000).
3. Serological pipet.
4. Clear flat bottom high-binding 96-well assay.
5. Plate washer.
6. Plate reader for high throughput absorbance measurements.

3 Methods

3.1 Production of EPS/GXM Standard

1. Grow 5 mL culture of standard strain (we use H99) in YPD at 30 °C, shaking, for 48 h.
2. Induce capsule growth in minimal media by diluting 1:500 from YPD culture into minimal media culture. Grow at 30 °C, shaking, for 3 days.
3. Spin down cells by centrifuging the culture at 2300 × *g*, 15 min.
4. Sterilize the filter supernatant through a 0.22 μm filter.

5. Remove media contaminants by filtering the supernatant through a 3 kDa molecular weight cut off (MWCO) Amicon filter unit.
6. Lyophilize ultrafiltration retentate (>3 kDa EPS) to absolute dryness (*see* **Note 2**).
7. Weigh dried EPS and make up a standard solution at 40 mg/mL in water (w/v). Store at −20 °C or −80 °C and avoid repeated freeze-thaw cycles.

3.1.1 Indirect ELISA for GXM-Specific Antibody Quantification

The indirect ELISA is the most facile ELISA experiment both in terms of reagent usage and experimental time. Indirect ELISA is an excellent choice for confirming the relative or absolute amount of specific antibody from mouse sera or hybridoma supernatant but is limited somewhat by a relative lack of sensitivity compared to other methods. This ELISA utilizes secreted exopolysaccharide (EPS) from *Cryptococcus neoformans* H99 strain, a mixed motif expressing strain, as the GXM standard. Recent observations, both published [23] and not, indicate that lyophilization may change the structure of the polysaccharides and that these polysaccharides have low levels of protein associated with them. We have discussed these issues further in Subheading 4 of this chapter.

Carry out all incubation steps at 37 °C for 1 h or 4 °C overnight.

3.2 Coating the Plate

1. Prepare a dilution of antigen—lyophilized GXM exopolysaccharide (EPS)—by diluting to 1 μg/mL (w/v) in PBS (total volume of 5 mL). Vortex the dilution well to homogenize the solution.
2. Pipet the diluted antigen into a reagent reservoir and aliquot 50 μL into each well of a clear, flat bottom high-binding 96-well assay plate to uniformly coat each well with antigen. Tap the plate on a flat surface to ensure even dispersion of the antigen across the surface area of the well (*see* **Note 3**). Incubate.
3. Add an additional 200 μL of blocking solution (*see* **Note 4**). Incubate.
4. Wash the plate in a plate washer or with a multichannel pipet three times with wash solution. Tap the plate dry on a solid surface to remove any residual buffer after the plate wash.

3.3 Primary Antibody Binding

1. Prepare a standard mAb with a previously determined concentration (of the same isotype as the unknown mAbs) by diluting to 5 μg/mL in blocking buffer (*see* **Note 5**).
2. Aliquot 100 μL of the diluted standard into wells A1 and A2.

3.4 Unknown Primary Antibody Binding

1. Unknown GXM-specific antibodies should be diluted to an estimated range of 1–10 μg/mL (*see* **Note 6**).
2. Aliquot 100 μL in duplicate to wells A3–A10 (depending on the number of samples you are testing) (*see* **Note 7**).
3. Aliquot 100 μL of blocking buffer to A11 and A12. Columns 11 and 12 will serve as blanks (*see* **Note 8**).
4. Aliquot 50 μL of blocking buffer to every well in the range of B1–H12 (not in row A). Then serially dilute every sample down the rows by pipetting 50 μL (half the volume in row A) with a 12-channel pipet into row B, mix slowly and thoroughly to obtain a two-fold dilution, then pipet 50 μL of that mixture to row C. Repeat the process of serially mixing and diluting two-fold each sample into the corresponding well in the successive row until you reach row H (Fig. 1). Mix carefully as before and remove 50 μL and discard it into liquid waste, leaving 50 μL in the final row. Incubate.
5. Wash the plate in a plate washer or with a multichannel pipet three times with wash solution. Tap the plate dry on a solid surface to remove any residual buffer after the plate wash.

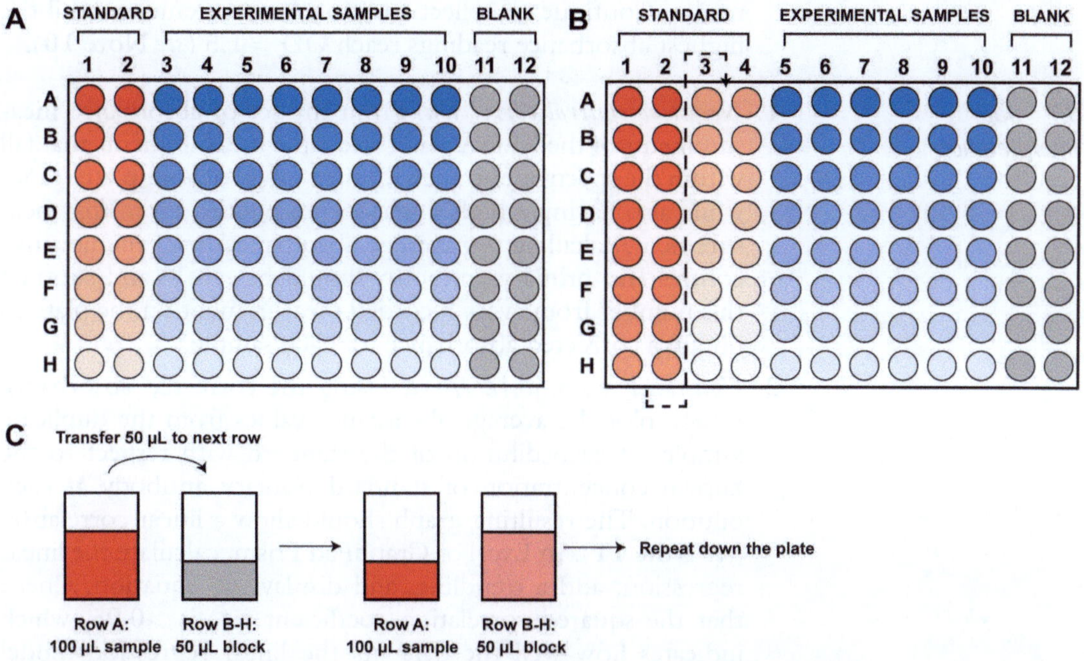

Fig. 1 Plate map for indirect GXM ELISA. Illustration mapping out sample locations and describing dilutions, as written in Subheading 3.4, **steps 2–4**. Conventional and alternative expanded dilution set for standards is displayed in panels **a** and **b**, respectively. (**c**) Dilution scheme for 1:1 serial dilution

3.5 Alkaline-Phosphatase Secondary Antibody Binding

1. Dilute alkaline phosphatase-conjugated secondary antibody recognizing isotype-specific constant domain epitopes of the primary antibody of interest to a concentration of 1 μg/mL in blocking solution in a volume of 5 mL. Vortex the dilution well to homogenize the solution.
2. Pipet the diluted secondary antibody into a reagent reservoir and aliquot 50 μL into each well of a clear flat bottom high-binding 96-well assay plate to uniformly coat each well with antibody solution. Incubate.
3. Wash the plate in a plate washer or with a multichannel pipet three times with wash solution. Tap the plate dry on a solid surface to remove any residual buffer after the plate wash.

3.6 Substrate Development and Detection

1. Prepare a 5 mL solution of 1 mg/mL alkaline phosphatase substrate (p-nitrophenyl phosphate, pNPP) in substrate buffer.
2. Pipet the diluted pNPP into a reagent reservoir and aliquot 50 μL into each well of a clear flat bottom high-binding 96-well assay plate to uniformly coat each well with substrate. Incubate at room temperature or 37 °C until a yellow color is visibly developed (*see* **Note 9**).
3. Measure the absorbance at 405 nm for every well using a plate reader. Continue to collect additional measurements until the highest absorbance readings reach OD = 1.5 (*see* **Note 10**).

3.7 Data Interpretation

1. *Negative control correction:* From the set of absorbance measurements of the ELISA plate, select the measurements that fall within the spectrophotometric linear range of your plate reader while also having a high signal-to-noise ratio. From this measurement, calculate the average absorbance from the negative control (no primary antibody, should be <0.1) and subtract this number from every recorded measurement on the plate to obtain a corrected absorbance for each sample.
2. *Construct a standard curve:* Using the corrected absorbance values, plot the average absorbance values from the duplicate samples at each dilution of the standard with respect to the known concentration of standard primary antibody at each dilution. The resulting graph should show a linear correlation (*see* **Note 11**). In Excel or GraphPad Prism, calculate the linear regression, add a trendline, and display the equation. Check that the squared correlation coefficient (r^2) is >0.95, which indicates how well the data fits the linear regression model represented with the equation (in the form of $y = mx + b$). If the r^2 value is acceptable, move on to interpolation of the unknown values. If not, repeat the assay and focus on pipetting accuracy and ensuring homogeneity of your samples or adjust the concentration gradient of your standard.

3. *Interpolate unknown sample absorbance values into your standard curve*: Using GraphPad Prism, unknown sample absorbance values may be quickly interpolated into the standard curve to yield calculated concentrations of *each unknown sample dilution* under the Analysis function. In Excel, absorbance values may be converted to concentrations by substitution of the A_{405} value into the equation displayed on the chart. Next, multiply the calculated concentration at each dilution by the dilution factor to give an estimation of the *total antibody concentration of your original solution*. Average all values of interpolable dilutions for a given sample to give a final estimate of antibody concentration (*see* **Note 12**).

3.8 Variations of the Indirect ELISA

1. Isotyping ELISA for monoclonal hybridomas. Follow the above procedure, but serially dilute the tissue culture supernatant (primary antibody) multiple times (in duplicate). For each set of duplicate rows, add a different secondary antibody specific to different isotypes to determine the isotype of the unknown monoclonal antibody.
2. Testing antigen specificity. Follow the above procedure, and coat with GXM exopolysaccharide antigen from different serotypes to determine the specificity of a given mAb to EPS secreted from *C. neoformans* strains of differing serotypes with unique GXM motif compositions.

3.9 Capture (Sandwich) ELISA for Quantification of GXM in EPS Samples

The capture GXM ELISA may be the most often used ELISA in our lab. Because of the low affinity of polysaccharides for the polystyrene of ELISA plates, and the signal amplification afforded by multivalency of the IgM capture antibody, capture ELISAs yield much better signal than conventional indirect ELISAs and may be used to measure much lower concentrations of GXM. Additionally, as the structure of GXM polymers has not been determined, how the antibody binding affinity is affected by interaction with the ELISA plates is unclear, but likely dominated by hydrophobic interactions with the polystyrene surface and may involve significant denaturation of the polysaccharide conformation at the solid-liquid interface. The capture and detection antibodies used must be different isotypes, which can be discriminated by the Fc-specific antibody in the final step to avoid background signal originating from capture antibody (Fig. 2a). As noted, we utilize the antibody pair IgM-2D10 and IgG1-18B7. Although other antibody pairs may be used, the 18B7-2D10 pair has been extensively validated.

Carry out all incubations for 1 h at 37 °C unless otherwise noted.

1. Prepare a 1 μg/mL solution of unlabeled goat anti-mouse (GAM) IgM in 1× PBS in a reagent reservoir (*see* **Note 13**). Mix well and then add 50 μL of the diluted GAM IgM solution

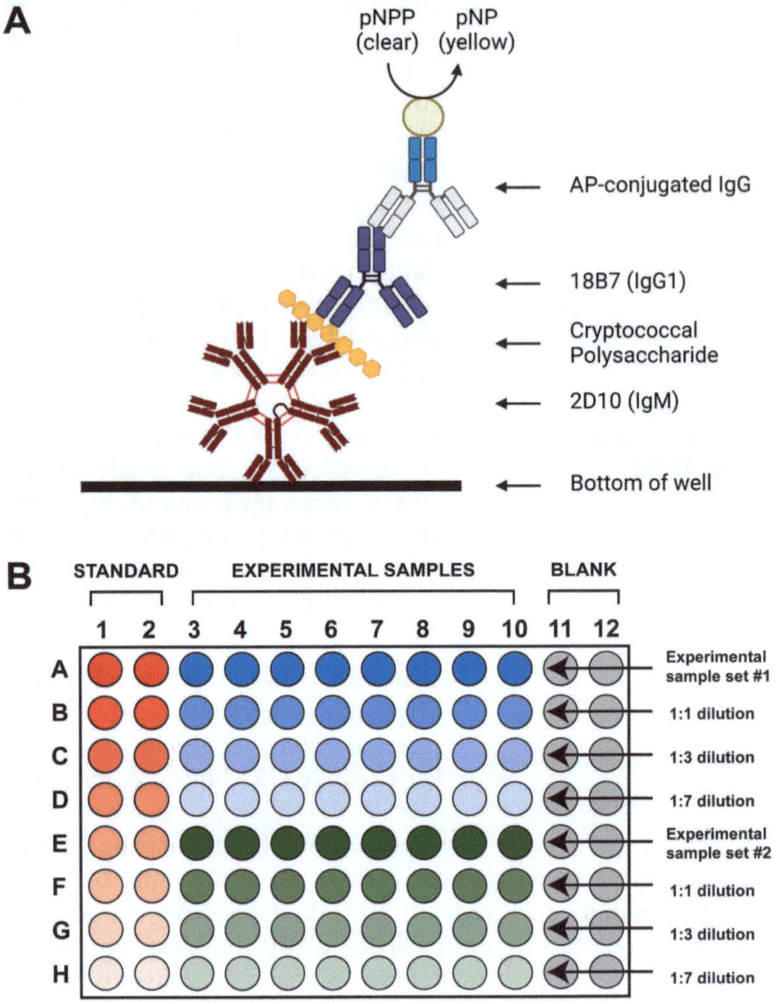

Fig. 2 Capture ELISA plate map and dilutions scheme. (**a**) Schematic of the antibody interactions in the GXM capture ELISA. (**b**) Schematic for the assay plate set up for measuring eight samples on a single plate with only three serial dilutions

to each well of a clear flat bottom high-binding 96-well ELISA plate using a multichannel pipette. Tap to ensure even distribution of the antibody across the wells. Incubate.

2. Add 200 μL of blocking solution to each well of the ELISA plate with a multichannel pipette. You do not need to remove the IgM solution before adding blocking solution. Incubate.
3. Empty the wells of the ELISA plate. Flip the ELISA plate and tap out any remaining liquid onto a paper towel or absorbent pad.
4. In a reagent reservoir, prepare a 2 μg/mL solution of one GXM-specific antibody in blocking buffer. Using a

multichannel pipette, add 50 μL of anti-GXM antibody solution to each well of the ELISA plate (*see* **Note 14**). Tap to ensure even distribution of solution and incubate (*see* **Note 15**).

5. During the first anti-GXM antibody incubation, prepare dilutions of GXM samples in blocking solution. Additionally, prepare a standard solution of GXM of known concentration (*see* **Note 16**). This will be used to generate a standard curve and quantify the concentration of the unknown GXM samples.
6. Remove the plate from incubation and empty the wells. Wash the plate three times with TBST using an automatic plate washer. Tap out the plate to remove excess wash buffer.
7. Prepare the ELISA plate for serial dilutions. Using a multichannel pipette, add 50 μL of blocking solution to each well of rows B–H (Fig. 2b) (*see* **Note 17**).
8. Add 100 μL of the GXM solution of known concentration to wells A1 and A2 (plus A3 if running samples in triplicate—*see* Fig. 2b). This will act as your standard dilution set with which a standard curve may be generated.
9. Add 100 μL of blocking solution to wells A11 and A12 (plus A10 if working in triplicate). This will act as your negative standard and be used to correct baseline absorbance measurements in the standard and unknown samples.
10. In duplicate (or triplicate), add 100 μL of unknown GXM dilutions to wells in row A. Then, with a multichannel pipette, remove 50 μL from the samples in Row A of the plate and add to the wells in Row B to create a two-fold dilution. Mix well and then continue to dilute by removing 50 μL of the volume in the wells of Row B and adding to the wells of Row C and continue until reaching row H. When at Row H, mix well, then remove 50 μL, and discard into liquid waste. Each well on the plate should contain 50 μL of liquid. Incubate (*see* **Note 15**).
11. Remove the plate from incubation and empty the wells. Tap out the plate to ensure liquid is removed. Wash the plate three times with TBST using an automatic plate washer. Tap again to remove any remaining wash buffer. Prepare a solution of 5 μg/mL anti-GXM detection antibody in blocking solution in a reagent reservoir. Add 50 μL of antibody solution to each well using a multichannel pipette. Gently shake plate to ensure even coverage of antibody solution. Incubate (*see* **Note 15**).
12. Remove plate from incubation and empty the wells. Tap out the plate to ensure liquid is removed. Wash the plate three times with TBST using an automatic plate washer. Tap again to remove any remaining wash buffer.

13. Prepare a 1 μg/mL AP-conjugated antibody solution in blocking solution in a reagent reservoir. Add 50 μL of AP-conjugated antibody solution to each well of the ELISA plate with a multichannel pipette. Gently tap and then incubate.
14. Remove the plate from incubation and empty the wells. Tap out the plate to ensure liquid is removed. Wash the plate three times with TBST using an automatic plate washer. Tap again to remove any remaining wash buffer.
15. Prepare a solution of alkaline phosphate substrate in substrate buffer. For each ELISA plate to be developed, add 5 mg of pNPP to 5 mL of substrate buffer. Mix these in a reagent reservoir. Add 50 μL to each well of the ELISA plate using a multichannel pipette. Incubate at room temperature or at 37 °C until yellow color is observed in wells.
16. Acquire absorbance readings at a wavelength of 405 nM using a plate reader. Collect measurements until the highest absorbance observed on the plate reaches OD = 1.5.
17. *See* Subheading 3.7 from the Indirect ELISA protocol (Subheading 3.7) for instructions to calculate concentrations of unknown GXM samples. Note that for capture ELISAs, the concentrations being determined are those of the unknown GXM solutions, rather than those of the unknown antibody solutions.

3.10 Competitive ELISA for GXM-Specific Antibody Quantification

A competition ELISA methodology (also known as "Competition/Inhibition Assays (C-ELISA/I-ELISA)" or "double sandwich competitive ELISA") establishes whether the epitope of one antibody overlaps with another. Simultaneous competitive binding of one monoclonal antibody (mAb) in the presence of a concentration gradient of a second competing mAb is measured to determine if the competing mAb interferes with binding of the mAb of interest. The inhibitory effect of the second mAb is measured by detecting interference in the ELISA assay signal. Similar to the capture ELISA, some method of discrimination of the two antibodies used must be employed to separate the signal from the mAb of interest from the competing mAb. Measuring the competition of mAbs with different isotypes may be accomplished with highly specific secondary antibodies. However, if antibodies to be tested are of the same isotype, a preliminary antigenic modification of the antibody of interest is required. This is traditionally accomplished by biotinylation and detection with a biotin-specific reporter.

Carry out all incubation steps at least 37 °C for 1 h or 4 °C overnight. Preferentially, wash plates using an automatic plate washer. After all the wash steps, tap out the plate to ensure liquid/wash buffer was removed before the next step.

The protocol below describes the assay as performed with only one set of reagents. It is suggested to assay each pair of mAbs of interest twice, inverting mAb X and mAb Y, to optimize competition conditions.

1. Uniformly coat one clear flat bottom high-binding 96-well assay plate with 1 μg/mL of GXM exopolysaccharide in PBS (50 μL/well). Incubate. Add 200 μL of blocking solution to each well and incubate once more. After the incubation time, wash plates three times.
2. Add 50 μL/well of blocking solution to all the wells in columns 2–12. For descriptive purposes, we name the pair of antibodies as mAb X and mAb Y. Designate that rows A–D and E–H will be developed with anti-Fc alkaline phosphatase reagents that recognize mAbs X and Y, respectively.
3. To wells 1A, 1B, 1C, 1D, 1E, and 1F, add 100 μL of 25 μg/mL mAb X diluted in blocking solution (Fig. 3a). To wells 1G and

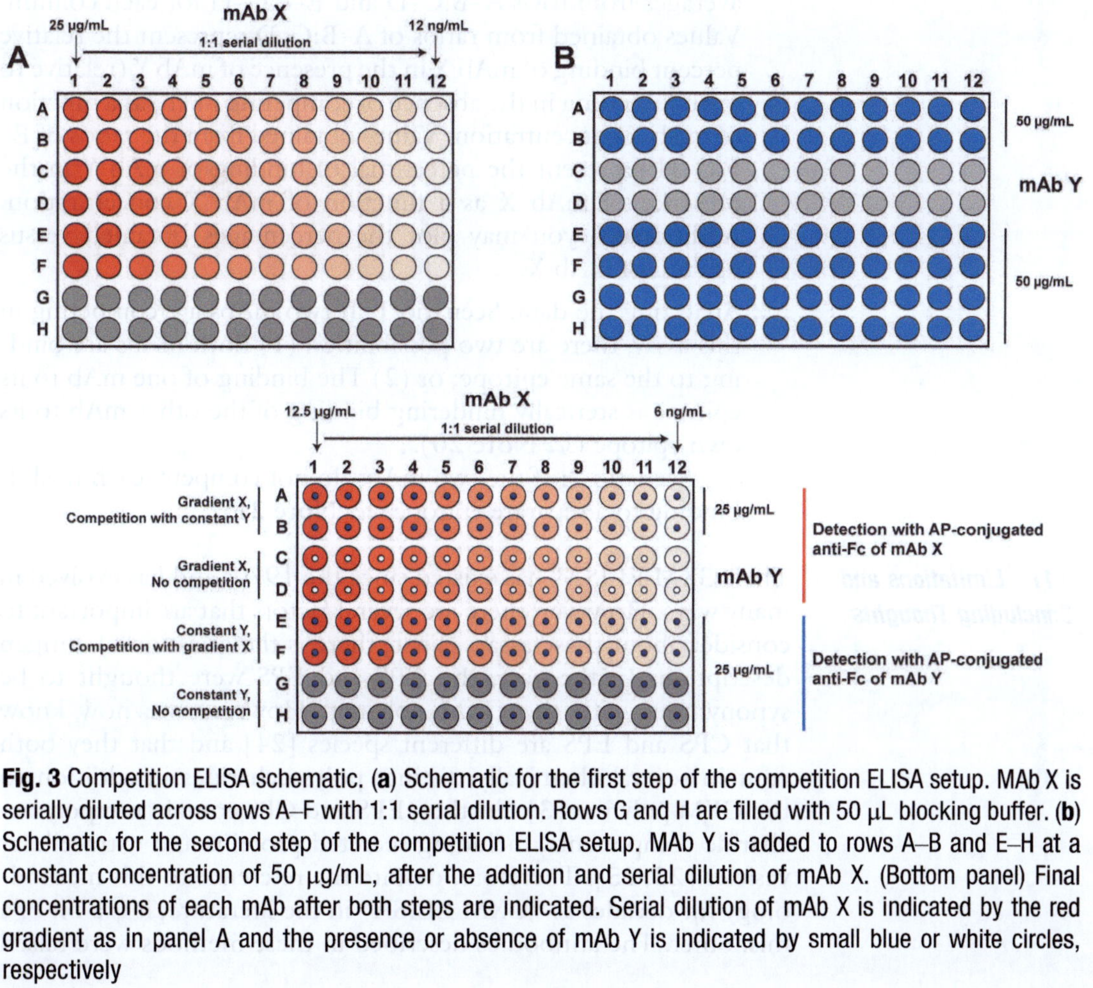

Fig. 3 Competition ELISA schematic. (**a**) Schematic for the first step of the competition ELISA setup. MAb X is serially diluted across rows A–F with 1:1 serial dilution. Rows G and H are filled with 50 μL blocking buffer. (**b**) Schematic for the second step of the competition ELISA setup. MAb Y is added to rows A–B and E–H at a constant concentration of 50 μg/mL, after the addition and serial dilution of mAb X. (Bottom panel) Final concentrations of each mAb after both steps are indicated. Serial dilution of mAb X is indicated by the red gradient as in panel A and the presence or absence of mAb Y is indicated by small blue or white circles, respectively

1H, add 100 μL of blocking solution (Fig. 3a). Serially dilute the solutions in column 1 across the plate using a 50 μL multichannel pipette (1:2 dilution). To each well in rows A–B, E–F, and G–H, add 50 μL of 5 μg/mL mAb Y (Fig. 3b). To each well in rows C–D, add 50 μL of blocking solution (Fig. 3b) (*see* **Notes 18** and **19**).

4. Incubate and wash plate three times. Then add 50 μL/well of alkaline phosphatase labeled anti-Fc of mAb X (1:1000 dilution in blocking solution) to wells in rows A–D (Fig. 3c). Add 50 μL/well of alkaline phosphatase labeled anti-Fc of mAb Y (1:1000 dilution in blocking solution) to wells in rows E–H (Fig. 3c).
5. Incubate and wash plate 3–5 times. Develop the entire plate by adding 50 μL/well of 1 mg/mL pNPP in substrate buffer. Incubate at room temperature or 37 °C until a yellow color is visibly developed.
6. Read the absorbance at 405 nm and calculate the ratio of the averages from rows A–B:C–D and E–F:G–H for each column. Values obtained from ratios of A–B:C–D represent the relative percent binding of mAb X in the presence of mAb Y (relative to mAb X binding in the absence of competing mAb) as a function of mAb X concentration. Values obtained from the ratios of E–F:G–H represent the percentual of binding of mAb Y in the presence of mAb X as a function of mAb X concentration. Additionally, you may plot the percentages obtained versus μg/mL of mAb X.
7. Analyzing the data. Scenario 1: If two mAbs are competing in this assay, there are two possibilities: (1) Both mAbs are binding to the same epitope; or (2) The binding of one mAb to its epitope is sterically hindering binding of the other mAb to its own epitope (*see* **Note 20**).

   Scenario 2: If the two mAbs do not compete, each mAb is binding to a separate epitope (*see* **Note 21**).

3.11 Limitations and Concluding Thoughts

The "GXM ELISA" has existed since the 1970s and has evolved in many ways. However, there are several factors that are important to consider about these assays. First, there is the question of antigen description. In the past, the CPS and EPS were thought to be synonymous with the GXM polymer. However, we now know that CPS and EPS are different species [24] and that they both contain several chemically distinct polysaccharides in addition to the GXM polymer [3]. Further, EPS and CPS contain nonpolysaccharide components, including secreted proteins and extracellular vesicles [23, 25]. Therefore, despite the reference to this methodology for decades as "GXM ELISA" in the literature, it is a bit of a misnomer. The antibodies described in these methods were raised against EPS, and the "GXM" standards used for absolute quantification are also EPS. Therefore, the ELISAs described in this chapter only reliably address the shed cryptococcal polysaccharides and not those attached to the cell and forming the capsule. Recent work by our group has described a new technique for the isolation of CPS without the use of detergents, chemicals, or lyophilization [26]. In the future, we hope to extend these protocols to include capsular polysaccharide.

Historically EPS isolated from cryptococcal strains required CTAB detergent isolation or ethanol precipitation followed by lyophilization for storage. However, these methods alter the polysaccharide structure [23]. Further, our preliminary data suggests that even modest lyophilization will cause noticeable aggregation of the polymers. This aggregation and altered polysaccharide structure could cause a host of problems including, but not limited to, differences in the size of polymers between standards and samples, availability of epitopes for antibody binding, and conformational epitope changes from lyophilized samples used in the generation of original hybridomas.

Our group uses the IgG1 mAb 18B7 prolifically, as do many in the cryptococcal field. 18B7 is particularly useful because of its cross-reactivity; it will bind polysaccharide from all four serotypes of *Cryptococcus*. Additionally, recent studies of 18B7 show that it possesses glycosidic activity and will slowly hydrolyze linkages within the GXM polymer [15, 21]. With these observations in mind, and noting that incubations occur at temperatures at which catalytic activity for 18B7 has been observed, we note that the timescale of this activity is slow enough as to not affect the assay. Further, we note that careful consideration of mAb reactivity profiles is crucial in the optimization of ELISA experiments targeting GXM antigen from different serotypes.

With the above issues and limitations in mind, we caution the reader that any quantification, especially of CPS, using these methods is relative. In the future, we hope a better understanding of cryptococcal capsule architecture, the structure of GXM polymers, and interactions between polysaccharides and proteins in both CPS and EPS will inform updates and improvements to these methods.

4 Notes

1. The pH of this solution is very important and should be checked frequently. The colorimetric conversion of pNPP to 4-Nitrophenol by alkaline phosphatase is dependent on an alkaline pH. 4-Nitrophenol is colorless at pH below 5.4 and increasingly yellow above 7.4. Additionally, the enzymatic activity of alkaline phosphatases increases three-fold from pH 7–9.

2. Dryness is defined as the point at which the mass no longer decreases over a span of 12–24 h.
3. Diluted antigen is first allowed to coat the well in the absence of blocking solution to ensure maximal antigen coating without competition with blocking proteins. The interaction between EPS and the 96-well plate is not as strong as other antigens, necessitating this step.
4. Blocking solution contains 1% BSA, which serves to coat the remaining surface area of the high-binding well bottom to prevent nonspecific adherence of mAbs to the well surface during later incubations.
5. mAb used for standard must be purified to a very high degree of homogeneity (>95%), as assessed by sodium dodecyl-sulfate polyacrylamide gel electrophoresis. After obtaining a suitably homogeneous standard mAb stock, total protein concentration may be determined by measuring the absorbance at 280 nM (if the primary sequence is known and a molar absorptivity coefficient can be estimated) or other total protein quantification methods (e.g., bicinchoninic acid or Bradford assays). The quantified standard can be stored at 4 °C in PBS (+0.02% sodium azide to prevent microbial growth, optionally) for months, but the homogeneity and concentration of the mAb standard should be assessed frequently to ensure appropriate correlation between measured ELISA absorbance values and mAb concentration. Very old stocks should be assessed for antigen binding relative to a freshly prepared antibody stock to ensure binding affinity is not diminished after extended storage.
6. Depending on the source of antibody, a rough estimation can be carried out. The concentration of polyclonal antibody mixtures from serum is generally ~10 mg/mL, but the concentration of polyclonal specific antibody is maximally ~1 mg/mL The concentration of monoclonal antibodies from tissue culture supernatant is ~50 μg/mL, but note that calf antibody will also be present if fetal bovine serum is used as an additive. Ascites generally produces 1–10 mg/mL antibody, including 10% contamination with nonspecific endogenous murine antibody [27]. However, because the sample will be serially diluted two-fold seven times, a rough estimate of the concentration should yield at least one dilution that falls within the linear range of the standard curve.
7. If the concentration of the unknown sample cannot be estimated, a second set of columns can be used to continue the serial dilution (i.e., serially dilute down from A3/A4 → H3/H4 and then continue with the serial dilution, using half the material from H3/H4 to dilute 50 μL of blocking buffer in

A5/A6, and continue down 5/6 column). This will give a very large range of concentrations (*see* Fig. 1).

8. ELISA blanks are important controls to rule out any nonspecific binding of the secondary antibody to antigen in the absence of primary antibody, or nonspecific degradation of the pNPP substrate. When properly controlled, these effects can be subtracted from the other absorbances in the experimental samples to obtain a net absorbance due only to primary antibody binding of antigen. Because GXM-specific antibodies are notoriously cross-reactive, it may also be necessary to add an additional control in which primary antibody is incubated with blocked wells that were not coated with antigen to rule out spurious binding of the primary antibody with the polystyrene well or BSA. This control experiment only needs to be conducted once per unique primary antibody.
9. The reaction may optionally be halted by the addition of 12.5 μL of 3 N NaOH.
10. Depending on the linear range and detection limits of your specific plate reader, plates can be read to higher OD values. Once outside of the linearity range of your plate reader, the intensity of the absorbance measurement is no longer correlated to the sample amount by a simple linear relationship, making the data much harder to fit.
11. If there are values on the extreme ends of the standard curve that clearly do not obey a linear relationship between concentration and absorbance, you may remove these values and re-calculate the linear regression to obtain a more accurate representation of the correlation inside the linear range.
12. Do not try to interpolate any A_{405} values from the unknown samples into the linear standard curve. Because they are outside of the absorbance range in which absorbance is linearly related to concentration, these must be excluded from this type of analysis. Usually there are several values along the dilution gradient that may be used.
13. Note the target species of the antibody to ensure proper binding during later antibody incubations. For our experiments, we use Goat anti-Mouse antibodies, as our GXM-specific antibodies were produced in mice.
14. Ensure that anti-GXM capture and detection antibodies are of different isotypes. For example, we use the anti-GXM IgM 2D10 as the capture antibody and the anti-GXM IgG1 18B7 as the detection antibody. This ensures only signal from the detection antibody is measured.
15. This is one of the three steps by which you may perform the incubation overnight at 4 °C. We recommend being consistent

with your incubation times and temperatures between replicates.

16. A concentration range of 5 μg/mL to 50 ng/mL for the EPS standard and dilution folds of 500–5000 for culture supernatants are routinely used in our group.
17. If hoping to conserve ELISA plates, you may also add blocking solution to rows B–D and rows F–H to double the number of samples analyzed per plate. However, this will result in a smaller range of concentrations per sample and may hinder concentration determination of the unknown samples.
18. As a result of adding two separate reagents to each well, the concentration of each component within the well is one half of that added to the well. Thus, the concentration of mAb Y is 2.5 μg/mL in each well to which it was added and the concentration of serially diluted mAb X in column 1 wells to which it was added is only 12.5 μg/mL.
19. In the top half of the plate, the binding of a decreasing gradient of mAb X (12.5 μg/mL–6 ng/mL) will be compared to the same gradient of mAb X in competition with a constant concentration of 2.5 μg/mL mAb Y. In the bottom half of the plate, the binding of mAb Y at a constant concentration of 2.5 μg/mL will be compared to competition with a decreasing gradient of mAb X (12.5 μg/mL–6 ng/mL).
20. Two sigmoidal curves that are mirror images of one another should be obtained if the mAbs compete with one another for the same binding site.
21. If the curves are not sigmoidal mirror images, the mAbs are indeed probably recognizing different binding sites.

References

1. van de Moer A, Salhi SL, Cherniak R et al (1990) An anti-*Cryptococcus neoformans* monoclonal antibody directed against galactoxylomannan. Res Immunol 141:33–42
2. Bose I, Reese AJ, Ory JJ et al (2003) A yeast under cover: the capsule of *Cryptococcus neoformans*. Eukaryot Cell 2:655–663
3. Jesus MD, Nicola AM, Chow S-K et al (2010) Glucuronoxylomannan, galactoxylomannan, and mannoprotein occupy spatially separate and discrete regions in the capsule of *Cryptococcus neoformans*. Virulence 1:500–508
4. Dromer F, Salamero J, Contrepois A et al (1987) Production, characterization, and antibody specificity of a mouse monoclonal antibody reactive with *Cryptococcus neoformans* capsular polysaccharide. Infect Immun 55: 742–748
5. Fromtling RA, Shadomy HJ (1982) Immunity in cryptococcosis: an overview. Mycopathologia 77:183–190
6. Dromer F, Charreire J, Contrepois A et al (1987) Protection of mice against experimental cryptococcosis by anti-*Cryptococcus neoformans* monoclonal antibody. Infect Immun 55:749–752
7. Casadevall A, Mukherjee J, Devi SJ et al (1992) Antibodies elicited by a *Cryptococcus neoformans*-tetanus toxoid conjugate vaccine have the same specificity as those elicited in infection. J Infect Dis 165:1086–1093
8. Cherniak R, Valafar H, Morris LC et al (1998) *Cryptococcus neoformans* chemotyping by quantitative analysis of 1H nuclear magnetic resonance spectra of glucuronoxylomannans

with a computer-simulated artificial neural network. Clin Diagn Lab Immunol 5:146–159

9. Guazzelli L, Ulc R, Bowen A et al (2020) A synthetic glycan array containing *Cryptococcus neoformans* glucuronoxylomannan capsular polysaccharide fragments allows the mapping of protective epitopes. Chem Sci 11:9209–9217
10. Crawford CJ, Guazzelli L, McConnell SA et al (2023) Synthetic glucuronoxylomannans enable molecular insights into antibody structure-function and fungal capsule architecture. ChemRxiv. Cambridge Open Engage, Cambridge
11. Todaro-Luck F, Reiss E, Cherniak R et al (1989) Characterization of *Cryptococcus neoformans* capsular glucuronoxylomannan polysaccharide with monoclonal antibodies. Infect Immun 57:3882–3887
12. Nussbaum G, Cleare W, Casadevall A et al (1997) Epitope location in the *Cryptococcus neoformans* capsule is a determinant of antibody efficacy. J Exp Med 185:685–694
13. Shapiro S, Beenhouwer DO, Feldmesser M et al (2002) Immunoglobulin G monoclonal antibodies to *Cryptococcus neoformans* protect mice deficient in complement component C3. Infect Immun 70:2598–2604
14. Ellerbroek P, Vecchiarelli A, Hoepelman A et al (2007) Immunology of infections with *Cryptococcus neoformans*. In: Brown GD, Netea MG (eds) Immunology of fungal infections. Springer Netherlands, Dordrecht, pp 383–407
15. Bowen A, Wear MP, Cordero RJB et al (2017) A monoclonal antibody to *Cryptococcus neoformans* glucuronoxylomannan manifests hydrolytic activity for both peptides and polysaccharides. J Biol Chem 292:417–434
16. Bowen A, Wear M, Casadevall A (2017) Antibody-mediated catalysis in infection and immunity. Infect Immun 85:e00202–e00217
17. Janda A, Eryilmaz E, Nakouzi A et al (2015) Variable region identical IgA and IgE to *Cryptococcus neoformans* capsular polysaccharide manifest specificity differences. J Biol Chem 290:12090–12100
18. Cordero RJB, Pontes B, Frases S et al (2013) Antibody binding to *Cryptococcus neoformans* impairs budding by altering capsular mechanical properties. J Immunol 190:317–323
19. McClelland EE, Nicola AM, Prados-Rosales R et al (2010) Ab binding alters gene expression in *Cryptococcus neoformans* and directly modulates fungal metabolism. J Clin Invest 120: 1355–1361
20. Jiang A, Bryan RA, Morgenstern A et al (2012) Treatment of early and established *Cryptococcus neoformans* infection with radiolabeled antibodies in immunocompetent mice. Antimicrob Agents Chemother 56:552–554
21. Crawford CJ, Wear MP, Smith DFQ et al (2021) A glycan FRET assay for detection and characterization of catalytic antibodies to the *Cryptococcus neoformans* capsule. Proc Natl Acad Sci U S A 118:e2016198118
22. Mukherjee J, Casadevall A, Scharff MD (1993) Molecular characterization of the humoral responses to *Cryptococcus neoformans* infection and glucuronoxylomannan-tetanus toxoid conjugate immunization. J Exp Med 177: 1105–1116
23. Wear MP, Hargett AA, Kelly JE et al (2022) Lyophilization induces alterations in cryptococcal exopolysaccharide resulting in reduced antibody binding. Carbohydr Polym 291: 119547
24. Frases S, Nimrichter L, Viana NB et al (2008) *Cryptococcus neoformans* capsular polysaccharide and exopolysaccharide fractions manifest physical, chemical, and antigenic differences. Eukaryot Cell 7:319–327
25. Rodrigues ML, Nimrichter L, Oliveira DL et al (2007) Vesicular polysaccharide export in *Cryptococcus neoformans* is a eukaryotic solution to the problem of fungal trans-cell wall transport. Eukaryot Cell 6:48–59
26. Wear MP, Jacobs E, Wang S et al (2021) *Cryptococcus neoformans* capsule regrowth experiments reveal dynamics of enlargement and architecture. J Biol Chem 298:101769
27. Harlow E, Lane D (1988) Antibodies: a laboratory manual. CSHL Press, Cold Spring Harbor

Chapter 17

Measuring Laccase Activity and Melanin Production in *Cryptococcus neoformans*

Patrícia Albuquerque, Herdson Renney de Sousa, Stefânia de Oliveira Frazão, Larissa Verginia do Nascimento Miranda, Hugo Costa Paes, Ildinete Silva Pereira, and André Moraes Nicola

Abstract

Melanin is a complex dark pigment synthetized by the phenoloxidase enzyme laccase in *Cryptococcus neoformans.* In vitro, this enzyme oxidizes exogenous catecholamines to produce melanin that may be secreted or incorporated into the fungal cell wall. This pigment has multiple roles in *C. neoformans* virulence during its interaction with different hosts and probably also in protecting fungal cells in the environment against predation and oxidative and radiation stresses, among others. However, it is important to note that laccase also has melanin-independent roles in *C. neoformans* interactions with host cells. In this chapter, we describe a quantitative laccase assay and a method for evaluating the kinetics of melanin production in *C. neoformans* colonies.

Key words *Cryptococcus neoformans*, Cryptococcosis, Melanin, Melanization, Densitometry, Photography

1 Introduction

Laccases are copper-dependent diphenol-oxidases present in many fungi, both pathogenic and non-pathogenic. Cryptococci encode two laccase genes Lac1 and Lac2, but only Lac1 seems to be implicated in virulence [1–3]. Cryptococcal laccase is a well-described virulence factor in these pathogens, first rising to prominence in the literature as the enzyme responsible for the synthesis of melanin. Melanin itself has long been known as a key factor contributing to the ability of these fungi to cause disease, be it by impairing phagocytosis [4], exerting immunomodulatory effects on host cells [5], or protecting yeast cells against oxidative damage [6]. The enzyme is known to be both secreted in extracellular vesicles [7], which may contribute to its effects on host cells, and to be anchored to the cryptococcal cell wall [8], allowing for the

Erin E. McClelland (ed.), *Cryptococcus neoformans: Methods and Protocols*, Methods in Molecular Biology, vol. 2775,
https://doi.org/10.1007/978-1-0716-3722-7_17,

synthesis of a melanin protective layer around it. Because the melanization rate has been strongly correlated with the pathogenicity of fungal isolates from patients with cryptococcal central nervous system (CNS) disease [9], quantitative measurement of melanization in vitro is of paramount importance for phenotypic characterization of isolates, as well as for the elucidation of molecular mechanisms that regulate melanin synthesis and its incorporation to the cell wall or release in the extracellular milieu. A key melanization assay uses medium containing L-3,4-dihydroxyphenylalanine (L-DOPA), a substrate that laccase uses directly to synthesize melanin [10]. It is a quick, inexpensive quantitative assay that yields a temporal melanization curve that may be used to classify isolates, which will be described in detail in this chapter.

Nonetheless, there is solid evidence in the literature that laccase activity has substrates other than L-DOPA and its corresponding products also contribute to the virulence of cryptococci. A well-known ability of laccase is the synthesis of prostaglandins E2 and F2α [11, 12], which influence the inflammatory milieu caused by the fungus [13]. There is also evidence that laccase detoxifies iron [4] in a murine model of lung infection with *C. neoformans.* Other possible substrates have been postulated [14], and laccase has roles in infection that have been investigated independently of melanization, such as lung breakout [15] and CNS invasion [16]. In fact, the correlation of laccase activity with severity of CNS disease seems to be even stronger than that of melanization [17]. Thus, cell-free assays of laccase activity using substrates other than L-DOPA may also be used to quantify enzyme secretion and thus add more phenotypical information on isolates. This chapter also describes a standard cell-free assay for measurement of secreted laccase activity using L-DOPA or 2′2-azino-bis(3′ethylbenzthiazoline-6-sulfonate (ABTS)) as substrates.

2 Materials

2.1 Solutions and Media for Handling and Growth of C. neoformans

1. Phosphate-buffered saline: 137 mM NaCl, 2.7 mM KCl, 10 mM Na_2HPO_4, 1.8 mM KH_2PO_4. Dissolve 8 g of NaCl, 0.2 g of KCl, 1.42 g of Na_2HPO_4, and 0.2 g of KH_2PO_4 in 1 L of distilled water, adjust the pH to 7.4. Autoclave and store at room temperature.
2. Sabouraud's dextrose broth and agar: 1% peptone, 2% D-glucose. Dissolve 10 g of peptone and 20 g of D-glucose in 1 L of distilled water. Autoclave and store at room temperature. For the solid medium, add agar at 1.5% to the solution prior to autoclaving.

3. YPD broth and agar: 1% Yeast extract, 2% peptone, 2% D-glucose. Dissolve 10 g of yeast extract, 20 g of peptone, and 20 g of glucose in 1 L of distilled water. Autoclave and store at room temperature. For the solid medium, add agar at 1.5% to the solution prior to autoclaving.
4. 30 mM thiamine hydrochloride stock solution (vitamin B1; 10,000×): Dissolve 101 mg of thiamine hydrochloride (molecular weight [MW] 337.27) in 10 mL of distilled water. Filter through a 0.22 μm membrane to sterilize, distribute in 1 mL aliquots, and store at −20 °C.
5. Minimal medium (MM): 15 mM D-Glucose, 10 mM $MgSO_4$, 29.4 mM KH_2PO_4, 13 mM glycine, 3 μM thiamine-HCl. Dissolve 2.7 g of D-glucose (MW 180), 2.5 g of $MgSO_4.7H_2O$ (MW 246.48), 4.0 g of KH_2PO_4 (MW 136.1), and 0.975 g of glycine (MW = 75.07) in 900 mL of distilled water. Adjust the pH to 5.5 using KOH, bring up the volume to 1 L, and autoclave. Let cool at room temperature and add 100 μL of 30 mM thiamine to the medium. Store at room temperature (*see* **Note 1**).

2.2 Evaluation of Laccase Activity

1. Asparagine salt medium with glucose: 0.3% D-glucose, 0.1% L-asparagine, 0.025% $MgSO_4$, 10 mM Na_2HPO_4 (pH 6.5), 3 μM thiamine, 10 μM $CuSO_4$ (*see* **Note 2**).
2. Asparagine salt medium without glucose: 0.1% L-asparagine, 0.025% $MgSO_4$, 10 mM Na_2HPO_4 (pH 6.5), 3 μM thiamine, 10 μM $CuSO_4$.
3. 10 mM L-DOPA solution: Dissolve 19.72 mg of L-DOPA (MW 197.2) in 10 mL of distilled water. Filter through a 0.2 μm syringe-filter to sterilize, distribute in 1 mL aliquots, and store at −20 °C protected from light (*see* **Notes 3**, **4**, and **5**).
4. 50 mM Na_2HPO_4 (pH 7.0): Dissolve 3.872 g of $Na_2HPO_4.7H_2O$ (MW 268.07) and 1.457 g of $NaH_2PO4.H_2O$ in 400 mL of distilled water. Check the pH and adjust the final volume to 500 mL with distilled water.
5. 10 mM ABTS: Make fresh in sterile distilled water or in 50 mM Na_2HPO_4 (pH 7.0). Keep the solution protected from light (*see* **Note 5**).

2.3 Kinetics of Melanin Production

1. 2× minimal medium (2× MM): 30 mM D-Glucose, 20 mM $MgSO_4$, 58.8 mM KH_2PO_4, 26 mM glycine, 6 μM thiamine-HCl. Dissolve 0.54 g of D-glucose (MW 180), 0.5 g of $MgSO_4.7H_2O$ (MW 246.48), 0.8 g of KH_2PO_4 (MW 136.1), and 0.195 g of glycine (MW 75.07) in 90 mL of distilled water. Adjust the pH to 5.5 using KOH, bring the volume up to 100 mL and autoclave. Let cool at room

Fig. 1 Evaluation of colony melanization kinetics. (**a** and **b**) Image capture apparatus consisting of a metal box with a mirror that allows visualization of the bottom of horizontal plates. (**c**) Example of the setup of the apparatus to image plates inside a biological safety cabinet

temperature and add 20 μL of 30 mM thiamine to the medium. Store at room temperature (*see* **Note 1**).

2. 3.75% Agar: Dissolve 3.75 g of agar in 100 mL of distilled water. Autoclave. Store at room temperature (*see* **Note 6**).
3. 10 mM L-DOPA solution: Dissolve 19.72 mg of L-DOPA (MW 197.2) in 10 mL of distilled water. Filter through a 0.2 μm syringe-filter to sterilize, distribute in 1 mL aliquots, and store at −20 °C protected from light (*see* **Notes 3**, **4**, and **5**).
4. Digital camera: A digital photographic camera will be used to image the plates with melanizing *C. neoformans* colonies. Although any camera may be used, we suggest using a digital

single-lens reflex (DSLR) camera. With a DSLR camera, the user may accurately control many aspects of the image capture to make it accurately reproducible at every time point. Macro lenses are the most appropriate. In our lab, we normally take the pictures in a biosafety cabinet placing the plates in the apparatus presented in Fig. 1.

5. Digital image processing software: To quantify melanization, you will need software to process the images and to extract quantitative data. We use Adobe Photoshop to process the images and the freely available ImageJ freeware (https://imagej.nih.gov/ij/download.html) to extract quantitative data, but other programs may be used.
6. Statistical analysis software: To quantify the kinetics of melanization, you will need to do a logistic regression. Several programs are available, but we use GraphPad Prism.

3 Methods

3.1 Evaluating Laccase Activity

1. Plate fungal cells from frozen stocks on Sabouraud's or YPD agar plates to obtain isolated colonies. Inoculate one colony of the desired strain into 5 mL of YPD and grow at 30 °C with 200 rpm shaking for 24 h (*see* **Note 7**).
2. The following day, harvest the cells by centrifugation at 1000× *g* for 3 min. Discard the supernatant and do two rounds of resuspending the pellet in 1 mL of PBS and centrifuging to wash the cells. Then, resuspend the cells in 5 mL of asparagine salt medium with glucose and incubate at 30 °C for 5 days (*see* **Note 8**).
3. Harvest the cells by centrifugation at 1000× *g* for 10 min and wash once with 5 mL of 50 mM Na_2HPO_4, pH 7.0, and once with 3 mL of asparagine salt medium without glucose (*see* **Notes 9** and **10**). Count the cells using a hemocytometer and adjust the cell density as needed to prepare a culture containing 1×10^8 cells/mL in 5 mL of asparagine salt medium without glucose (*see* **Note 11**). Incubate at 30 °C for 72 h at 200 rpm to induce laccase expression.
4. After incubation, estimate the density of each culture by counting the cells using a hemocytometer. This number will be used later to normalize the secreted laccase activity in the supernatant by the number of cells present at the end of the culture. Centrifuge the cultures at 2000× *g* for 5 min to harvest the supernatants.
5. To evaluate laccase activity in the supernatant, add 180 μL of supernatant from each culture and 20 μL of 10 mM L-DOPA (1 mM final concentration) to the wells of a 96-well plate

[18]. Prepare a blank by adding to the respective wells 180 μL 50 mM Na_2HPO_4, pH 7.0, plus 20 μL 10 mM L-DOPA (*see* **Note 12**).

6. Incubate at 30 °C. After that, laccase activity may be evaluated by a single endpoint measurement of A_{480nm} 10–30 min after the addition of L-DOPA. Alternatively, a kinetic assay may be performed by measuring the A_{480nm} every 30 min for 24 h using an automatic plate reader with incubation.
7. Alternatively, ABTS may be used as the substrate to evaluate laccase activity. For this, prepare a 10 mM solution of ABTS and substitute L-DOPA by ABTS in the above steps. After being oxidized by the enzyme, this compound will generate a green pigment and laccase activity may be followed in a spectrophotometer at 419 nm (*see* **Note 13**).

3.2 Kinetic Evaluation of Melanin Production by C. neoformans Colonies

1. Inoculate an isolated *C. neoformans* colony in 5 mL of Sabouraud or YPD medium and incubate the culture at 30 °C for 16–24 h. Centrifuge the culture at 1000× *g* for 2 min and discard the supernatant. Add an equal volume of PBS to resuspend the cells and centrifuge at 1000× *g* for 2 min. Repeat this wash step twice more (*see* **Note 10**). Count the cells using a hemocytometer and prepare suspensions of 2×10^7 cells/mL (*see* **Note 14**).
2. Prepare the plates containing L-DOPA agar MM. As thiamine and L-DOPA may be degraded at high temperatures, the 3.75% agar solution should be heated separately to 60 °C (*see* **Note 15**). For a 24-well plate, you may prepare 30 mL of medium by mixing 12 mL of 3.75% agar, 15 mL of 2× MM, and 3 mL of L-DOPA. Mix well and promptly distribute 1 mL in each well, avoiding bubble formation.
3. Inoculate 5 μL of the cell suspension of each isolate in the center of each well of the 24-well plate. Let the drop dry thoroughly before flipping the plate over and moving it to the incubator. The plates should be incubated at 30 °C protected from light and photographed at regular intervals for 7 days (Fig. 2).
4. After 24 h of incubation, take the first photograph of the plates. The colonies should be very small and not melanized at this point, which will be a baseline. Afterward, photograph each plate every 12 h for a total of 7 days (*see* **Note 16**).
5. The following six suggestions should help obtain more consistent results:
6. We photograph plates inside a biosafety cabinet, using the bespoke apparatus shown in Fig. 1. This allows removing the lids of the 24-well plates to collect better photos without

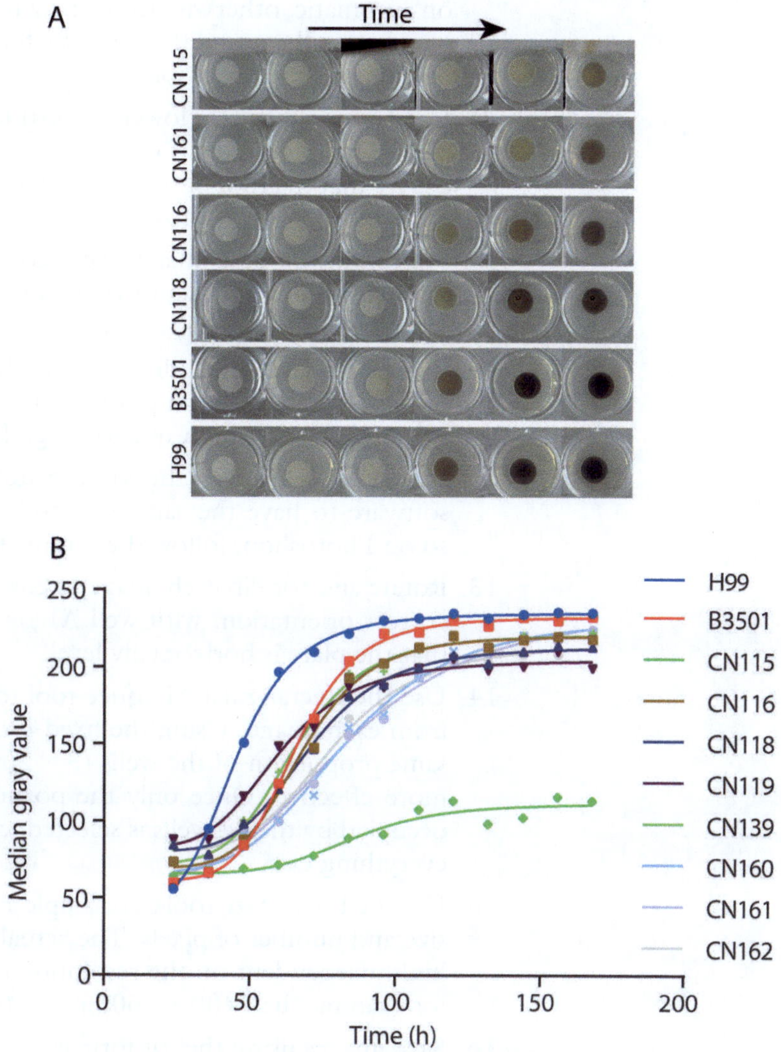

Fig. 2 Representative results of an experiment performed using this protocol. (**a**) Photographs of colonies through time. (**b**) Melanization curves of different clinical isolates of *C. neoformans.* (Adapted from [24])

generating a biohazard for the experimenter and without contaminating the plate with environmental microorganisms.

7. It is important to standardize the conditions used to take the photographs at each time point. To do so, always take photos in the same biosafety cabinet, using the same illumination setup (i.e., remember to be consistent with what lights are turned on in the room and in the biosafety cabinet).
8. Using a DSLR camera, you may also more easily standardize the photographs. For the first image, determine exposure time, aperture (f stop), white balance, and zoom. For all other time points, use the exact same settings. Do not leave these settings

on automatic, otherwise the camera might capture photos with parameters that are different at each time point, and the comparisons will be unreliable.

9. Use larger apertures (lower f stops) to minimize the depth of field. This makes focusing slightly more challenging but will ensure that the roof of the biosafety cabinet (or the room) does not appear in the photos.
10. We found the best images were collected with a macro lens, although the experiment may be done with other more common lenses.
11. We saw no discernible difference in the results when collecting photos using a RAW output or a high-quality JPEG output, so we suggest using high-quality, large JPEG output.
12. Adjust all photos of the same plate in an image processing software to have the same size and number of pixels. To do so on Photoshop, follow these four steps:
13. Rotate and/or flip each image to ensure that the plate is in the correct orientation, with well A1 on the top left corner, and that the plate is horizontally level.
14. Use the Rectangular Marquee tool to select only the 24 wells from each image. Using the fixed-proportion option with the same proportion of the wells (3 × 2, width × length) makes it more effective. Once only the portion of each image that is occupied by the 24 wells is selected, crop the image to remove everything else.
15. Use the Image Size tool to resample all images to have the same size and number of pixels. The actual number of pixels will be highly dependent on the resolution of the camera, but could, for example, be 2400 × 1600 pixels (width × height).
16. Save images using the .tif format.
17. Open the photographs on ImageJ. Using menu Image > Type, convert the image to 8-bit grayscale. Use Edit > Invert to invert the image.
18. For each colony at each time point, draw a circular region of interest (ROI) around the colony. Use the tool on the Analyze > Measure menu to quantify the mean gray level of the ROI. Organize the mean gray level for each colony at each time point on a spreadsheet.
19. Arrange the incubation times and the mean gray levels for each colony on GraphPad Prism (or other statistical analysis software) to perform a nonlinear regression (four-parameter logistic curve). In GraphPad Prism, we specifically used the "[Agonist] vs. response – Variable slope (four parameters) equation". The regression will result in four parameters:

20. Bottom: the mean gray level of the colony at the first time-point, which reflects how dark the colonies were at the beginning of the experiment. In our experience, this parameter is not very useful.
21. Hill slope: the slope of the curve at the inflection point, a reflection of how fast melanin is produced while the colony is melanizing. We have not observed much variation in this parameter between different *C. neoformans* strains.
22. Top: the mean gray level of the colony at the end of the experiment, which is proportional to the actual amount of melanin the colony has produced by the end of the experiment. This parameter is very useful to differentiate strains or conditions based on the amount of pigment being produced.
23. EC50: the time it took for the colony to get to the inflection point of the sigmoidal melanization curve, or the time it took for the colony to reach half-maximum melanization. This parameter is very useful to compare how fast different colonies melanize.

4 Notes

1. Nitrogen sources are important in determining how fast *C. neoformans* produces melanin and how much of the pigment it produces [19]. This classic recipe of *Cryptococcus* minimal medium uses glycine. If asparagine is used instead, fungal growth and melanization are faster and fungal colonies get even darker.
2. Laccase has identifiable copper binding sites in its sequence and copper has been shown to be an important inducer of laccase transcription [20, 21].
3. *C. neoformans* cannot synthesize its own melanin precursors, so it only produces melanin in the presence of exogenous substrates, such as L-DOPA. Other substances containing diphenol and diamino groups, such as dopamine, epinephrine, catechol and hydroquinone, may also be used [1].
4. Although L-DOPA solutions may be frozen for storage, it is better to prepare them fresh because L-DOPA may self-polymerize upon exposure to air and light. If you absolutely need to store them, store them in aliquots and protect from light.
5. Avoid preparing solutions with higher concentrations, as both compounds are hard to dissolve. If freezing L-DOPA or ABTS solutions, make small aliquots to avoid multiple freeze-thaw cycles.

6. It is better to prepare the medium right before using it and to keep it in a 60 °C water bath. You may also prepare it beforehand and melt the medium in a microwave oven. In this case, prepare smaller volumes to avoid multiple melting cycles that would lead to water evaporation and increased agar concentration.
7. Some strains have been reported to melanize poorly in vitro when grown at 37 °C [21, 22], so this assay is normally performed at 30 °C.
8. The main objective here is to obtain a high number of cells for the next steps, as melanization rates are dependent on fungal cell density.
9. This is a condition that induces an increase in capsule thickness. Consequently, the centrifugation time must be increased to avoid losing cells during this step.
10. We have previously observed that a quorum sensing molecule that is secreted by *C. neoformans* induces melanization, so this molecule must be removed [23]. This is the reason why washing the cells to remove molecules that have accumulated in the medium is so important.
11. Glucose strongly inhibits laccase expression in *C. neoformans* [1]; therefore, after obtaining a high number of cells, using a medium without glucose will stimulate laccase expression.
12. Instead of measuring secreted laccase in the supernatant, you may discard the supernatant and use the cell pellet to measure the whole cell laccase activity. Just prepare suspensions of 1×10^8 cells/mL in 50 mM Na_2HPO_4, pH 7.0, and follow the same steps described for measuring the activity in the supernatant.
13. When using ABTS, keep in mind that the kinetics are a lot faster than when using L-DOPA. Therefore, it is recommended to set up a timer right after adding the substrate to accurately determine the time to measure the absorbance in endpoint assays.
14. It is crucial to accurately count the cells, so you spot a precise number of cryptococci in each well. The reason is that quorum sensing regulates melanization, so colonies with different numbers of fungi will melanize at different times because of cell density-regulated melanization [23]. Thus, other means of inoculating cells onto the plates, such as an inoculation loop, a needle, or sterile toothpick will lead to unreliable and irreproducible results.
15. We recommend keeping a 60 °C water bath and a vortex at hand to avoid agar solidification before plating.

16. This time may be optimized according to the characteristics of the experiment being performed. Common laboratory strains such as H99 and B3501 melanize very fast, completing the process between 2 and 4 days. Other strains may take much longer.

Acknowledgments

This work was supported by FAPDF-Brazil Grant numbers 00193.00000888/2021-67, 00193.0000142/2019-39, and 0193.001.200/2016 and scholarships from CAPES/Brazil and CNPq/Brazil.

References

1. Williamson PR (1994) Biochemical and molecular characterization of the diphenol oxidase of *Cryptococcus neoformans*: identification as a laccase. J Bacteriol 176(3):656–664
2. Missall TA, Moran JM, Corbett JA et al (2005) Distinct stress responses of two functional laccases in *Cryptococcus neoformans* are revealed in the absence of the thiol-specific antioxidant Tsa1. Eukaryot Cell 4(1):202–208
3. Pukkila-Worley R, Gerrald QD, Kraus PR et al (2005) Transcriptional network of multiple capsule and melanin genes governed by the *Cryptococcus neoformans* cyclic AMP cascade. Eukaryot Cell 4(1):190–201
4. Liu L, Tewari RP, Williamson PR (1999) Laccase protects *Cryptococcus neoformans* from antifungal activity of alveolar macrophages. Infect Immun 67(11):6034–6039
5. Mednick AJ, Nosanchuk JD, Casadevall A (2005) Melanization of *Cryptococcus neoformans* affects lung inflammatory responses during cryptococcal infection. Infect Immun 73(4):2012–2019
6. Cordero RJB, Camacho E, Casadevall A (2020) Melanization in *Cryptococcus neoformans* requires complex regulation. MBio 11(1):e03313-19
7. Rodrigues ML, Nakayasu ES, Oliveira DL et al (2008) Extracellular vesicles produced by *Cryptococcus neoformans* contain protein components associated with virulence. Eukaryot Cell 7(1):58–67
8. Zhu X, Gibbons J, Garcia-Rivera J et al (2001) Laccase of *Cryptococcus neoformans* is a cell wall-associated virulence factor. Infect Immun 69(9):5589–5596
9. de Sousa HR, de Oliveira GP Jr, Frazão SO et al (2022) Faster *Cryptococcus* melanization increases virulence in experimental and human cryptococcosis. J Fungi (Basel) 8(4):393
10. Salas SD, Bennett JE, Kwon-Chung KJ et al (1996) Effect of the laccase gene CNLAC1, on virulence of *Cryptococcus neoformans*. J Exp Med 184(2):377–386
11. Erb-Downward JR, Huffnagle GB (2007) *Cryptococcus neoformans* produces authentic prostaglandin E2 without a cyclooxygenase. Eukaryot Cell 6(2):346–350
12. Erb-Downward JR, Noggle RM, Williamson PR et al (2008) The role of laccase in prostaglandin production by *Cryptococcus neoformans*. Mol Microbiol 68(6):1428–1437
13. Shen L, Liu Y (2015) Prostaglandin E2 blockade enhances the pulmonary anti-*Cryptococcus neoformans* immune reaction via the induction of TLR-4. Int Immunopharmacol 28(1): 376–381
14. Eisenman HC, Mues M, Weber SE et al (2007) *Cryptococcus neoformans* laccase catalyses melanin synthesis from both D- and L-DOPA. Microbiology (Reading) 153(Pt 12): 3954–3962
15. Noverr MC, Williamson PR, Fajardo RS et al (2004) CNLAC1 is required for extrapulmonary dissemination of *Cryptococcus neoformans* but not pulmonary persistence. Infect Immun 72(3):1693–1699
16. Qiu Y, Davis MJ, Dayrit JK et al (2012) Immune modulation mediated by cryptococcal laccase promotes pulmonary growth and brain dissemination of virulent *Cryptococcus neoformans* in mice. PLoS One 7(10):e47853
17. Sabiiti W, Robertson E, Beale MA et al (2014) Efficient phagocytosis and laccase activity affect the outcome of HIV-associated cryptococcosis. J Clin Invest 124(5):2000–2008

18. Bach CE, Warnock DD, Van Horn DJ et al (2013) Measuring phenol oxidase and peroxidase activities with pyrogallol, L-DOPA, and ABTS: effect of assay conditions and soil type. Soil Biol Biochem 67:183–191
19. Chaskes S, Tyndall RL (1975) Pigment production by *Cryptococcus neoformans* from para- and ortho-diphenols: effect of the nitrogen source. J Clin Microbiol 1(6):509–514
20. Zhu X, Gibbons J, Zhang S et al (2003) Copper-mediated reversal of defective laccase in a Deltavph1 avirulent mutant of *Cryptococcus neoformans*. Mol Microbiol 47(4):1007–1014
21. Zhu X, Williamson PR (2004) Role of laccase in the biology and virulence of *Cryptococcus neoformans*. FEMS Yeast Res 5(1):1–10
22. Jacobson ES, Emery HS (1991) Temperature regulation of the cryptococcal phenoloxidase. J Med Vet Mycol 29(2):121–124
23. Albuquerque P, Nicola AM, Nieves E et al (2013) Quorum sensing-mediated, cell density-dependent regulation of growth and virulence in *Cryptococcus neoformans*. MBio 5(1):e00986–e00913
24. Frazao SO, de Sousa HR, da Silva LG et al (2020) Laccase affects the rate of *Cryptococcus neoformans* nonlytic exocytosis from macrophages. MBio 11(5):e02085–e02020

Chapter 18

Measuring Urease and Phospholipase Secretion in *Cryptococcus neoformans*

Hunter J. Hoffman and Erin E. McClelland

Abstract

Urease and phospholipase are enzymes that are important virulence factors for *Cryptococcus neoformans*. These are two of the most studied enzymes involved in how *C. neoformans* breaches the blood-brain barrier. Additionally, phospholipase secretion also supports dissemination from the lungs. This chapter describes the methods used to measure the secretion of these enzymes, which may be used to characterize strain invasiveness and virulence.

Key words Urease, Phospholipase, Virulence, *Cryptococcus*

1 Introduction

Cryptococcus neoformans (*Cn*) is an encapsulated fungus that causes meningitis in patients with an immunocompromised immune system and has a mortality rate of 74% [1], even with treatment. Due to a combination of climate change where the world is getting hotter and more humid, and an increasing number of patients that are immunocompromised due to infection with HIV, cancer chemotherapy, or organ transplantation, there are increasing numbers of fungal infections [2]. In fact, *Cn* was just listed by the World Health Organization as one of the top four fungi that are a critical threat to public health [3]. Part of the reason *Cn* is a successful pathogen is due to the production of a myriad number of virulence factors that enhance its survivability and infectivity, including capsule production [4, 5] and secretion [6], melanin production [7], and secretion of a variety of enzymes, such as urease [8] and phospholipase [9], which are both involved in crossing the blood-brain barrier [10, 11].

Cn uses urease to catalyze the hydrolysis of urea into ammonia [12]. This creates a more alkaline environment that also increases the amount of melanization, conferring protection against host

Erin E. McClelland (ed.), *Cryptococcus neoformans: Methods and Protocols*, Methods in Molecular Biology, vol. 2775,
https://doi.org/10.1007/978-1-0716-3722-7_18,

immunity in nearby cells [12]. Urease has also been shown to increase *Cn*'s ability to enter the central nervous system (CNS) by allowing it to pass more easily through cerebral capillary beds [10]. While this exact mechanism is not fully understood, urease has been shown to increase the number of sites by which *Cn* crosses the blood-brain barrier [13]. By measuring the amount of urease secretion in different *Cn* strains, we may better understand the complementary functions of its virulence factors, as well as its ability to disseminate into the CNS.

Phospholipases are extracellular enzymes used by some fungal pathogens to hydrolyze ester linkages of cell membrane phospholipids [14]. Phospholipase B has three different enzyme activities, a lipase, subtilisin protease aspartate, and phospholipase [15]. These enzymatic activities act to hydrolyze glycerophospholipids, release fatty acids, and synthesize phospholipids from lyso-phospholipids [16]. Hydrolysis of phospholipids results in release of fatty acids, diacylglycerol, free fatty acids, phosphatidic acid, and lysophospholipids, all of which are important in various signaling pathways [17]. In *Cn*, phospholipase B is important for survival within macrophages [18], as well as within the CNS [19], and helps the fungus disseminate from the lungs [20] and cross the blood-brain barrier [11]. Thus, it is important to test for phospholipase B activity to determine the virulence of a strain. This chapter describes detailed methods for measuring urease and phospholipase activity in *Cryptococcus*.

2 Materials

2.1 Urease

Prepare all solutions with deionized water.

1. Yeast peptone dextrose (YPD) agar plates. After autoclaving premixed YPD medium, allow to cool and then pour into 100 × 20 mm tissue culture plates, filling approximately halfway. Allow plates to solidify overnight, then store at 4 °C.
2. Sabouraud dextrose agar (SDA) plates. After autoclaving premixed SDA, allow to cool and then pour into 100 × 20 mm tissue culture plates, filling approximately halfway. Allow plates to solidify overnight, then store at 4 °C.
3. 2× Roberts urea broth (*see* **Note 1**): Urea: 4 g, Yeast extract: 0.02 g, Phenol red: 0.002 g (or 0.4 mL in 0.5% phenol red solution), KH_2PO_4 0.273 g, Na_2HPO_4: 0.285 g. Filter sterilize and store at 4 °C.
4. Phosphate-buffered saline (PBS): 137 mM sodium chloride, 2.7 mM potassium chloride, 1.5 mM potassium phosphate, and 8.5 mM disodium hydrogen phosphate in deionized water. Autoclave to sterilize and store at room temperature.

5. YPD broth (*see* **Note 2**).
6. Sabouraud dextrose broth (SDB) (*see* **Note 3**).
7. 15 mL conical centrifuge tubes.
8. 50 mL conical centrifuge tubes.
9. 100 × 20 mm tissue culture plates.
10. 250 mL filter units with 0.1 μL aPES membrane.
11. Autoclave.
12. Class II A2 Biosafety cabinet.
13. Centrifuge, including a rotor for 15 mL conical tubes.
14. Oscillating incubator that maintains 37 °C at 150 RPM.
15. Spectrophotometer.
16. Vortex.

2.2 Phospholipase

1. YPD or SDB.
2. PBS.
3. Malt egg yolk agar plates: For 500 mL (approximately 10 plates), add the following to a flask: 22.5 g malt agar powder, 29.22 g NaCl, and 0.277 $CaCl_2$. Add 400 mL of diH_20, and QC to 500 mL. Autoclave at 121 °C, for 20 min and cool. Add 2% 50% egg yolk and immediately pour plates. Let plates solidify overnight at room temperature and then store, inverted in a closed bag, at 4 °C.
4. 50 mL conical centrifuge tubes.
5. 100 × 20 mm tissue culture plates.
6. Hemocytometer.
7. 30 °C incubator.
8. Digital imaging station for capturing images of plates, such as an Alfa Imager, with a color filter.
9. Image J, FIJI, or Adobe Photoshop software.

3 Methods

Since *Cn* is a human pathogen, all yeast preparation should take place in a BSL-II certified A2 Biosafety cabinet. All cell waste should be disposed of in a beaker containing 10% bleach for 24 h to ensure death of yeast.

3.1 Urease

Work at room temperature with solutions on ice. All centrifugation steps are at room temperature.

1. Working inside a Biosafety cabinet, streak *Cn* out onto SDA or YPD plates and grow for 48 h at 30 °C or 37 °C.

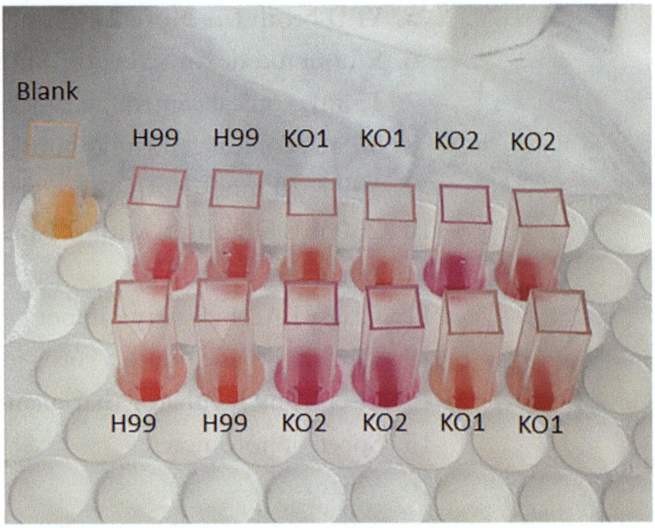

Fig. 1 Image of urease supernatants after 4 h of incubation at 37 °C. H99, wild type; KO1, knockout strain 1; KO2, knockout strain 2. Image generated by Ronak Patel

2. Pick colonies and suspend in 2.0 mL of sterile PBS to an optical density of 0.8–1.0 at 600 nm (*see* **Note 4**). If possible, pick at least three colonies and perform the following steps in triplicate.
3. Vortex cell suspension and add 1.0 mL of cell suspension to 1.0 mL of 2× Roberts Urea Broth in a 15 mL conical tube.
4. Mix 1.0 mL of PBS with 1.0 mL of Roberts Urea Broth in a 15 mL conical tube to create the control.
5. Place tubes in an oscillating incubator at 37 °C and 150 RPM for 4 h.
6. After 4 h, remove tubes and centrifuge for 5 min at 710 × *g*.
7. After centrifugation, transfer 1.0 mL of supernatant from each tube into separate cuvettes. If cells are secreting urease, the supernatant is pink in color (Fig. 1).
8. Blank the spectrophotometer using the PBS:Roberts Urea Broth control (Subheading 3.1, **step 4**).
9. Place samples into spectrophotometer and read absorbance at 560 nm.

3.2 Phospholipase (see Note 5)

1. Start an overnight culture of *Cn* in YPD or SDB and grow to log-phase at 30 °C or 37 °C.
2. Collect cells by centrifugation at 710 × g, 5 min and wash 3× with PBS.

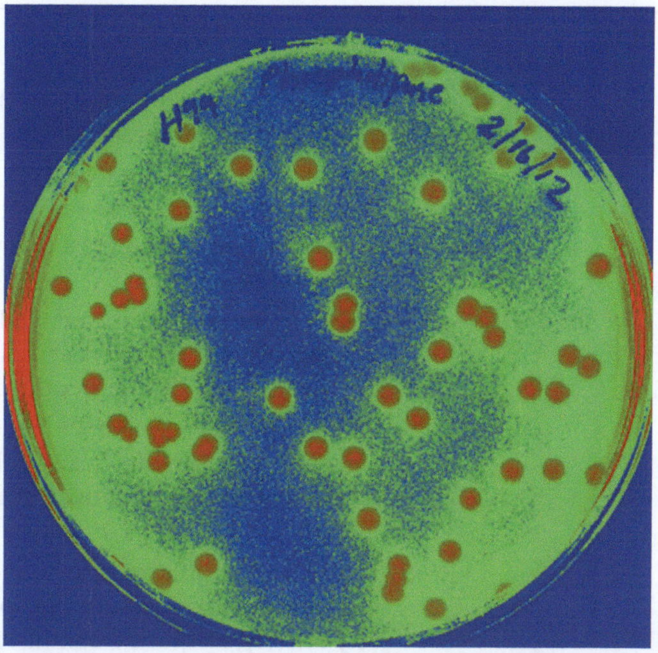

Fig. 2 Image of malt egg yolk agar plate of the wild type H99 strain after 10 days at 30 °C. Image was taken using an Alfa Imager with a color filter. The color of the plate background was changed to highlight the precipitation around the colonies. Image generated by Ronak Patel

3. Resuspend cells in PBS such that 50–200 yeast are counted for each outer quadrant on a hemocytometer and then calculate the number of cells/mL (average of four outer quadrants × dilution factor × 10^4/mL).
4. Dilute cells to 1 × 10^3 cells/mL in 1 mL of PBS.
5. Plate 50 μL of cells on malt egg yolk agar plates.
6. Incubate at 30 °C for 10 days.
7. After incubation, take pictures of plates using an Imager with a color filter and change the color of the plate to highlight the precipitation around the colony (Fig. 2).
8. Using Image J, FIJI, or Adobe Photoshop, measure the colony diameter, and the diameter of the colony plus the zone of precipitation around the colony.
9. Calculate the phospholipase index: the ratio of the colony diameter to the diameter of the colony plus the precipitation zone. There is a high phospholipase enzyme activity when the ratio is close to 0.

4 Notes

1. Dissolve solutes in 100 mL of diH_2O and filter sterilize. Store at 4 °C.
2. Dissolve 50 g of premixed YPD in 1000 mL of diH_2O. Mix thoroughly and autoclave at 121 °C for 20 min. Broth may also be made using 10.0 g yeast extract, 20.0 g peptone, and 20.0 g dextrose.
3. Dissolve 30 g of premixed SDB in 1000 mL of diH_2O. Mix thoroughly and autoclave at 121 °C for 15 min. Broth may also be made using 5.0 g peptic digest of animal tissue, 5.0 g pancreatic digest of casein, and 20.0 g dextrose.
4. Previous experiments using YPD in these conditions have found that an optical density of 0.8–1.0 corresponds to approximately 1×10^7–5×10^7 cells of *Cn*.
5. It is also possible to use western blotting or a radiometric enzyme assay to measure phospholipase secretion and activity, respectively, in specific conditions [21–23]. However, an antibody to phospholipase B is required and may need to be made.

References

1. Rajasingham R, Govender NP, Jordan A et al (2022) The global burden of HIV-associated cryptococcal infection in adults in 2020: a modelling analysis. Lancet Infect Dis 22(12): 1748–1755. https://doi.org/10.1016/S1473-3099(22)00499-6
2. Christensen J (2023) The fungal threat to human health is growing in a warmer, wetter, sicker world. CNN. https://www.cnn.com/2023/02/07/health/fungus-health-threat-scn/index.html#:~:text=Discovery.warmer%2C%20wetter%20and%20sicker%20world.&text=%E2%80%9CWe're%20always%20surrounded%20by,even%20evolved%20into%20modern%20humans. Accessed 7 Feb 2023
3. World Health Organization (2022) WHO fungal priority pathogens list to guide research, development and public health action. https://www.who.int/publications/i/item/9789240060241. Accessed 23 Feb 2023
4. Doering TL (2009) How sweet it is! Cell wall biogenesis and polysaccharide capsule formation in *Cryptococcus neoformans*. Annu Rev Microbiol 63:223–247. https://doi.org/10.1146/annurev.micro.62.081307.162753
5. Casadevall A, Coelho C, Cordero RJB et al (2019) The capsule of *Cryptococcus neoformans*. Virulence 10(1):822–831. https://doi.org/10.1080/21505594.2018.1431087
6. Vecchiarelli A, Pericolini E, Gabrielli E et al (2013) Elucidating the immunological function of the *Cryptococcus neoformans* capsule. Future Microbiol 8(9):1107–1116. https://doi.org/10.2217/fmb.13.84
7. Williamson PR (1994) Biochemical and molecular characterization of the diphenol oxidase of *Cryptococcus neoformans*: identification as a laccase. J Bacteriol 176(3):656–664. https://doi.org/10.1128/jb.176.3.656-664.1994
8. Cox GM, Mukherjee J, Cole GT et al (2000) Urease as a virulence factor in experimental cryptococcosis. Infect Immun 68(2): 443–448. https://doi.org/10.1128/IAI.68.2.443-448.2000
9. Cox GM, McDade HC, Chen SC et al (2001) Extracellular phospholipase activity is a virulence factor for *Cryptococcus neoformans*. Mol Microbiol 39(1):166–175. https://doi.org/10.1046/j.1365-2958.2001.02236.x
10. Olszewski MA, Noverr MC, Chen GH et al (2004) Urease expression by *Cryptococcus neoformans* promotes microvascular sequestration, thereby enhancing central nervous system invasion. Am J Pathol 164(5):1761–1771. https://doi.org/10.1016/S0002-9440(10)63734-0
11. Maruvada R, Zhu L, Pearce D et al (2012) *Cryptococcus neoformans* phospholipase B1 activates host cell Rac1 for traversal across the

blood-brain barrier. Cell Microbiol 14(10): 1544–1553. https://doi.org/10.1111/j.1462-5822.2012.01819.x

12. Baker RP, Casadevall A (2023) Reciprocal modulation of ammonia and melanin production has implications for cryptococcal virulence. Nat Commun 14(1):849. https://doi.org/10.1038/s41467-023-36552-7
13. Shi M, Li SS, Zheng C et al (2010) Real-time imaging of trapping and urease-dependent transmigration of Cryptococcus neoformans in mouse brain. J Clin Invest 120(5): 1683–1693. https://doi.org/10.1172/JCI41963
14. Barman A, Gohain D, Bora U et al (2018) Phospholipases play multiple cellular roles including growth, stress tolerance, sexual development, and virulence in fungi. Microbiol Res 209:55–69. https://doi.org/10.1016/j.micres.2017.12.012
15. Jones PM, Turner KM, Djordjevic JT et al (2007) Role of conserved active site residues in catalysis by phospholipase B1 from *Cryptococcus neoformans.* Biochemistry 46(35): 10024–10032. https://doi.org/10.1021/bi7009508
16. Ghannoum MA (2000) Potential role of phospholipases in virulence and fungal pathogenesis. Clin Microbiol Rev 13(1):122–143. https://doi.org/10.1128/CMR.13.1.122
17. Kohler GA, Brenot A, Haas-Stapleton E et al (2006) Phospholipase A2 and phospholipase B activities in fungi. Biochim Biophys Acta 1761(11):1391–1399. https://doi.org/10.1016/j.bbalip.2006.09.011
18. Evans RJ, Li Z, Hughes WS et al (2015) Cryptococcal phospholipase B1 is required for intracellular proliferation and control of titan cell morphology during macrophage infection. Infect Immun 83(4):1296–1304. https://doi.org/10.1128/IAI.03104-14
19. Hamed MF, Araujo GRS, Munzen ME et al (2023) Phospholipase B is critical for *Cryptococcus neoformans* survival in the central nervous system. MBio 14(2):e0264022. https://doi.org/10.1128/mbio.02640-22
20. Santangelo R, Zoellner H, Sorrell T et al (2004) Role of extracellular phospholipases and mononuclear phagocytes in dissemination of cryptococcosis in a murine model. Infect Immun 72(4):2229–2239. https://doi.org/10.1128/IAI.72.4.2229-2239.2004
21. Siafakas AR, Wright LC, Sorrell TC et al (2006) Lipid rafts in *Cryptococcus neoformans* concentrate the virulence determinants phospholipase B1 and Cu/Zn superoxide dismutase. Eukaryot Cell 5(3):488–498. https://doi.org/10.1128/EC.5.3.488-498.2006
22. Siafakas AR, Sorrell TC, Wright LC et al (2007) Cell wall-linked cryptococcal phospholipase B1 is a source of secreted enzyme and a determinant of cell wall integrity. J Biol Chem 282(52):37508–37514. https://doi.org/10.1074/jbc.M707913200
23. Chayakulkeeree M, Johnston SA, Oei JB et al (2011) SEC14 is a specific requirement for secretion of phospholipase B1 and pathogenicity of *Cryptococcus neoformans.* Mol Microbiol 80(4):1088–1101. https://doi.org/10.1111/j.1365-2958.2011.07632.x

Chapter 19

Measuring Stress Phenotypes in *Cryptococcus neoformans*

Rajendra Upadhya, Corinna Probst, J. Andrew Alspaugh, and Jennifer K. Lodge

Abstract

Cryptococcus neoformans is an opportunistic human fungal pathogen capable of surviving in a wide range of environments and hosts. It has been developed as a model organism to study fungal pathogenesis due to its fully sequenced haploid genome and optimized gene deletion and mutagenesis protocols. These methods have greatly aided in determining the relationship between *Cryptococcus* genotype and phenotype. Furthermore, the presence of congenic mata and matα strains associated with a defined sexual cycle has helped further understand cryptococcal biology. Several in vitro stress conditions have been optimized to closely mimic the stress that yeast encounter in the environment or within the infected host. These conditions have proven to be extremely useful in elucidating the role of several genes in allowing yeast to adapt and survive in hostile external environments. This chapter describes various in vitro stress conditions that could be used to test the sensitivity of different mutant strains, as well as the protocol for preparing them. We have also included a list of mutants that could be used as a positive control strain when testing the sensitivity of the desired strain to a specific stress.

Key words *Cryptococcus* growth, Cell wall stress, Oxidative stress, Nitrosative stress, Alkaline stress, Temperature stress, Capsule, Melanin

1 Introduction

Cryptococcus neoformans and closely related species are basidiomycetes fungi that grow in a wide variety of environments and infect a broad range of host species [1–5]. They also grow and proliferate in multiple organs and tissue environments even within a single host. This is because, to survive, they have an exceptional ability to sense, respond to, and adapt to their surroundings. *Cryptococcus* species have evolved a number of signaling pathways that are activated by specific cell wall sensors and are critical for the organism's survival when subjected to a specific stress [6, 7]. The high osmolarity glycerol (HOG) kinase, the cAMP/protein kinase A, the Ras signaling, the protein kinase C cell wall integrity (CWI), the Rim signaling, the calcium calcineurin signaling, and the unfolded

Erin E. McClelland (ed.), *Cryptococcus neoformans: Methods and Protocols*, Methods in Molecular Biology, vol. 2775, https://doi.org/10.1007/978-1-0716-3722-7_19,

protein response (UPR) signaling pathways are some of these signaling pathways [8]. Deleting or inactivating a single gene or a family of genes causes the fungus to be more sensitive to a specific type of external stress [9–12]. Furthermore, *Cryptococcus* expresses several virulence-related traits that are critical for its virulence within the mammalian host [12, 13]. Several in vitro culture conditions that closely mimic the mammalian host have been identified, and phenotypic analysis of yeast strains grown under these conditions has been useful to understand the mechanism of fungal pathogenesis [14, 15]. These stress phenotypes are important for understanding *C. neoformans*' survival strategies in a variety of environments, and they may also provide new antifungal targets.

Stress response phenotypes in *C. neoformans* include changes in fungal growth, viability, morphogenesis (cell size, shape, or colony morphology), pigmentation, and changes in capsule production or shedding and are accompanied by gene expression changes [6]. In *Cryptococcus*, several transformation methods, including biolistic transformation and a recently optimized CRISPR CAS9-based system, have allowed the generation of gene deletions and other mutations. The potential importance of a specific gene or its product in mediating a specific stress response will be revealed by subjecting these mutant strains to a range of phenotypic assays under numerous stress conditions. The type of the culture media and growth conditions employed for each stress reagent depends on the nature of the stressor being investigated. One common phenotypic assay involves spotting serially diluted cell suspension onto a stress plate and monitoring the cell growth over time. This growth assay may also be carried out in liquid media with varying concentrations of the desired stressor and monitoring the growth of the yeast by measuring the absorbance at 600 nm. A particular stressor or a stress condition may be fungistatic or fungicidal. This can be assessed by monitoring the viability of the yeast cells at different times either by using Alamar Blue or by plating the serially diluted cell suspension onto a permissive growth medium, such as a yeast extract-peptone-dextrose (YPD) plate and enumerating the colony forming units. The disc diffusion assay is widely used to determine yeast sensitivity to common antifungals. The sensitivity of yeast strains to various antifungals may be compared more precisely by determining their minimal inhibitory concentration (MIC) values. In this chapter, we describe protocols for preparing various stress plates/media and procedures to evaluate the phenotypes of the fungal strains to be tested. The various types of stress conditions include thermotolerance, cell wall stress, oxidative and nitrosative stress, alkaline pH stress, hypoxia, endoplasmic reticulum (ER) stress, mitochondrial stress, and metal stress. Some of the phenotypes associated with host interaction include capsule expression, melanin production, Titan cell formation, and a macrophage killing assay.

2 Materials

Cryptococcus is a Biosafety Level 2 (BSL-2) organism. Personnel working in the BSL2 lab must have general training in handling pathogenic agents. It is necessary to have Institutional Biosafety Committee (IBC) approval to study *Cryptococcus.* As a BSL2 pathogen, work with *Cryptococcus* may be done on an open bench top as long as aerosols are not generated. However, if the procedure involves the potential release of aerosols, working in a biosafety hood is preferable. All culture-related materials are autoclaved or incinerated prior to disposal.

We recommend double-distilled or Milli-Q water (18 megaohm) for making solutions and media.

Room temperature (22–25 °C) is abbreviated as RT.

2.1 General Equipment Needed

1. Incubators at different temperatures (25 °C, 30 °C, 37 °C, and 39 °C) for incubating growth plates at different temperatures.
2. Incubator for shaking cultures capable of maintaining 300 rpm and 30 °C and 37 °C.
3. Polypropylene round-bottom tubes (14 mL) with cap for 4 mL cultures.
4. Sterile culture tubes (15 mL Falcon tubes).
5. Sterile microcentrifuge tubes.
6. Sterile petri dishes.
7. Sterile inoculation loop.
8. Sterile glass beads (to spread cells).
9. Filter sterilizer.
10. Phosphate-buffered saline (PBS), 1×.
11. Clear bottom 96 well plates.
12. Multichannel pipettor.
13. Spectrophotometer capable of measuring absorbance at 600 nm.
14. pH-Meter or pH strips.
15. Camera fixed to a photo station to capture the images of the plate.
16. Light microscope.

2.2 Preparation of Growth Media

1. Yeast extract.
2. Bacto-peptone.
3. Dextrose/glucose.
4. Yeast Nitogen Base (YNB) without amino acids.
5. Synthetic complete (SC) supplement mixture.

6. Roswell Park Memorial Institute (RPMI)-1640 medium with L-Glutamine, without glucose and sodium bicarbonate.
7. Agar, bacteriological.
8. 3-(N-morpholino) propanesulfonic acid (MOPS).
9. 4-(2-hydroxyethyl)-1-piperazineethanesulfonic acid (HEPES).
10. NaOH and HCl solution to adjust pH.
11. Amino acids, purine, and pyrimidine supplements (adenine, uracil, histidine, leucine, and lysine).
12. Potassium dihydrogen phosphate (KH_2PO_4).
13. Magnesium sulfate heptahydrate ($MgSO_47H_2O$).
14. L-3,4-Dihydroxyphenylalanine (L-DOPA).
15. Urea agar base.

2.3 Preparation of Cell Wall Stress Plates

1. Calcofluor white (CFW), also called fluorescent brightener 28.
2. Congo red.
3. Caffeine.
4. Sodium dodecyl sulfate (SDS).
5. Sodium chloride.
6. Sorbitol.
7. Cobalt chloride.
8. Tunicamycin.
9. Dithiothreitol (DTT).
10. Rotenone.
11. Salicylhydroxamic acid (SHAM).
12. Carboxin.
13. Antimycin A.
14. Sodium azide.
15. Calcium chloride.
16. Copper sulfate.
17. Bathocuproine disulphonate (BCS).
18. Bathophenanthrolinedisulfonate (BPS).

2.4 Stress Assays

1. 6 mm blank filter discs (e.g., BD 231039).
2. MIC strips (e.g. Biomerieux E-test).
3. Sterile tweezers.
4. Alamar Blue.

2.5 Analysis of Host Stress Associated Phenotypes

1. India Ink.
2. Light microscope with a minimum objective of 40× magnification for visualizing capsule.
3. Tissue culture incubator to induce capsule at 37 °C with 5% CO_2.
4. Dulbecco's Modified Eagle Medium (DMEM)/RPMI medium (e.g., Gibco DMEM 11995065, Sigma RPMI-1640 R6504-10L).
5. Heat inactivated fetal bovine serum (FBS).
6. CO_2-independent medium.
7. Bovine serum albumin (BSA).
8. Capsule specific antibody clone mAB18B7.
9. Paraformaldehyde.
10. Fluorescently conjugated mouse secondary antibody.
11. Fluorescent microscope.
12. Azoalbumin.
13. J774A.1 macrophage culture.
14. Penicillium-streptomycin.
15. Nonessential amino acid mix.
16. Tissue culture flasks and 96-well plates.
17. Trypan blue.
18. Interferon γ (INF-γ).
19. Lipopolysaccharide (LPS).
20. Hemocytometer: Standard glass or disposable hemocytometer or a TC20 automated cell counter and counting slides.
21. Light microscope to view hemocytometer.

2.6 Preparation of Growth Media

1. YPD agar/liquid: Solution A: 1% (w/v) yeast extract, 2% (w/v) bacto-peptone, and 2% (w/v) agar in H_2O (subtract the volume of solution B which is added to achieve the final 2% glucose concentration from volume of solution A). Solution B: 2.0% (w/v) dextrose/glucose in H2O. Autoclave both solutions A and B separately at 121 °C for 30 min. After autoclaving, mix solutions A and B together to make YPD agar plates with a final dextrose/glucose concentration of 2%. Pour into plates (~20–30 mL per plate). YPD liquid media is made as described above, without the addition of agar (*see* **Note 1**).
2. Buffered YPD agar: Solution A: 1% (w/v) yeast extract, 2% (w/v) bacto-peptone, and 150 mM HEPES in H_2O. Adjust pH to designated pH using a NaOH or HCl solution (HEPES pH range 6.8–8.2). Use a different buffer system in case the designated pH is outside the pH range of HEPES, for example, 150 mM 2-(N-morpholino)ethanesulfonic acid for

pH 5.8–6.5. Once pH is set, add 2% (w/v) agar. Solution B: 20% (w/v) dextrose/glucose in H_2O. Autoclave both solutions A and B separately at 121 °C for 30 min. After autoclaving mix, solutions A and B together, as described in Subheading 2.6, **item 1** and pour into plates. Buffered YPD liquid media is made as described above, without the addition of agar.

3. Yeast Nitrogen Base (YNB) agar: Make 2× YNB 1.34% (w/v) yeast nitrogen base without amino acids and 4% (w/v) dextrose and filter sterilize. Autoclave 4% (w/v) agar at 121 °C for 30 min. To make YNB agar plates mix an equal volume of 2× YNB solution and 4% agar solution and pour into plates. YNB liquid media is made as described above, without the addition of agar.
4. YNB, pH 4.0: 0.67% (w/v) yeast nitrogen base without amino acids, 2% dextrose, and 25 mM sodium succinate at pH 4.0 (*see* **Note 2**). Filter sterilize the media. To make plates follow instructions in Subheading 2.6, **item 3**. YNB, pH 4.0 liquid media is made as described above, without the addition of agar.
5. YNB, pH 7.0: 0.67% (w/v) yeast nitrogen base without amino acids, 2% dextrose and 25 mM MOPS at pH 7.0 (*see* **Note 3**). Filter sterilize the media. To make plates follow instructions in Subheading 2.6, **item 3**. YNB, pH 7.0 liquid media is made as described above, without the addition of agar.
6. Yeast Extract with Supplements (YES) agar: Dissolve 3% (w/v) dextrose/glucose, 0.5% (w/v) yeast extract, 0.0 2% (w/v) adenine, 0.02% (w/v) uracil, 0.02% (w/v) histidine, 0.02% (w/v) leucine, and 0.02% (w/v) lysine in H_2O. Add 2% (w/v) agar and autoclave. Pour into plates.
7. Synthetic Complete (SC) agar: Make 2× SC solution in H_2O (0.4% (w/v) SC mix, 1.42% (w/v) YNB without amino acids, and 4% dextrose/glucose). Dissolve completely, adjust pH to 5.6, and filter sterilize the solution. Autoclave 4% (w/v) agar at 121 °C for 30 min. To make SC agar plates, mix an equal volume of 2× SC solution and 4% agar solution and pour into plates.
8. RPMI agar: Make 2× RPMI-1640 (2.08% (w/v) RPMI-1640, 6.9% (w/v) MOPS, and 4% Glucose), adjust pH to 7.0, and filter sterilize. Autoclave 4% (w/v) agar at 121 °C for 30 min. To make RPMI agar plates, mix an equal volume of 2× RPMI-1640 solution and 4% agar solution and pour into plates.
9. L-DOPA agar: For 500 mL L-DOPA medium: Dissolve 0.57 g L-asparagine (7.6 mM), 0.5 g glucose (5.6 mM), 1.5 g KH_2PO_4 (22 mM), 0.123 g $MgSO_4$ $7H_2O$ (1 mM), and 50 mg L-DOPA (0.5 mM) in 300 mL of H_2O. Dissolve components completely and adjust the pH to 5.6. Fill up to 500 mL with H_2O and add 10 g of agar. Autoclave media for 20 min at

121 °C. Cool to ~50–60 °C and add 50 μL of 3 mM Thiamine stock solution (sterile filtered in H_2O, final concentration 0.3 μM) and 10 μL of 1 mM Biotin stock solution (sterile filtered in H_2O, final concentration 20 nM). Pour into plates. L-DOPA agar plates may be stored for 2 weeks at 4 °C, protected from light.

10. Urease Agar: For 500 mL Urease agar: Dissolve 14.5 g of Urea agar base in 50 mL of H_2O and filter sterilize. Mix 7.5 g agar with 450 mL of H_2O. Autoclave the agar mix for 30 min at 121 °C. Cool to ~50–60 °C and add the sterile filtered Urea agar base solution. Mix well and pour into plates.

2.7 Thermotolerance of C. neoformans

YPD agar plates are used or RPMI-agar to mimic the in vitro host environment.

2.8 Oxidative and Nitrosative Stress

YNB pH 4.0 plates containing different concentrations of H_2O_2 (1 mM and 2 mM; *see* **Note 4**) or $NaNO_2$ (0.75 mM and 1 mM; *see* **Notes 5** and **6**). Alternatively, a disc diffusion assay may be performed using YPD (*see* Subheading 3.3 for disc diffusion assay).

2.9 Alkaline Stress

Buffered YPD agar plates from pH 8 to pH 8.2.

2.10 Cell Wall Stress

1. Calcofluor (CFW) plates: Dissolve 5–15 mg/mL of CFW stock in 20% Glucose and filter sterilize. YPD plates containing 0.5 mg/mL (make 5 mg/mL CFW stock), 1.0 mg/mL (make 10 mg/mL CFW stock), and 1.5 mg/mL (make 15 mg/mL CFW stock) of CFW may be used. Make YPD agar solution A as described in Subheading 2.6, **item 1**. Autoclave at 121 °C for 30 min. After autoclaving, cool the solution while stirring at RT. When solution is cool to the touch (~50–60 °C), add 1 part of the sterile filtered CFW-glucose solution to 9 parts of YPD-agar solution (*see* **Note 7**). Pour into plates.
2. YPD + Congo red (0.5% w/v): Make YPD agar solution A as described in Subheading 2.6, **item 1** and add 0.5% (w/v) Congo red. Make solution B as described in Subheading 2.6, **item 1**. Autoclave both solutions at 121 °C for 30 min. Mix solution A and B together and pour into plates as described in Subheading 2.6, **item 1**.
3. YPD+ Caffeine: Make a 10 mg/mL stock of caffeine in H_2O and filter sterilize the solution. Make and autoclave YPD agar medium by following the steps in Subheading 2.6, **item 1**. Combine Solution A and B and cool to ~50–60 °C. Add Caffeine stock to a final concentration of 0.5 mg/mL or 1 mg/mL and pour into plates.

4. YPD+ SDS: Make a 10% (w/v) stock solution of SDS in H_2O, filter sterilize, and store at RT. Make and autoclave YPD agar medium by following the steps in Subheading 2.6, **item 1**. Combine Solution A and B and cool to ~50–60 °C. Add SDS stock to a final concentration of either 0.01%, 0.03%, or 0.06%. Pour into plates.
5. YPD+ 1.5 M NaCl: Make YPD agar solution A as described in Subheading 2.6, **item 1** and add NaCl to a final concentration of 1.5 M. Make solution B as described in Subheading 2.6, **item 1**. Autoclave both solutions at 121 °C for 30 min. Mix solution A and B together and pour into plates as described in Subheading 2.6, **item 1**.
6. YPD+ 1 M Sorbitol (*see* **Note 8**): Make YPD agar solution A as described in Subheading 2.6, **item 1** and add sorbitol to a final concentration of 1 M. Make solution B as described in Subheading 2.6, **item 1**. Autoclave both solutions at 121 °C for 30 min. Mix solution A and B together and pour into plates as described in Subheading 2.6, **item 1**.

2.11 Hypoxia Stress

Make a stock of 1 M $CoCl_2$ in H_2O and filter sterilize. Store at RT. Make YES agar as described in Subheading 2.6, **item 5**. Autoclave YES agar medium as described in Subheading 2.6, **item 5**. Cool autoclaved YES agar medium and add $CoCl_2$ to a final concentration of 0.7 mM $CoCl_2$. Pour into plates.

2.12 ER Stress

1. Tunicamycin stress plates: Make 5 mg/mL Tunicamycin stock solution in dimethyl sulfoxide (DMSO). Store at 4 °C. Make YPD agar solution A as described in Subheading 2.6, **item 1**. Make solution B as described in Subheading 2.6, **item 1**. Autoclave both solutions at 121 °C for 30 min. Mix solution A and B together as described in Subheading 2.6, **item 1**. When solution is cool to the touch (~50–60 °C), add Tunicamycin to a final concentration of 5 μg/mL and pour into plates.
2. 1,4-Dithiothreitol (DTT) stress plates: Make 1 M DTT stock solution in H_2O and filter sterilize. Store at −20 °C. Make YPD agar solution A as described in Subheading 2.6, **item 1**. Make solution B as described in Subheading 2.6, **item 1**. Autoclave both solutions at 121 °C for 30 min. Mix solution A and B together as described in Subheading 2.6, **item 1**. When solution is cool to the touch (~50–60 °C), add DTT to a final concentration of 1 mM and pour into plates.

2.13 Mitochondrion Stress

1. Make stock solution and filter sterilize.

Compound	Inhibition	Stock	Storage
Rotenone	Complex I	50 mg/mL in DMSO	−20 °C
Salicylhydroxamic acid (SHAM)	AOX	250 mM in DMSO	−20 °C

(continued)

Compound	Inhibition	Stock	Storage
Carboxin	Complex II	125 mg/mL in Acetone	−20 °C
Antimycin A	Complex III	50 mg/mL in Ethanol	−20 °C
Sodium azide	Complex IV	1% (=154 mM) in H_2O	RT

Make YPD agar solution A as described in Subheading 2.6, **item 1**. Make solution B as described in Subheading 2.6, **item 1**. Autoclave both solutions at 121 °C for 30 min. Mix solution A and B together as described in Subheading 2.6, **item 1**. When solution is cool to the touch (~50–60 °C), add stressor to final concentration as follows:

Rotenone – 0.5 mg/mL.

SHAM – 2.5 mM.

Carboxin – 0.1 mg/mL.

Antimycin A – 3 μg/mL.

Sodium azide – 0.5 mM.

Pour into plates.

2.14 Calcineurin

Make YPD agar solution A as described in Subheading 2.6, **item 1** and add $CaCl_2$ to a final concentration of 0.35 M. Make solution B as described in Subheading 2.6, **item 1**. Autoclave both solutions at 121 °C for 30 min. Mix solution A and B together and pour into plates as described in Subheading 2.6, **item 1**.

2.15 Metal Stress

1. Copper toxicity (*see* **Note 9**): Make 1 M $CuSO_4$ stock and filter sterilize. Make and autoclave SC agar medium as described in Subheading 2.6, **item 6**. Add $CuSO_4$ to the SC agar to a final concentration ranging from 0.25 mM to 1 mM. Pour into plates.
2. Copper starvation: Make 200 mM BCS stock in H_2O and filter sterilize. Make and autoclave SC agar medium as described in Subheading 2.6, **item 6**. Cool medium to ~50–60 °C and add BCS to the SC agar to a final concentration ranging from 0.5 mM to 3 mM. Pour into plates.
3. Iron starvation: Make 100 mM BPS stock in H_2O and filter sterilize. Make and autoclave YNB agar medium as described in Subheading 2.6, **item 6**. Cool medium to ~50–60 °C and add BPS to the YNB agar to a final concentration of 100 μM. Pour into plates.

3 Methods

3.1 Inoculation of Plates from Glycerol Stock and Precultures

1. Streak out the designated *Cryptococcus* strain from glycerol onto an YPD or YNB agar plate using a sterile inoculation loop or autoclaved wood stick.
2. Incubate for 2–3 days at 30 °C or RT or until cells are grown.
3. Start overnight culture by transferring yeast cells into 5–20 mL in YPD liquid media using a sterile inoculation loop.
4. Grow for ~16 h at 30 °C, shaking (*see* **Note 10**).

3.2 Spot Dilution Assay

1. Grow designated strains on an agar plate as described in Subheading 3.1. Make sure to include the corresponding wild type (WT) strain and if possible a control knockout strain (*see* Table 1), which demonstrates the stress phenotype analyzed in the assay.
2. Start and grow overnight culture as described in Subheading 3.1. For the spot dilution assay, a 5 mL culture (in 12 mL sterile PP tubes) in YPD, YNB, or SC is suitable.
3. Measure cell density of the overnight culture. Dilute the cell suspension appropriately (either 1:10 or 1: 20 depending on the sensitivity of the spectrophotometer being used) and determine the density of the culture by measuring absorbance of the culture at 600 nm.
4. Adjust the OD_{600} of the cell suspension to 1.0. Make 5 more ten-fold serial dilutions of the culture in a 96-well plate using this dilution. To perform serial dilution for multiple strains, use a multichannel pipettor.
5. Using a multichannel pipettor, aspirate 3–5 μL of cell suspension and spot onto a suitable plate (*see* **Note 11**).
6. Incubate the plates at the recommended temperatures. Take images of the plate on days 2, 3, 4, and 5. Ensure that the image resolution is at least 300 dpi.

3.3 Disc Diffusion Assay

The disc diffusion assay is a plate-based assay that can be used to analyze the inhibitory effects of soluble drugs/compounds. Sterile filters are soaked with defined drug or compound concentrations and then placed on an inoculated yeast plate. Up to 4 filter discs with different drugs or containing the same drug at increasing temperatures may be analyzed per plate. If the drug or compound is dissolved in a solvent other than H_2O, the inhibitory effect of the solvent only has to be tested.

Table 1
Selected *Cryptococcus neoformans* gene knockout strains with known defects in stress assays

Stress assays	Knockout strain	Pathway/function	Additional stress phenotypes	References
High temp (37–39 °C)	*tps1Δ*	Trehalose biosynthesis	Temperature defect restored by sorbitol addition	[17]
	cna1Δ	Calcineurin signaling	Impaired growth on 0.35 M $CaCl_2$, 0.03% SDS, 5 mg/mL CFW and 1% Congo red	[18]
	ras1Δ	Ras-signaling		[19]
Oxidative stress				
ROS stress induced by menadione	*sod2Δ*	Superoxide dismutase 2, ROS detoxification (mitochondrial)	Impaired growth during high temperature (37–39 °C)	[20]
ROS stress induced by peroxides	*gpx1Δ*	Glutathione peroxidase 1		[21]
ROS stress induced by H_2O_2	*hog1Δ*	MAPK of the high osmolarity and glycerol (HOG) pathway	Impaired growth on 1 M KCl, and during high temperature (40 °C)	[22]
NOS stress induced by sodium nitrite	*grl1Δ*	Glutathione reductase	WT-like growth in the presence of ROS stress (peroxides)	[23]
NOS stress induced by sodium nitrite	*tsa1Δ*	Thiol peroxidase	Impaired growth also in presence of ROS stress (peroxides)	[10]
Alkaline pH stress	*rim101Δ*	pH sensing transcription factor	Impaired growth on 1.5 M NaCl	[24]
	sre1Δ	Sterol sensing transcription factor	Impaired growth on 0.03% SDS, azole drugs, 1.5 M NaCl, and hypoxia ($CoCl_2$)	[25]
Cell wall stress				
Calcofluor white (CFW)	*chs3Δ*	Chitin synthase 3, chitosan biosynthesis	Impaired growth on 0.03% SDS, 1.5 M NaCl, 0.5 mg/mL caffeine, and 0.5% Congo red	[26]
Congo red	*chs3Δ*	Chitin synthase 3, chitosan biosynthesis	Impaired growth on 0.03% SDS, 1.5 M NaCl, 0.5 mg/mL caffeine, and 1.5 mg/mL CFW	[26]
Caffeine	*chs3Δ*	Chitin synthase 3, chitosan biosynthesis	Impaired growth on 0.03% SDS, 1.5 M NaCl, 0.5% Congo red, and 1.5 mg/mL CFW	[26]
Membrane stress				
SDS	*sre1Δ*	Sterol-sensing transcription factor	Impaired growth on azole drugs, 1.5 M NaCl, hypoxia ($CoCl_2$), as well as alkaline pH	[25]
Azole drugs	*sre1Δ*	Sterol-sensing transcription factor	Impaired growth on 0.03% SDS, 1.5 M NaCl, hypoxia ($CoCl_2$), as well as alkaline pH	[27]

(continued)

Table 1
(continued)

Stress assays	Knockout strain	Pathway/function	Additional stress phenotypes	References
Salt stress				
NaCl/KCl	*hog1Δ*	MAPK of the HOG pathway	Impaired growth on 1 M KCl, 3 mM H_2O_2, and during high temperature (40 °C)	[22, 28]
NaCl	*rim101Δ*	pH sensing transcription factor	Impaired growth on alkaline pH	[24]
Sorbitol stress	*nrg1Δ*	cAMP regulated transcription factor	Impaired growth on 1.5 M NaCl and 0.01% SDS	[29]
Hypoxia	*sre1Δ*	Sterol-sensing transcription factor	Impaired growth on azole drugs, 0.03% SDS, 1.5 M NaCl, as well as alkaline pH	[25, 27]
ER stress				
Tunicamycin	*ire1Δ*	Unfolded protein response (UPR) pathway	Impaired growth on 5 mg/mL Congo red, 1.5 mg/mL CFW, as well as during high temperature growth (37 °C)	[30]
DTT	*ire1Δ*	UPR pathway	Impaired growth on 5 mg/mL Congo red, 1.5 mg/mL CFW, as well as during high temperature growth (37 °C)	[30]
Mitochondrion stress				
Rotenone	*mrj1Δ*	Mitochondrial respiration J-domain protein	Impaired growth at 30 °C and 37 °C, as well as reduced growth on 150 mM BPS	[31]
Salicylhydroxamic acid	*mrj1Δ*	Mitochondrial respiration J-domain protein	Impaired growth at 30 °C and 37 °C, as well as reduced growth on 150 mM BPS	[31]
Carboxin	*mar1Δ*	Mitochondrial metabolic activity	Impaired growth at 37 °C, as well as reduced growth on 1.5 M NaCl, 0.5% Congo red, and 0.06% SDS	[32]
Antimycin A	*mar1Δ*	Mitochondrial metabolic activity	Impaired growth at 37 °C, as well as reduced growth on 1.5 M NaCl, 0.5% Congo red, and 0.06% SDS	[32]
Sodium azide	*mar1Δ*	Mitochondrial metabolic activity	Impaired growth at 37 °C, as well as reduced growth on 1.5 M NaCl, 0.5% Congo red, and 0.06% SDS	[32]
Calcineurin/ $CaCl_2$	*cna1Δ*	Calcineurin signaling	Impaired growth at high temperature (37 °C), as well as on 0.03% SDS, 5 mg/mL CFW and 1% Congo red	[18]

(continued)

Table 1
(continued)

Stress assays	Knockout strain	Pathway/function	Additional stress phenotypes	References
Copper toxicity	*mt1Δ* *mt2Δ*	Copper detoxification		[33]
Copper starvation	*ctr1Δ* *ctr4Δ*	Copper uptake		[34]
Iron starvation	*cfo1Δ*	Iron ion uptake	Impaired growth on 1 μg/mL amphotericin B, as well as 10 μg/mL fluconazole	[35]

CFW calcofluor white, *ROS* reactive oxygen species, *MAPK* mitogen-associated protein kinase, *HOG* high osmolarity and glycerol, *NOS* nitric oxide synthase, *SDS* sodium dodecyl sulfate, *cAMP* cyclic adenosine monophosphate, *DTT* 1,4-Dithiothreitol, *UPR* unfolded protein response, *BPS* bathophenanthrolinedisulfonate

1. Grow designated strains on an agar plate as described in Subheading 3.1. Make sure to include the corresponding WT strain.
2. Start and grow overnight culture as described in Subheading 3.1. For the disc diffusion assay, use a 25 mL culture (in sterile 250 mL flasks) in YPD grown for ~16 h at 30 °C, shaking.
3. Measure cell density of the overnight culture as described in Subheading 3.2.
4. Bring all cultures to OD_{600} of ~16 in 10 mL of YPD.
5. Dilute sample 1:1000 in 10 mL of YPD.
6. Spread 500 μL of the diluted sample onto YPD agar plates (*see* Subheading 2.6, **item 1**) using sterile glass beads.
7. Once agar plates are dry, apply drug or compound-soaked filter discs (~15 μL/filter) onto the inoculated agar plates using sterile tweezers. Use sterile 6 mm blank filter discs.
8. Incubate at designated temperatures (RT or 30–39 °C) for 72 h. Image once per day.
9. Determine the area of inhibition (= area around the disc with no cell growth) using the measurement tool in Fiji/ImageJ (*see* **Note 12**).

3.4 Determination of the MIC of a Soluble Compound Using Premade MIC Strips

Common antifungal drugs, such as Fluconazole or Amphotericine B, are available as premade MIC strips (e.g., Biomerieux E-test). These may be used for a quick assessment of drug efficacy changes in different strains compared to WT or in different growth conditions.

1. Prepare cell cultures and prepare plates as described in Subheading 3.1.

2. Once agar plates are dried, apply the MIC strip to the middle of the dried plate using sterile tweezers. Make sure to only touch the top of the MIC strips and to not move the strip once applied to the plate.
3. Incubate at designated temperatures (RT or 30 °C to 39 °C) for 72 h. Image once per day.
4. Determine the MIC by reading the MIC value at complete inhibition of all growth.

3.5 Determination of the MIC in Liquid Cultures: Preparation of Strains

1. Grow designated strains on an agar plate as described in Subheading 3.1. Make sure to include the corresponding WT strain.
2. Start and grow overnight culture as described in Subheading 3.1. For the MIC assay, use a 25 mL culture (in sterile 250 mL flasks) in YPD grown for ~16 h at 30 °C, shaking.
3. Harvest cells and wash once in 25 mL of PBS.
4. Resuspend cells in 25 mL of PBS and measure the OD_{600} as described in Subheading 3.2.
5. Make Suspension I: Dilute overnight cultures to approximately 0.23–0.27 (1–5 × 10^6 cells/mL) in PBS or the medium of choice.
6. Make Suspension II: 1:100 dilution of Suspension I in PBS or the medium of choice (~10^4 cells/mL).
7. Make Suspension III: 1:20 dilution of Suspension II. This will be the working suspension for the MIC tests.
8. Plate 100 μL of Suspension III on YPD to ensure that the cell density is correct (should yield ~100 colonies).

3.6 Determination of the MIC in Liquid Cultures: Preparation of the 10× Drug Plate

Many drugs or compounds are not soluble in H_2O but are soluble in solvents, such as DMSO. These solvents alone may affect growth of *Cryptococcus neoformans*. To make sure that the observed growth inhibitions are not caused by the solvent, a drug-free, solvent-only MIC test must be performed. High concentrations of DMSO inhibit *Cryptococcus neoformans* growth, especially when grown in RPMI or at higher temperatures. Therefore, the final DMSO concentration in the test plate should not exceed 1%. Furthermore, DMSO containing plates may have to be incubated longer than non-DMSO containing plates to achieve proper cell density for measurement. To minimize the influence of DMSO on the MIC data, the DMSO concentration must be kept constant in each sample including the no drug control. To not exceed a DMSO concentration of 1% for drugs or compounds dissolved in DMSO, a 100× drug plate is prepared.

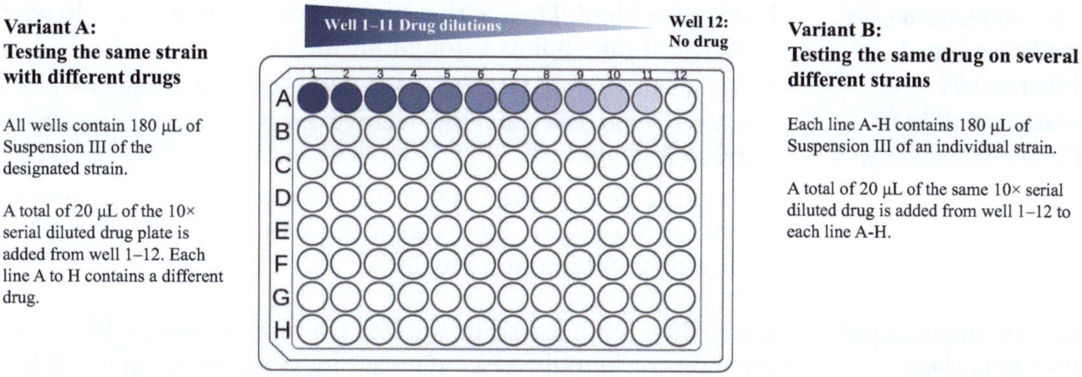

Fig. 1 MIC analysis in liquid cultures. Schematic outline of the test plate for the two described MIC test variant methods

1. Make 10× drug stock(s), minimum 500 μL volume.
2. Aliquot 200 μL of the corresponding solvents into wells 2–12 of a 96-well plate (Variant A) or 11 centrifuge tubes labelled 2–11 (Variant B).
3. Pipet 400 μL of your 10× drug stock solution in well 1 (Variant A only).
4. Make 1:2 serial dilutions:

   Variant A: Transfer 200 μL from well 1 into well 2 and mix by pipetting. Proceed with the 1:2 serial dilution using 200 μL for each well through well 11. Do not add drug to well 12.

   Variant B: Transfer 200 μL of your 10× stock solution into tube 2 and mix by pipetting. Proceed with the 1:2 serial dilution using 200 μL for each tube through tube 11. Do not add drug to tube 12.

3.7 Determination of the MIC in Liquid Cultures: Preparation of Test Plate (Fig. 1)

1. Aliquot suspension III into the 96-well plate:

   Variant A: Transfer 180 μL of Suspension III of the same strain into all wells.

   Variant B: Transfer 180 μL of Suspension III of a different strain in each line (Strain A: wells A1–12, Strain B, wells B1–12).
2. Add 20 μL of the 10× drug serial dilution:

   Variant A: Using a multi-channel pipet, transfer 20 μL from each line of the 10× drug plate to the corresponding line of the test plate and mix by pipetting.

   Variant B: Transfer 20 uL from the 10× drug dilution tubes into the corresponding numbered well and mix by pipetting (Tube 1 into A1, B1...H1; Tube 2 into A2, B2, ... H2 etc.).
3. Incubate at the designated temperature 48–72 h. Observe growth in each well.

3.8 Determination of the MIC in Liquid Cultures: MIC Measurement Using Cell Density at OD_{600}

1. Measure the OD_{600} with a plate reader. This may be done at different time points throughout the experiment.
2. To calculate the MIC, set the no drug control sample to 100% growth and calculate the relative growth rate of the drug-treated samples.
3. Plot the growth data against its corresponding drug concentration to calculate 50% of growth inhibition (MIC50) or 80% of growth inhibition (MIC80).

3.9 Determination of the MIC in Liquid Cultures: MIC Measurement of Viability Using Alamar Blue

Alamar Blue is a fluorescent viability dye that measures the reduction of resazurin to fluorescent resourfin. It works well at 37 °C but may be inconsistent at 30 °C. It does not work well at low pH or in YPD. Since Alamar Blue is measured by fluorescence, it is best to set up the test plate using 96-well black plates that are clear on the bottom.

1. After 48 h of incubation of the MIC experiment, add 20 μL of Alamar Blue to each well and mix by pipetting.
2. Incubate for 24 h at the designated temperature (the optimum temperature is 37 °C).
3. Measure fluorescence using a plate reader (Excitation 530–560 nm, Emission 590 nm).
4. To calculate the MIC, set your no drug control sample to 100% growth and calculate the relative growth rate of the drug-treated samples.
5. Plot the growth data against its corresponding drug concentration to calculate MIC50 or MIC80.

3.10 Plate-Based Melanin Assay

Melanin is a brown pigment formed by *Cryptococcus neoformans* using host catecholamines (e.g., L-Dopamine or L-DOPA or other diphenols) as substrate. Melanin formation is indicated by a brown coloration of the colony.

1. Make L-DOPA agar plates as described in Subheading 2.6, **item 8**.
2. Grow designated strains on an agar plate as described in Subheading 3.1. Make sure to include the corresponding WT strain and a control knockout strain (Table 2).
3. Start and grow overnight culture as described in Subheading 3.1. For the Melanin assay, a 5 mL culture (in 12 mL sterile PP tubes) in YPD or SC is suitable.
4. Harvest overnight culture and wash once with 5 mL of PBS.
5. Resuspend in 1 mL of PBS and measure cell density as described in Subheading 3.2.
6. Dilute to an OD_{600} of 2.5 in PBS.

Table 2
Selected *Cryptococcus neoformans* gene knockout strains with known defects in stress phenotypes

Stress phenotype	Knockout strain	Pathway/function	References
Melanin			
No melanin formation	*lac1Δ*	Melanin biosynthesis	[36]
Leaky melanin	*chs3Δ*	Chitosan biosynthesis	[26]
Capsule			
Reduced capsule (India ink)	*cap59Δ*		[37]
Increased capsule (India ink)	*hog1Δ*	MAPK of the HOG pathway	[22]
Urease activity	*ure1Δ*	Urease 1	[38]
Protease activity			[39]
Titan cells			
No Titan cell formation	*rim101Δ*	pH sensing transcription factor	[40]
Increased formation of Titan cells	*cln1Δ*	Cyclin	[41]

MAPK mitogen-associated protein kinase, *HOG* high osmolarity and glycerol

7. Spot 5 μL of the diluted sample on the L-DOPA agar. Up to four strains per plate may be analyzed.
8. Incubate at either 30 °C or 37 °C for 72 h. Image once per day.

3.11 Melanin Assay in Liquid Culture

1. In 14 mL culture tubes, inoculate the WT and mutant strains in 4 mL of YPD. Grow them at the desired temperature, with shaking.
2. Collect the cells by centrifugation at 3330× *g* for 10 min at 25 °C at the end of the growth period (24–36 h).
3. Wash the cell pellet twice with PBS, pH 7.4.
4. Add 4 mL of L-DOPA liquid medium and incubate the cells for 36–48 h at 30 °C.
5. Examine the color change in both the cells and the supernatant. The absorption at 400 nm may be used to quantify the color of the supernatant.

3.12 Capsule Assay

In response to host environmental stresses, *Cryptococcus neoformans* produces a polysaccharide capsule, which is attached to its outer cell wall.

Quantification via India Ink Staining

India ink is a negative stain and visualizes the capsule as a light ring surrounding the cell body.

1. Grow designated strains on an agar plate as described in Subheading 3.1. Make sure to include the corresponding WT strain and a control knockout strain (Table 2).

3.13 Capsule Induction Using CO_2-Independent Medium (No CO_2-Incubator Required)

2. Start and grow overnight cultures as described in Subheading 3.1. For the capsule assay, a 5 mL culture (in 12 mL sterile PP tubes) in YPD or SC is suitable.
3. Harvest overnight culture and wash twice with 5 mL of PBS.
4. Resuspend in 1 mL of CO_2-independent medium (CIM).
5. Measure cell density as described in Subheading 3.2.
6. Inoculate 5 mL of CIM (in sterile 25 mL flasks; alternatively, sterile 12 mL PP tubes may be used) to an OD_{600} of 0.1 with the PBS-washed cells.
7. Incubate for 3 days at 37 °C, shaking.
8. Harvest cells and resuspend in ~100 μL of PBS.
9. Mix cell suspension 1:4 with an India ink solution.
10. Analyze microscopically using the differential interference contrast (DIC) channel.
11. Images taken using the microscope are used for quantification of the capsule thickness by measuring the cell body diameter, as well as the cell diameter including the capsule using the measurement tool in Fiji/ImageJ. Capsule thickness is calculated as follows:

$$\text{Capsule thickness} = \frac{(\text{Cell diameter including capsule} - \text{Cell body diameter})}{2}$$

3.14 Capsule Induction Using RPMI Medium +10% FBS (CO_2-Incubator Required)

1. Grow designated strains on an agar plate as described in Subheading 3.1. Make sure to include the corresponding WT strain and a control knockout strain.
2. Start and grow overnight culture as described in Subheading 3.1. For the capsule assay, a 5 mL culture (in 12 mL sterile PP tubes) in YPD or SC is suitable.
3. Harvest overnight culture and wash twice with 5 mL of PBS.
4. Measure cell density using a hemocytometer or automated TC cell counter.
5. Suspend cells in RPMI medium containing 10% FBS at a cell density of 500 cells/μL and incubate at 37 °C in the presence of 5% CO_2 for 4 days.
6. Prepare India ink sample by mixing 8 μL of the cell suspension with 1:4 India ink solution (in H_2O).
7. Image and quantify capsule thickness as described in Subheading 3.8, **step 11**.

3.15 Immuno-staining of Capsule

1. Grow the WT and the corresponding mutant, as well as the positive control strain (e.g., *cap59Δ*) at the desired media and growth conditions.
2. Adjust the cell suspension to an OD_{600} of 1.0.
3. Transfer 1 mL of the $OD_{600} = 1.0$ cells to a 1.5 mL Eppendorf tube. Fix the cells by incubating them in PBS containing 4% paraformaldehyde at 4 °C for 30 min.
4. Wash the cells three times in PBS at 4 °C. Resuspend the cells in 1 mL of PBS containing 2% BSA. Block the cells for 30 min at RT.
5. Collect the cells by centrifugation (8000× *g* for 10 min at 4 °C). Resuspend the cells in PBS containing capsule-specific antibody (18B7 or 3C2) at a concentration of 1 μg/mL. Incubate at 37 °C for 30 min with mixing.
6. Collect the cells by centrifugation (8000× *g* for 10 min at 4 °C) and wash them three times with PBS.
7. Incubate the cells with a fluorescently conjugated mouse secondary antibody of choice for 30 min at 37 °C in PBS.
8. After the incubation, wash the cells three times with PBS.
9. Spot 5–8 μL of the cell suspension onto a microscope slide and image using a fluorescent microscope.

3.16 Urease Assay

1. Make Urease agar plates as described in Subheading 2.6, **item 9**.
2. Grow designated strains on an agar plate as described in Subheading 3.1. Make sure to include the corresponding WT strain and a control knockout strain.
3. Start and grow overnight culture as described in Subheading 3.1. For the Urease assay, a 5 mL culture (in 12 mL sterile PP tubes) in YPD or SC is suitable.
4. Harvest overnight culture and wash once with 5 mL of PBS.
5. Resuspend in 1 mL of PBS and measure the cell density as described in Subheading 3.2.
6. Dilute to an OD_{600} of 2.5 in PBS.
7. Spot 5 μL of the diluted sample on a Urease agar plate. Up to four strains per plate may be analyzed.
8. Incubate at either 30 °C or 37 °C for 72 h. Image once per day.
9. Urease positive strains show a pink zone around the colony. The size of the zone may be quantified using the measurement tool of Fiji/ImageJ.

3.17 Protease Assay

1. Prepare an azoalbumin plate in YNB agar with 0.1% (w/v) azoalbumin.
2. Grow the cells as described in Subheading 3.16, spot them on an azoalbumin plate, and incubate at 30 °C for 7 days.
3. The presence of secreted protease is indicated by the appearance of a clear zone around the colony.

3.18 In Vitro Titan Cell Formation

Titan cells are enlarged yeast cells that were first observed in cells harvested from lungs (16). Several groups have established in vitro conditions to induce and produce Titan cells outside of the lung environment. Nutritional starvation and low cell density are a common feature of each of these induction protocols. This method is adapted from [16].

1. Grow designated strains on an agar plate as described in Subheading 3.1. Make sure to include the corresponding WT strain and a control knockout strain (Table 2).
2. Start and grow overnight culture as described in Subheading 3.1. Overnight culture must be performed in YNB medium (20 mL in a sterile 250 mL flask).
3. Harvest overnight culture and wash once with 20 mL of PBS.
4. Resuspend cells in 20 mL of PBS and measure the OD_{600} as described in Subheading 3.2.
5. Inoculate 10 mL of PBS + 10% heat-inactivated FBS to an OD_{600} of 0.001. Aliquot 10 mL of cell suspension into a well of a 6-well plate.
6. Incubate plate for 24–72 h at 37 °C and 5% CO_2.
7. To analyze Titan cell formation, harvest cells at the designated time point, resuspend in ~20 μL of PBS, and visualize cells using a microscope on the DIC channel. For better contrast, an India ink stain may be performed (*see* Subheading 3.12).
8. Cell size may be measured from images taken at the microscope using the measurement tool in Fiji/ImageJ.

3.19 Macrophage Killing of C. neoformans: General Remarks for Handling Macrophages

In this section, yeast-macrophage interaction methods are outlined using J774A.1 macrophage-like cells (Fig. 2). If other macrophage cell lines or primary cells are used, the growth conditions will need to be adapted to accommodate the different growth requirements.

1. Cells are incubated in a humidified 37 °C incubator with 5% CO_2.
2. Tissue culture flasks require 20–25 mL of medium; never let cells dry out.
3. Washing cells requires a gentle flow. Vigorous washing disrupts the monolayer.
4. Growth medium: DMEM +10% heat inactivated FBS, 1× penicillium-streptomycin, 1% nonessential amino acid mix.

5. Cells require passage twice weekly (or when 80% confluent). To passage cells, remove old medium and wash twice with prewarmed PBS (37 °C). Add 10 mL of prewarmed fresh growth medium and scrape cells from flask using a cell scraper. Add 500 μL of the cell suspension into a new flask with 15 mL of prewarmed growth medium.

3.20 Macrophage Killing of C. neoformans: Preparing Activated Macrophages for Co-culture

1. When cells are confluent, remove old medium and wash twice with prewarmed PBS.
2. Add 10 mL of prewarmed fresh growth medium and scrape cells from flask using a cell scraper.
3. Add 500 μL of the cell suspension into a new flask with 15 mL of prewarmed medium to maintain the culture.
4. Transfer the residual cell suspension into a 50 mL reaction tube and centrifuge at 900× *g* for 10 min at 4 °C.
5. Gently resuspend in 2 mL of prewarmed growth medium.
6. Measure cell density: Mix 20 μL of the cell suspension with 180 μL of trypan blue (dead cells are blue) and count viable cells using a hemocytometer. Make 1 mL of a 1:100 dilution (or higher dilution depending on the initial cell density) in PBS. Apply 10–12 μL of diluted cell suspension between the cover slip and hemocytometer. Count cells located in the four outer quadrants of the hemocytometer (*n*) and calculate the average cell number (*n*/4). Total cell number = average cell number (*n*) × dilution factor (e.g., 100) × 10^4/mL.
7. Add 1 × 10^5 viable cells in 100 μL of culture medium to each well of a 96-well tissue culture plate.
8. Activate macrophages by adding 100 u/mL INF-γ and 0.6 μg/mL LPS.
9. Incubate overnight at 37 °C with 5% CO_2.

3.21 Macrophage Killing of C. neoformans: Preparing Yeast Cells for Co-culture

1. Grow designated strains on an agar plate as described in Subheading 3.1. Make sure to include the corresponding WT strain.
2. Start and grow overnight culture as described in Subheading 3.1 in 20 mL of YPD at 30 °C. The overnight culture has to be started on the same day the macrophages are activated.
3. Wash the overnight culture twice with PBS.
4. Resuspend yeast cells in 10–20 mL of macrophage growth medium.
5. Measure cell density of a 1:100 dilution using a hemocytometer.
6. Set the cell density to 1 × 10^6 cells/mL (for a 1:1 co-culture).

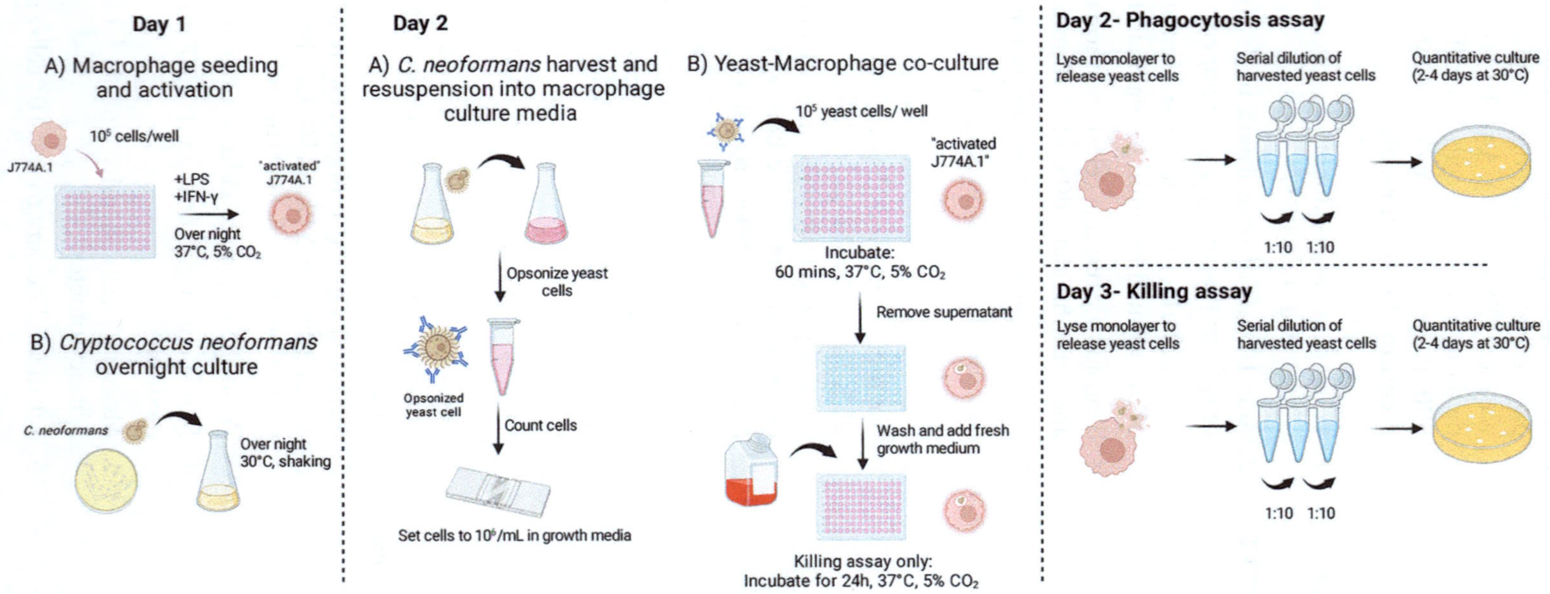

Fig. 2 Macrophage assay. Experimental outline of the phagocytosis and killing assay using J744A.1 macrophages and opsonized *Cryptococcus neoformans* cells

7. Opsonize yeast cells for 1 h at RT with mAb 18B7. Use 0.5 μL of 1.98 mg/mL stock for 1 mL yeast cell suspension (1 μg/mL 18B7).
8. Spread 100 μL of the yeast culture on to an YPD plate for quantitative culture to verify cell count. Incubate plate for 2–3 days at 30 °C. Count the colonies.

3.22 Macrophage Killing of C. neoformans: Macrophage Killing Assay

1. Remove the old macrophage growth medium from the previously prepared 96-well plate.
2. Add 100 μL of previously prepared yeast cell culture.
3. Incubate plate for 1 h at 37 °C with 5% CO_2.
4. Following incubation, gently wash wells once with PBS to remove extracellular yeast without disrupting the monolayer at the bottom of the wells.
5. Add fresh growth medium to each well.
6. Incubate plate for 24 h at 37 °C with 5% CO_2.
7. Add 200 μL of sterile, distilled H_2O to disrupt the macrophage monolayer. Incubate 5 min at RT.
8. Transfer the macrophage cell suspension into a 1.5 mL reaction tube, scrape the bottom of the well, and pipette up and down to further disrupt the monolayer.
9. Repeat **step** 7 twice and combine all washes from one well into a single reaction tube. Add 400 μL sterile H_2O to bring sample to a total volume of 1 mL.
10. Perform a 2-step 1:10 serial dilution (final dilution will be 1: 100).
11. Plate 100 μL onto an YPD plate for quantitative culture. Incubate plate for 2–3 days at 30 °C.
12. Once colonies are visible, count cells to calculate colony forming units for each co-culture.

3.23 Phagocytosis Efficiency

1. Perform co-culture as described in Subheadings 3.20 and 3.21.
2. Following incubation (Subheading 3.22, **step 3**), wash wells twice with PBS to remove extracellular yeast without disrupting the monolayer on the bottom of the well.
3. Phagocytosed yeast cells are immediately harvested and plated for quantitative culture as described in Subheading 3.22, **steps 10–12**.

4 Notes

1. For 500 mL of YPD agar, components of solution A are dissolved in 450 mL of H_2O. After autoclaving, 50 mL of solution B is added to 450 mL of solution A to achieve YPD agar containing 2% glucose.

2. For 1 M succinate, add 27 g of sodium succinate dibasic hexahydrate ($C_4H_4Na_2O_4.6H_2O$) to 90 mL of H_2O. Adjust the pH to 4.0 using HCl. Bring the volume to 100 mL with deionized H_2O. Store at RT.
3. For 1 M MOPS, add 20.9 g of MOPS ($C_7H_{15}NO_4S$) to 75 mL of H_2O. Adjust the pH to 7.0 using NaOH. Bring the volume to 100 mL with deionized H_2O. Store at 4 °C up to 4 weeks.
4. Hydrogen peroxide: Make a 200 mM stock of H_2O_2 by adding 230 μL of 30% H_2O_2 to 10 mL of H_2O. Store at 4 °C up to 7 days.
5. Sodium Nitrite: Make a stock of 100 mM $NaNO_2$ in H_2O. Make a fresh solution of $NaNO_2$ each time.
6. Oxidative and nitrosative stress tests may also be performed on YPD agar or in liquid media. However, depending on the culture density, cells grown in different media have varying capacities to degrade exogenous H_2O_2/$NaNO_2$. Higher stressor concentrations are recommended when using YPD media.
7. Example: 50 mL of 15 mg/mL CFW in 20% glucose is added to 450 mL of YPD-agar solution to achieve YPD + 1.5 mg/mL CFW agar plates.
8. 1 M sorbitol may also be added to different cell wall stress plates to determine if the cell wall stress sensitivity phenotype of a specific mutant strain may be rescued, indicating that the sensitivity phenotype is specific to the compromised cell wall integrity.
9. YPD medium is restrictive in copper. If metal stress assays are conducted in YPD medium, higher copper concentrations have to be used to induce a copper toxicity phenotype (2.5 mM $CuSO_4$) and lower metal chelator concentrations (BCS: 0.1–0.5 mM).
10. Other growth medium or temperatures may be used for certain growth or phenotypic analysis. Certain knockout strains may require longer growth times.
11. A metal prong replicator pin may also be used for spotting the serially diluted cell suspension onto the stress plate for a high throughput type of screen. In this case, three serially diluted suspensions are sufficient to determine whether there is a difference in the strain's sensitivity to a stressor compared to that of the WT.
12. If RPMI agar plates (*see* Subheading 2.6, **item** 7) are used instead of YPD agar plates, wash cells once with 25 mL of PBS prior to measurement at OD_{600} and perform the dilution in PBS instead of YPD.

Acknowledgments

This work was supported by NIH-NIAID grants R01AI072195 and R01AI125045 to JKL.

References

1. Idnurm A, Bahn YS, Nielsen K et al (2005) Deciphering the model pathogenic fungus *Cryptococcus neoformans*. Nat Rev Microbiol 3(10):753–764. https://doi.org/10.1038/nrmicro1245
2. Lin X, Heitman J (2006) The biology of the *Cryptococcus neoformans* species complex. Annu Rev Microbiol 60:69–105. https://doi.org/10.1146/annurev.micro.60.080805.142102
3. Kronstad JW, Attarian R, Cadieux B et al (2011) Expanding fungal pathogenesis: *Cryptococcus* breaks out of the opportunistic box. Nat Rev Microbiol 9(3):193–203. https://doi.org/10.1038/nrmicro2522
4. Hull CM, Heitman J (2002) Genetics of *Cryptococcus neoformans*. Annu Rev Genet 36:557–615. https://doi.org/10.1146/annurev.genet.36.052402.152652
5. May RC, Stone NR, Wiesner DL et al (2016) *Cryptococcus*: from environmental saprophyte to global pathogen. Nat Rev Microbiol 14(2):106–117. https://doi.org/10.1038/nrmicro.2015.6
6. Bahn YS, Jung KW (2013) Stress signaling pathways for the pathogenicity of *Cryptococcus*. Eukaryot Cell 12(12):1564–1577. https://doi.org/10.1128/EC.00218-13
7. Thak EJ, Son YJ, Lee DJ et al (2022) Extension of O-linked mannosylation in the golgi apparatus is critical for cell wall integrity signaling and interaction with host cells in *Cryptococcus neoformans* pathogenesis. MBio 13(6):e0211222. https://doi.org/10.1128/mbio.02112-22
8. Lengeler KB, Davidson RC, D'Souza C et al (2000) Signal transduction cascades regulating fungal development and virulence. Microbiol Mol Biol Rev 64(4):746–785. https://doi.org/10.1128/MMBR.64.4.746-785.2000
9. Gerik KJ, Donlin MJ, Soto CE et al (2005) Cell wall integrity is dependent on the PKC1 signal transduction pathway in *Cryptococcus neoformans*. Mol Microbiol 58(2):393–408. https://doi.org/10.1111/j.1365-2958.2005.04843.x
10. Missall TA, Pusateri ME, Lodge JK (2004) Thiol peroxidase is critical for virulence and resistance to nitric oxide and peroxide in the fungal pathogen, *Cryptococcus neoformans*. Mol Microbiol 51(5):1447–1458. https://doi.org/10.1111/j.1365-2958.2004.03921.x
11. Rutherford JC, Bahn YS, van den Berg B et al (2019) Nutrient and stress sensing in pathogenic yeasts. Front Microbiol 10:442. https://doi.org/10.3389/fmicb.2019.00442
12. Kronstad J, Saikia S, Nielson ED et al (2012) Adaptation of *Cryptococcus neoformans* to mammalian hosts: integrated regulation of metabolism and virulence. Eukaryot Cell 11(2):109–118. https://doi.org/10.1128/EC.05273-11
13. Fan W, Kraus PR, Boily MJ et al (2005) *Cryptococcus neoformans* gene expression during murine macrophage infection. Eukaryot Cell 4(8):1420–1433. https://doi.org/10.1128/EC.4.8.1420-1433.2005
14. O'Meara TR, Holmer SM, Selvig K et al (2013) *Cryptococcus neoformans* Rim101 is associated with cell wall remodeling and evasion of the host immune responses. MBio 4(1):e00522-12. https://doi.org/10.1128/mBio.00522-12
15. Upadhya R, Baker LG, Lam WC et al (2018) *Cryptococcus neoformans* Cda1 and its chitin deacetylase activity are required for fungal pathogenesis. MBio 9(6):e02087-18. https://doi.org/10.1128/mBio.02087-18
16. Dambuza IM, Drake T, Chapuis A et al (2018) The *Cryptococcus neoformans* Titan cell is an inducible and regulated morphotype underlying pathogenesis. PLoS Pathog 14(5):e1006978. https://doi.org/10.1371/journal.ppat.1006978
17. Petzold EW, Himmelreich U, Mylonakis E et al (2006) Characterization and regulation of the trehalose synthesis pathway and its importance in the pathogenicity of *Cryptococcus neoformans*. Infect Immun 74(10):5877–5887. https://doi.org/10.1128/iai.00624-06
18. Chow EWL, Clancey SA, Billmyre RB et al (2017) Elucidation of the calcineurin-Crz1 stress response transcriptional network in the human fungal pathogen *Cryptococcus neoformans*. PLoS Genet 13(4):e1006667. https://doi.org/10.1371/journal.pgen.1006667
19. Ballou ER, Kozubowski L, Nichols CB et al (2013) Ras1 acts through duplicated Cdc42 and Rac proteins to regulate morphogenesis

and pathogenesis in the human fungal pathogen *Cryptococcus neoformans*. PLoS Genet 9(8):e1003687. https://doi.org/10.1371/journal.pgen.1003687

20. Kim D, Song M, Do E et al (2021) Oxidative stress causes vacuolar fragmentation in the human fungal pathogen *Cryptococcus neoformans*. J Fungi 7(7):523. https://doi.org/10.3390/jof7070523
21. Missall TA, Cherry-Harris JF, Lodge JK (2005) Two glutathione peroxidases in the fungal pathogen *Cryptococcus neoformans* are expressed in the presence of specific substrates. Microbiology 151(Pt 8):2573–2581. https://doi.org/10.1099/mic.0.28132-0
22. Bahn YS, Kojima K, Cox GM et al (2005) Specialization of the HOG pathway and its impact on differentiation and virulence of *Cryptococcus neoformans*. Mol Biol Cell 16(5):2285–2300. https://doi.org/10.1091/mbc.e04-11-0987
23. Missall TA, Pusateri ME, Donlin MJ et al (2006) Posttranslational, translational, and transcriptional responses to nitric oxide stress in *Cryptococcus neoformans*: implications for virulence. Eukaryot Cell 5(3):518–529. https://doi.org/10.1128/ec.5.3.518-529.2006
24. O'Meara TR, Xu W, Selvig KM et al (2014) The *Cryptococcus neoformans* Rim101 transcription factor directly regulates genes required for adaptation to the host. Mol Cell Biol 34(4):673–684. https://doi.org/10.1128/MCB.01359-13
25. Brown HE, Telzrow CL, Saelens JW et al (2020) Sterol-response pathways mediate alkaline survival in diverse fungi. MBio 11(3):e00719-20. https://doi.org/10.1128/mBio.00719-20
26. Baker LG, Specht CA, Donlin MJ et al (2007) Chitosan, the deacetylated form of chitin, is necessary for cell wall integrity in *Cryptococcus neoformans*. Eukaryot Cell 6(5):855–867. https://doi.org/10.1128/ec.00399-06
27. Chang YC, Bien CM, Lee H et al (2007) Sre1p, a regulator of oxygen sensing and sterol homeostasis, is required for virulence in *Cryptococcus neoformans*. Mol Microbiol 64(3):614–629. https://doi.org/10.1111/j.1365-2958.2007.05676.x
28. Kim SY, Ko YJ, Jung KW et al (2011) Hrk1 plays both Hog1-dependent and -independent roles in controlling stress response and antifungal drug resistance in *Cryptococcus neoformans*. PLoS One 6(4):e18769. https://doi.org/10.1371/journal.pone.0018769
29. Cramer KL, Gerrald QD, Nichols CB et al (2006) Transcription factor Nrg1 mediates capsule formation, stress response, and pathogenesis in *Cryptococcus neoformans*. Eukaryot Cell 5(7):1147–1156. https://doi.org/10.1128/EC.00145-06
30. Cheon SA, Jung K-W, Chen Y-L et al (2011) Unique evolution of the UPR pathway with a novel bZIP transcription factor, Hxl1, for controlling pathogenicity of *Cryptococcus neoformans*. PLoS Pathog 7(8):e1002177. https://doi.org/10.1371/journal.ppat.1002177
31. Horianopoulos LC, Hu G, Caza M et al (2020) The novel J-domain protein Mrj1 is required for mitochondrial respiration and virulence in *Cryptococcus neoformans*. MBio 11(3):e01127-20. https://doi.org/10.1128/mBio.01127-20
32. Telzrow CL, Esher Righi S, Cathey JM et al (2023) *Cryptococcus neoformans* Mar1 function links mitochondrial metabolism, oxidative stress, and antifungal tolerance. Front Physiol 14:1150272. https://doi.org/10.3389/fphys.2023.1150272
33. Ding C, Festa RA, Chen YL et al (2013) *Cryptococcus neoformans* copper detoxification machinery is critical for fungal virulence. Cell Host Microbe 13(3):265–276. https://doi.org/10.1016/j.chom.2013.02.002
34. Sun T-S, Ju X, Gao H-L et al (2014) Reciprocal functions of *Cryptococcus neoformans* copper homeostasis machinery during pulmonary infection and meningoencephalitis. Nat Commun 5(1):5550. https://doi.org/10.1038/ncomms6550
35. Jung WH, Hu G, Kuo W et al (2009) Role of ferroxidases in iron uptake and virulence of *Cryptococcus neoformans*. Eurkaryot Cell 8(10):1511–1520. https://doi.org/10.1128/EC.00166-09
36. Walton FJ, Idnurm A, Heitman J (2005) Novel gene functions required for melanization of the human pathogen *Cryptococcus neoformans*. Mol Microbiol 57(5):1381–1396. https://doi.org/10.1111/j.1365-2958.2005.04779.x
37. García-Rivera J, Chang YC, Kwon-Chung KJ et al (2004) *Cryptococcus neoformans* CAP59 (or Cap59p) is involved in the extracellular trafficking of capsular glucuronoxylomannan. Eukaryot Cell 3(2):385–392. https://doi.org/10.1128/ec.3.2.385-392.2004
38. Singh A, Panting RJ, Varma A et al (2013) Factors required for activation of urease as a virulence determinant in *Cryptococcus*

neoformans. MBio 4(3):e00220-13. https://doi.org/10.1128/mBio.00220-13

39. Chen LC, Blank ES, Casadevall A (1996) Extracellular proteinase activity of *Cryptococcus neoformans*. Clin Diagn Lab Immunol 3(5): 570–574. https://doi.org/10.1128/cdli.3.5.570-574.1996
40. Okagaki LH, Wang Y, Ballou ER et al (2011) Cryptococcal titan cell formation is regulated by G-protein signaling in response to multiple stimuli. Eukaryot Cell 10(10):1306–1316. https://doi.org/10.1128/ec.05179-11
41. Altamirano S, Li Z, Fu MS et al (2021) The cyclin Cln1 controls polyploid titan cell formation following a stress-induced G(2) arrest in *Cryptococcus*. MBio 12(5):e0250921. https://doi.org/10.1128/mBio.02509-21

Part VII

Measuring Specific Phenotypes

Chapter 20

Antibody Isolation in *C. neoformans*

Susana Ruiz Mendoza, Marina da Silva Ferreira, Michele Ramos Valente, and Allan Jefferson Guimarães

Abstract

The importance of humoral immunity to fungal infections remains to be elucidated. In cryptococcosis, patients that fail to generate antibodies against antigens of the fungus *Cryptococcus neoformans* are more susceptible to the disease, demonstrating the importance of these molecules to the antifungal immune response. Historically, antibodies against *C. neoformans* have been applied in diagnosis, therapeutics, and as important research tools to elucidate fungal biology. Throughout the process of generating monoclonal antibodies (mAbs) from a single B-cell clone and targeting a single epitope, several immunization steps might be required for the detection of responsive antibodies to the antigen of interest in the serum. This complex mixture of antibodies comprises the polyclonal antibodies. To obtain mAbs, B-lymphocytes are harvested (from spleen or peripheral blood) and fused with tumor myeloma cells, to generate hybridomas that are individually cloned and specifically screened for mAb production. In this chapter, we describe all the necessary steps, from the immunization to polyclonal antibody harvesting, hybridoma generation, and mAb production and purification. Additionally, we discuss new cutting-edge approaches for generating interspecies mAbs, such as humanized mAbs, or for similar species in distinct host backgrounds.

Key words Antibodies, *Cryptococcus neoformans*, Capsule, Melanin, Polyclonal antibodies, Monoclonal antibodies

1 Introduction

Cryptococcus neoformans is an opportunistic pathogenic fungus of worldwide distribution that commonly causes cryptococcosis in immunocompromised hosts. Upon inhalation of fungal basidiospores or desiccated viable yeasts, fungi reach the lung where it may cause pneumonia and often disseminates to other organs, including the central nervous system, resulting in meningoencephalitis [1].

The host immune response starts with innate immune cell recognition of yeasts through pattern recognition receptors (PRRs), including Dectin-1, Toll-like receptors (TLRs), mannose receptors, and also CD14 [2, 3]. However, infectious propagules promptly express a polysaccharide capsule mainly composed of

Erin E. McClelland (ed.), *Cryptococcus neoformans: Methods and Protocols*, Methods in Molecular Biology, vol. 2775, https://doi.org/10.1007/978-1-0716-3722-7_20,

glucuronoxylomannan (GXM), glucuronoxylomanogalactan (GXMGal), and mannoproteins that hinder fungal recognition by alveolar macrophages and inhibit proinflammatory cell recruitment and the expression of costimulatory molecules and T-cell differentiation (Th1), as well as stimulate regulatory T cells (Tregs) to secrete anti-inflammatory cytokines. Thus, opsonization with antibodies or complement proteins might be required for successful phagocytosis by alveolar macrophages, which recruit monocytes and dendritic cells (DCs) through the secretion of macrophage chemotactic protein (MCP-1). DCs are the main antigen presenting cells (APCs) and present cryptococcal antigens through major histocompatibility complex (MHC) class II. This initiates the adaptive immune response through $CD4^+$ T-cell stimulation, which further differentiates into effector subsets, such as T helper 1 (Th1), Th2, and Th17 cells, depending on the cytokine profile [4]. T-helper cells in turn also induce B cell maturation and clonal expansion, resulting in antibody production, as well as activate inflammatory macrophages for infection resolution [1, 4].

Antibodies (Abs) are secreted versions of the antigenic receptors of B lymphocytes and are important glycoproteins. They are structurally composed of two symmetric axes, made of two identical heavy- and two light-chains. Heavy and light-chains are combined, forming a variable region bearing the antigen binding site and a constant region (Fc), which is the effector domain and determines the classification into different isotypes (or subclasses). These isotypes determine antibody effector function, that is, how antibodies interact with the wide variety of immune cells, and therefore enable antigen recognition (Fig. 1a) [5].

Classically, antibody preparations may be divided into polyclonal and monoclonal antibodies (mAbs). Polyclonal antibodies, or hyper immune serum, are a mixture of antibodies generated by several B cell clones and are thus composed of a wide range of antibodies with distinct recognition capacities to multiple epitopes, or distinct specificities. In contrast, mAbs are originated from a single B cell clone, having unique epitope recognition and specificity [6]. The technology for the generation of mAbs was described by Kohler and Milstein in 1975, providing the possibility to create specific antibodies for immunotherapeutic evaluations against several infectious diseases [7]. Briefly, B lymphocytes from immunized mice are fused with immortal myeloma cells to obtain a cell culture with the capacity to produce antibodies with a single specificity (mAbs) and to endlessly proliferate [7]. Starting in 1980, many murine mAbs, most from the γ-immunoglobulin (IgG) class, were clinically developed (suffix -omab) [8] and were often related to allergic reactions and the induction of anti-drug antibodies (ADAs) (Fig. 1b) [9]. Later, these limitations were overcome with the initial development of chimeric mouse-human antibodies (suffix -ximab), where the variable domain of a mouse antibody has been fully

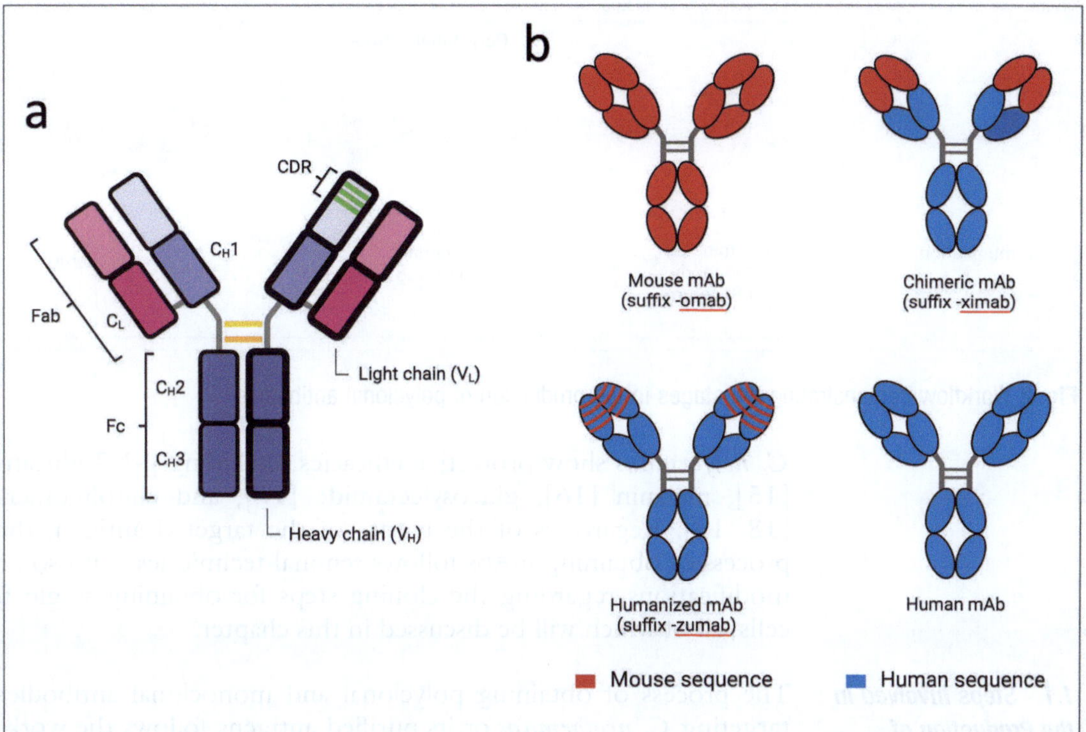

Fig. 1 (**a**) Antibody structure. An immunoglobulin G (IgG) composed of two pairs of light chain and heavy chain connected by disulfide bonds. The heavy chain contains the variable domains (V_H) and three constant domains (C_H1, C_H2, and C_H3). The light chain contains a variable domain (V_L) and a constant domain (C_L). Overall, the quaternary structure of the immunoglobulin may be divided in fragment antigen-binding (F_{AB}) and fragment crystallizable (F_C) regions. The variable domains may be subdivided into hypervariable regions (or complementary determining regions, CDR). (**b**) Nomenclature and types of monoclonal antibodies (mAb). Therapeutic mAbs are classified into mouse mAb (−omab), chimerics (−ximab), humanized (−zumab), and human (−umab). (Modified from [9])

inserted in an expression vector system containing the constant domain of a human antibody [10] (Fig. 1b). Despite the extended half-life in humans and lower immunogenicity, considerable ADAs were observed. For this reason, humanized mAbs (suffix -zumab) have been proposed, where the hypervariable regions (complementarity determining regions (CDRs)) of the mouse antibody have been inserted into new human antibody vectors, resulting in the expression of antibodies containing 95% human content, and with equal specificity to the parental murine antibody [11].

In the setting of cryptococcosis, humoral responses have been correlated with protection [12]. More specifically, mAbs against GXM, the major capsular antigen and virulence factor of *C. neoformans*, have offered protective, nonprotective, and disease-enhancing activities in animal models [13, 14]. In addition, mAbs to other important antigens and virulence factors of

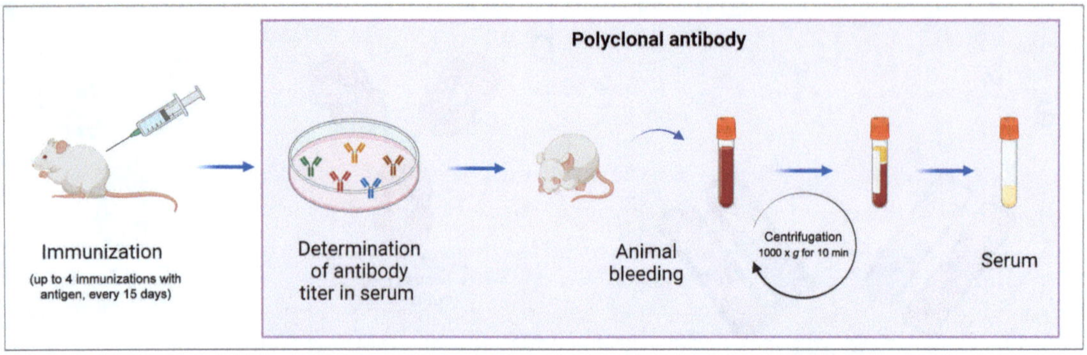

Fig. 2 Workflow demonstrating the stages in the production of polyclonal antibodies

C. neoformans show protective efficacies, including β-1,3-glucans [15], melanin [16], glucosylceramide [17], and chitoligomers [18, 19]. Regardless of the nature of the targeted antigen, the process of obtaining mAbs follows seminal techniques, with some modifications regarding the cloning steps for obtaining single B cells, all of which will be discussed in this chapter.

1.1 Steps Involved in the Production of Antibodies Against C. neoformans

The process of obtaining polyclonal and monoclonal antibodies targeting *C. neoformans* or its purified antigens follows the workflows shown in Figs. 2 and 3, respectively. It usually starts with immunization steps where the antibody response in serum may be evaluated by several protocols, among which the enzyme-linked immunosorbent assay (ELISA) is the most widely used. Upon detection of antibodies in the serum that recognize the target antigen, polyclonal antibodies may be obtained by simply bleeding the animals (Fig. 2), after a centrifugation step to separate red blood cells and collecting polyclonal serum. Additional steps are required to produce mAbs (Fig. 3), most often following classic protocols for obtaining hybridoma cells, which further include cell fusions, hybridoma-growth selection, screening, clone expansion, mAb production, purification, and characterization. Recent developments of mAb production include the cloning of selected B lymphocytes (from immunized animals or even from cryptococcosis patients), amplification of the V(D)J sequences of the heavy and light chains, cloning into mouse (or human antibody) expression cassettes, and vector transfection to mammalian cells, such as KEK293 or CHO cells, for mAb expression.

2 Materials

All solutions must be prepared using ultrapure water (MilliQ grade) and analytical grade reagents. Culture mediums and solutions that will be in contact with cells should be sterilized by filtration with a

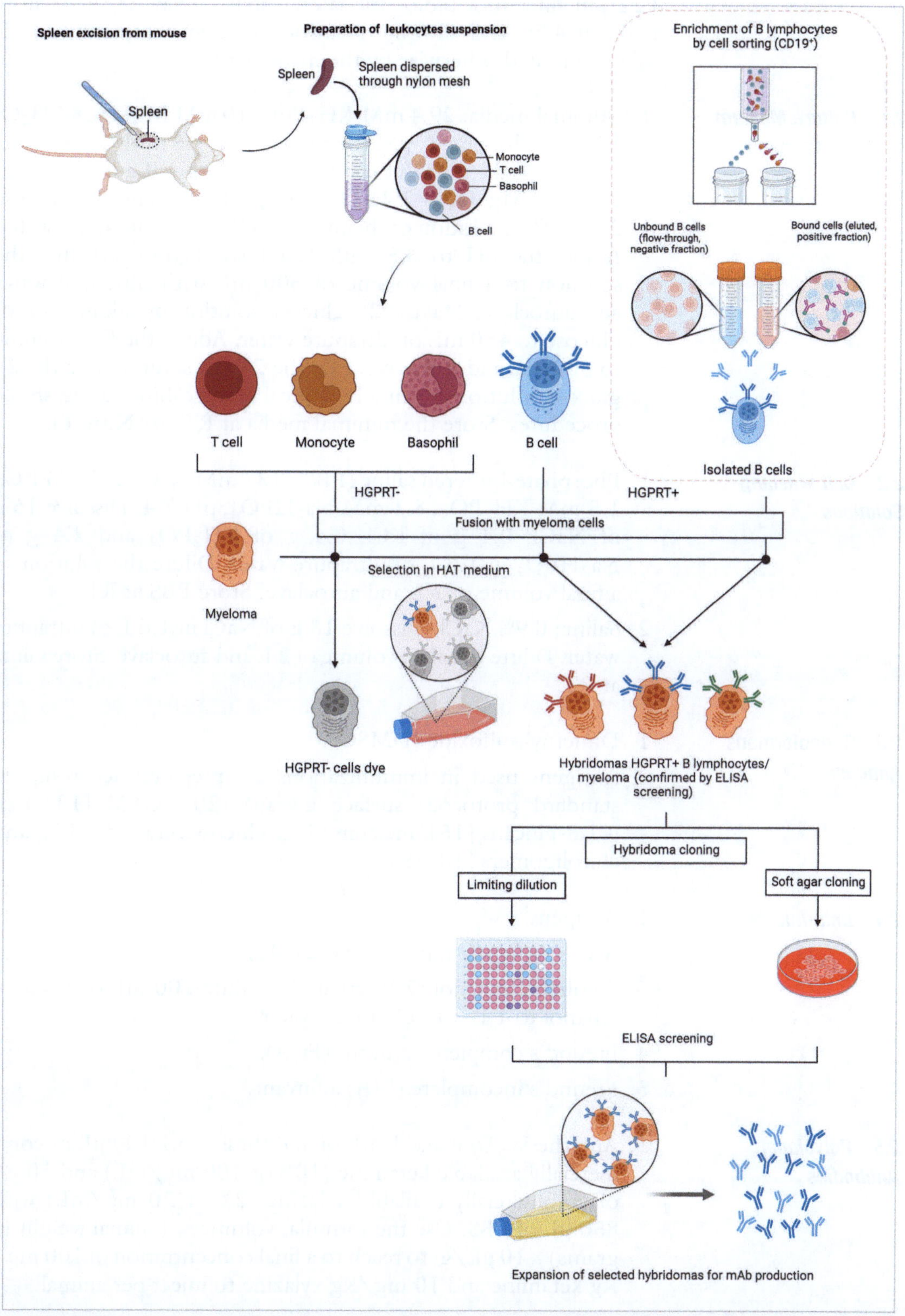

Fig. 3 Workflow demonstrating the stages in the production of monoclonal antibodies (mAbs)

0.22 μm filter in a biological safety cabinet (class II/A2). Store media at 4 °C and all other solutions at room temperature (RT, unless indicated otherwise by the manufacturer).

2.1 Culture Medium for C. neoformans

1. Minimal media: 29.4 mM KH_2PO_4, 10 mM $MgSO_4 \times 7H_2O$, 13 mM glycine, 3 μM thiamine, and 15 mM D-Glucose, pH 5.5. Make a 2× salt solution by adding 4.0 g of KH_2PO_4, 2.5 g of $MgSO_4 \times 7\ H_2O$, 0.98 g glycine, and 51 μL of a 20 mg/mL solution of thiamine to 450 mL of ultrapure water. Adjust the pH to 5.5 with KOH (or H_3PO_4). Dilute the solution to a final volume of 500 mL with ultrapure water and autoclave. Make a 2× glucose solution by adding 3 g of glucose to 450 mL of ultrapure water. Adjust the final volume to 500 mL and autoclave. Mix the 2× salt solution and the 2× glucose solution 1:1 in a biological safety cabinet using sterile procedures. Store the minimal media at RT (*see* **Note 1**).

2.2 Cell Washing Solutions

1. Phosphate-buffered saline (PBS): 137 mM NaCl, 2.7 mM KCl, 1.5 mM KH_2PO_4, 8.1 mM Na_2HPO_4, pH 7.4. Dissolve 16 g of NaCl, 0.4 g of KCl, 0.4 g of KH_2PO_4 and 2.4 g of Na_2HPO_4 in 1.6 L of ultrapure water. Dilute the solution to a final volume of 2 L and autoclave. Store PBS at RT.
2. Saline: 0.9% NaCl. Dissolve 18 g of NaCl in 1.6 L of ultrapure water. Dilute to a final volume of 2 L and autoclave. Store saline at RT.

2.3 C. neoformans Antigens

1. Dimethyl sulfoxide (DMSO).
2. Antigens used in immunizations are prepared according to standard protocols: surface glycans [20], GXM [13, 14], β-1,3-glucans [15], melanin [16], glucosylceramide [17], and chitoligomers [18, 19].

2.4 Immunizations

1. Antigens.
2. BALB/c female mice (~6 weeks old).
3. Sterilizing ethanol (70% ethanol): Dilute 700 mL of absolute ethanol to 1 L with ultrapure water.
4. Freund's complete adjuvant (FCA).
5. Freund's incomplete (FIA) adjuvant.

2.5 Polyclonal Antibodies

1. Anesthesia. To make 1 mL of anesthesia, mix 100 μL of commercially available ketamine (10% or 100 mg/mL) and 50 μL of commercially available xylazine (2% or 20 mg/mL) with 850 μL of PBS. Use the formula, volume = (animal weight in grams) × 10 μL/g, to reach to a final concentration of 100 mg/Kg ketamine and 10 mg/Kg xylazine to inject per animal.

2. Blood drawn from immunized animals (or cryptococcosis patients, *see* **Note 2**).

2.6 Indirect ELISA

1. Blood drawn from immunized animals or cryptococcosis patients.
2. Carbonate buffer (protein coating): 0.015 M Na_2CO_3, 0.035 M $NaHCO_3$, and 0.003 M NaN_3, pH 9.6. Dilute 1.59 g of Na_2CO_3, 2.94 g of $NaHCO_3$, and 0.2 g of NaN_3 to 0.8 L with ultrapure water. Adjust the pH to 9.6 with HCl and dilute to a final volume of 1 L. Carbonate buffer can be stored at 4 °C.
3. Washing buffer (TBS-Tween, TBS-T): 0.01 M Tris–HCl, 0.15 M NaCl, 0.001 M NaN_3, and 0.1% Tween 20. To prepare 10× TBS, dilute 12.11 g of Tris–HCl, 87.7 g of NaCl, and 0.65 g of NaN_3 to 800 mL ultrapure water. Adjust the pH to 7.2 with concentrated HCl and then dilute to a final volume of 1 L. To make 1× TBS, dilute 100 mL of 10× TBS to a final volume of 1 L. TBS buffer can be autoclaved. To make the final TBS-T, add 1 mL of Tween 20. TBS-T washing buffer can be stored at RT.
4. Blocking buffer: 1% bovine serum albumin (BSA) in TBS-T. Add 1 g of BSA to every 100 mL of TBS-T buffer. Store at 4 ° C.
5. Anti-mouse alkaline phosphatase conjugate (α-mouse AP conjugate): Dilute the α-mouse AP conjugate to 1 μg/mL (usually dilute 1 μL of a 1 mg/mL conjugate stock in blocking buffer to a final volume of 1 mL).
6. Bicarbonate substrate buffer: 0.001 M $MgCl_2 \times 6\ H_2O$ and 0.05 M Na_2CO_3, pH 9.6–9.8. Dissolve 0.2 g of $MgCl_2 \times 6\ H_2O$ and 5.3 g of Na_2CO_3 in 800 mL of ultrapure water. Adjust to pH to 9.6–9.8 with concentrated HCl and dilute to a final volume of 1 L. Store at RT.
7. Substrate solution: para-Nitrophenylphosphate (pNPP). Dissolve one tablet (~5 mg) of pNPP to every 5 mL of bicarbonate substrate buffer to make a 1 mg/mL solution.

2.7 Media for Cell Culture

1. Dulbeco's Modified Eagles Medium (DMEM). DMEM can be stored at 4 °C.
2. Hybridoma-tested fetal bovine serum (FBS). FBS can be stored at 4 °C.
3. 1× penicillin/streptomycin (100 U/mL). Penicillin/streptomycin antibiotic stock solution is commercially available 100× concentrated (10,000 U/mL). To supplement the medium with antibiotics, add 10 mL of the stock solution to a final volume of 1 L of medium.

4. 1× nonessential amino acids. Amino acid stock solution is commercially available 100× concentrated (10,000 U/mL). To supplement the medium, add 10 mL of the stock solution to a final volume of 1 L.
5. DMEM 10% FBS + amino acids + antibiotics: To make 1 L of medium, mix 880 mL of plain DMEM, 100 mL of FBS, 10 mL of 100× penicillin/streptomycin stock, and 10 mL of 100× non-essential amino acids stock. Sterilize by 0.22 μm filtration in a biological safety cabinet. Store all media at 4 °C.
6. 50× hypoxanthine, aminopterin, and thymidine (HAT) stock: 5 mM hypoxanthine, 0.02 mM aminopterin, and 0.8 mM thymidine. To prepare 100 mL of 100× hypoxanthine and thymidine (HT), dissolve 136 mg of hypoxanthine and 38 mg of thymidine in 80 mL of H_2O ultrapure water (*see* **Note 3**). Dilute to a final volume of 100 mL. To prepare 100 mL of 100× aminopterin, dissolve 1.76 mg of aminopterin in 80 mL of 1 N NaOH. Dilute to a final volume of 100 mL with 1 N NaOH (*see* **Note 4**). To make 200 mL of 50× HAT stock, mix 100 mL of 100× HT with 100 mL of 100× aminopterin stock.
7. DMEM conditioned medium: DMEM 20% FBS + 10% macrophage conditioned medium + 2× HAT + amino acids + antibiotics. To make 1 L of DMEM conditioned medium, mix 660 mL of plain DMEM, 200 mL of FBS, 100 mL of macrophage conditioned medium (*see* **Note 5**), 20 mL of 50× HAT stock, 10 mL of 100× penicillin/streptomycin stock and 10 mL of 100× non-essential amino acids stock. Sterilize by 0.22 μm filtration in a biological safety cabinet. Store all media at 4 °C.
8. DMEM HAT medium: DMEM 10% FBS + 2× HAT + amino acids + antibiotics. To make 1 L of DMEM conditioned medium, mix 860 mL of plain DMEM, 100 mL of FBS, 20 mL of 50× HAT stock, 10 mL of 100× penicillin/streptomycin stock, and 10 mL of 100× nonessential amino acids stock.

2.8 Splenocyte Preparation

1. Ammonium-Chloride-Potassium (ACK) lysis buffer: 0.15 M NH_4Cl, 0.01 M $KHCO_3$, and 0.0001 M disodium ethylenediaminetetraacetic acid (EDTA). Dissolve 4.0 g of NH_4Cl, 0.5 g of $KHCO_3$, and 0.019 g of disodium EDTA in 450 mL of ultrapure water. Adjust the pH to 7.2–7.4 and add ultrapure water to 500 mL. Sterilize the solution in an autoclave or by filtration (0.22 μm).

2.9 Cell Fusion

1. Polyethylene glycol (PEG 1450).
2. 50% PEG solution. Weigh 5 g of PEG and slowly add PBS. Adjust the solution to a final volume of 10 mL.

2.10 Cell Preservation

1. Freezing medium: 70% v/v DMEM, 20% v/v FBS, 10% v/v DMSO. For every 10 mL of freezing medium, mix 7 mL of DMEM, 2 mL of FBS, and 1 mL of DMSO.

2.11 Hybridoma Cloning

1. Hank's balanced salt solution (HBSS): 0.15 M NaCl, 0.05 M KCl, 0.001 M $CaCl_2$, 0.0004 $Mg(SO_4) \times 7\ H_2O$, 0.0005 M $MgCl_2 \times 6\ H_2O$, 0.0003 M $Na_2HPO_4 \times 2\ H_2O$, 0.0004 M KH_2PO_4, 0.006 M D-Glucose (dextrose), and 0.004 $NaHCO_3$. Dissolve 8.87 g of NaCl, 0.4 g of KCl, 0.14 g of $CaCl_2$, 0.1 g $Mg(SO_4) \times 7\ H_2O$, 0.1 g of $MgCl_2 \times 6\ H_2O$, 0.06 g of $Na_2HPO_4 \times 2\ H_2O$, 0.06 g of KH_2PO_4, 1 g of dextrose, and 0.35 g of $NaHCO_3$ in a beaker containing 800 mL of ultrapure water. Adjust the pH to 7.2 with HCl and dilute to a final volume of 1 L. Sterilize the solution in an autoclave or by filtration (0.22 μm).
2. 5% low-melt agarose: Mix 5 g of low melt agarose with 100 mL of HBSS solution. Heat the solution and homogenize in a magnetic stirrer. Autoclave the solution.
3. 0.5% low-melt agarose + DMEM HAT medium: Warm the 5% low melt agarose in a microwave until it completely dissolves. To make 500 mL of 0.5% low melt agarose DMEM HAT medium, aseptically transfer 450 mL of DMEM HAT medium to a sterile bottle, keeping it warm in a water bath at 50 °C. Add 50 mL of 5% low melt agarose, and gently swirl the bottle to mix the contents (*see* **Note 6**). Keep the solution at 50 °C until use.
4. 0.5% low melt agarose + DMEM HAT medium plates: Aseptically plate 9 mL of the above solution onto 50 mm cell culture plates. Allow plates to dry in a biosafety cabinet (*see* **Note** 7).

2.12 MAb Precipitation

1. Saturated ammonium sulfate: To make 500 mL of saturated ammonium sulfate ($(NH_4)_2SO_4$), transfer 500 mL of ultrapure water to a beaker and slowly add ammonium persulfate, while stirring the solution. Keep adding the ammonium persulfate powder until the liquid becomes turbid (*see* **Note 8**). Allow the powder to precipitate and carefully transfer the solution to a new bottle to avoid carriage of the precipitated ammonium sulfate. Store at RT.
2. 12,000–14,000 Da dialysis bag.
3. Elution buffer: 0.1 M glycine (pH 2.5). Dissolve 3.75 g of glycine in 450 mL of ultrapure water. Adjust the pH to 2.5 with HCl and dilute to a final volume of 500 mL. Store at RT.

3 Methods

3.1 Animal Immunizations

An antigen to be used in the immunization steps should be immunogenic enough to stimulate a humoral response in the animal model used. In the case of *C. neoformans*, several antigen preparations to induce a humoral response have been described in the literature, including GXM [21], GXMGal [22], glycans [20], β-1,3-glucan [15], chitooligomers [18], and melanin [23]. As T-independent antigens, these polysaccharides or complex phenolic structures, such as melanin, induce low titers and short-lived antibodies [24, 25]. Therefore, they usually require conjugation with proteins, such as tetanus toxoid or BSA, which are presented by B-lymphocytes via MHC-II to $CD4^+$ T-cells, which, in turn, induce the responsive B cell differentiation into plasmocytes [21, 26], resulting in an overall high-titer antibody response.

Antigen injections (in the scale of micrograms) to an animal (usually mice or rabbit) are performed peritoneally in a 1:1 emulsification with FCA. Repeat the injections (in 14 days intervals, usually 2–4 times) with antigen and FIA, for additional stimulation of responsive B-lymphocytes to the antigen used. Assay the titer of antibodies in the serum before and after each immunization step (as described in Subheading 3.9, **steps 1–15**, but using serially diluted serum in **step** 7 for determination of titer of antibodies), which is expected to increase during the process.

1. Make an antigen solution using up to 50 μg/animal (usually in 100 μL) to be immunized in PBS or saline solutions.
2. Mix the same volume of antigen preparation with the resuspended FCA into a microtube.
3. Mix the antigen and FCA vigorously in a vortex to make an emulsion. At this stage, a thickened mixture is formed.
4. Draw up the emulsion into a disposable sterile 1 mL syringe.
5. Attach a sterile 20-G needle to the syringe containing the antigen and FCA emulsion.
6. Clean the abdominal area of the animal with a cotton swab embedded with 70% ethanol.
7. Inject the emulsion intraperitoneally into the animal.
8. For additional boosters, repeat the procedures described in **steps 1–6**, using the same dose of antigen emulsified (1:1) in FIA.

3.1.1 Polyclonal Antibodies

Upon immunization, as described above, when the desired serum concentration of the antibodies is reached, polyclonal antibodies may be obtained by bleeding the animals, followed by centrifugation (Fig. 2). Usually, this consists of a mixture of antibodies from different isotypes and distinct specificities, but all targeting the same antigen used in the immunization steps. Also, polyclonal antibodies may be obtained from cryptococcosis patients, upon centrifugation of blood samples. However, in this case, we expect the serum to contain antibodies against the distinct antigens found in *C. neoformans* (polysaccharides, proteins, and lipids), as we are dealing with the context of infection/disease.

3.2 Mouse Bleeding from Tail Vein (Up to 0.2 mL)

1. Hold the mouse by the base of its tail and place it on a surface where it can grasp the surface.
2. Firmly place your free hand on its shoulders, and using your thumb and forefinger, grasp the scruff of its neck.
3. Restrain the tail using your little finger.
4. Clean the base of its tail with 70% alcohol.
5. Locate the lateral tail vein.
6. Using a 25–27 G needle, collect the blood from the lateral tail vein.
7. Cease blood flow by applying pressure with a sterile gauze.
8. Centrifuge blood at 1000× *g* for 10 min and collect the serum (upper layer).

3.3 Mouse Bleeding from Retro-Orbital Sinus (Up to 0.2 mL)

1. Anesthetize the mouse using an appropriate method (approved by the Institutional Ethics Committee; usually 100 mg/Kg ketamine and 10 mg/Kg xylazine, intraperitoneally).
2. Gently scruff the neck of the animal, which will cause the eye to bulge.
3. Insert a capillary tube or a Pasteur pipette laterally into the eye.
4. Allow the blood to flow by capillary action.
5. Cease blood flow by applying pressure with a sterile gauze.
6. Centrifuge blood at 1000× *g* for 10 min and collect the serum (upper layer).

3.4 Mouse Bleeding by Cardiac Puncture (Up to 1 mL)

1. Anesthetize the mouse using an appropriate method (approved by the Institutional Ethics Committee; usually 100 mg/Kg ketamine and 10 mg/Kg xylazine, intraperitoneally).
2. The animal may also be sacrificed using an overdose of anesthetics.
3. Locate the left ventricle (about the middle part of its chest).
4. Using a 25–27 G needle, slowly withdraw blood.
5. Centrifuge the blood at 1000× *g* for 10 min and collect the upper liquid (serum).

3.5 Human Blood Withdrawal

1. Ask the person to expose one arm.
2. Place a tight elastic band (tourniquet) around the limb to partially back up the blood flow in the veins.
3. The vein can be easily identified and accessed.
4. Clean the target skin area with 70% alcohol.
5. Connect an 18 G needle to a special vacuum tube or a syringe.
6. Collect the blood (by gentle aspiration in the case of a syringe).
7. Release the tourniquet and gently remove the needle from the vein.
8. Using a gauze, gently apply pressure to stop the bleeding.
9. Cover the punctured site with a bandage.
10. Centrifuge the blood at 1000× *g* for 10 min and collect the serum (upper layer).

3.5.1 Monoclonal Antibodies

The desired antigen to be used in the immunization steps should be immunogenic enough to stimulate a humoral response in the animal model used. Upon immunizations and certification of the presence of antibodies in serum, animals may be additionally boosted 3 days before splenocyte collection (Fig. 3). Splenocytes may then be used in fusion steps with myeloma cells. The desired hybridomas may be cloned and screened by assaying for the presence of antibodies to the target antigen using immunoassays.

3.6 Myeloma Cell Culture

Selecting the desired myeloma cells as a fusion partner is a required and key step. For obtaining mouse hybridomas, BALB/c-derived hybridomas are often used, such as the sp2/0 cells. Rat and human myeloma cells are also available for fusion, as needed. Desirable property of such cells is the absence of antibody secretion and the sensitivity to HAT, due to the lack of the hypoxanthine-guanine phosphoribosyl transferase (HGPRT) enzyme. Myeloma cells as fusion partners are prepared using the following steps:

1. Retrieve a vial of myeloma cells from liquid nitrogen.
2. Quickly thaw the cells at 37 °C in a water bath.
3. Transfer the cells to a sterile 50 mL conical tube containing 10 mL of prewarmed DMEM plain medium.
4. Centrifuge the cells at 800× *g* for 10 min.
5. Add 10 mL of DMEM 10% FBS + amino acids + antibiotics.
6. Subculture the myeloma cells every 48 h.
7. Prior to fusion, subculture the myeloma cells for 24 h to achieve highest viability.

3.7 Preparation of Splenocytes for Fusion

Three days after the last booster with the antigen, spleen lymphocytes from immunized mice are obtained as followed:

1. Anesthetize the mouse using an appropriate method (approved by the Institutional Ethics Committee; usually ketamine 100 mg/Kg and xylazine 10 mg/Kg, intraperitoneally).
2. The animal may also be sacrificed using an overdose of anesthetics.
3. Perform splenectomy by aseptically opening the peritoneal cavity of the animal and collecting the spleen with a scissor/forceps.
4. Wash the spleen in plain DMEM medium in a Petri dish.
5. Disrupt the spleen mechanically by rubbing it against the nylon mesh of a 70 μm cell strainer using a syringe plunger.
6. Alternatively, the spleen membrane may be disrupted using scissors or by perforating it with an 18 G needle. Flush the spleen internally with plain DMEM medium several times in Petri dishes containing medium.
7. Wash splenocytes 3× using DMEM medium and centrifugation.
8. Disrupt red blood cells by resuspending the pellet in ACK lysis buffer.
9. Wash the cells twice in DMEM 10% FBS + amino acids + antibiotics.

3.8 Myeloma-Splenocyte Fusion

The fusion step of splenocytes and myeloma cells, followed by selection in HAT medium, are crucial steps for hybridoma generation and mAb production. Additives, such as PEG, may interfere with cell membrane structure and permeation by the production of pores that enhance membrane hydrophilicity and interconnectivity [27, 28]. Myeloma cells and splenocytes are fused by chance and require the correct mixing ratios. Among the many fusion possibilities, myeloma-myeloma, splenocyte-splenocyte, and myeloma-splenocyte may be observed, as well as unfused splenocytes or myeloma cells. The aminopterin in the HAT medium blocks the de novo nucleotide synthesis pathway. Only cells expressing the HGPRT enzyme, such as B cells, perform the salvage pathway using hypoxanthine to produce guanine. Unfused myelomas or myeloma-myeloma fusions cannot grow because they lack HGPRT (HGPRT-). Unfused splenocytes or splenocyte-splenocyte fusions cannot grow indefinitely because of their short life span. Therefore, from all the fusion possibilities, only B cell (HGPRT+)-myelomas survive in HAT medium and grow indefinitely and are therefore positively selected. The steps for myeloma-splenocyte fusion and selection of B-lymphocyte-myeloma hybridomas are performed as follows:

1. Wash splenocytes and myeloma cells 3× with plain DMEM medium by centrifugation at 1000× *g* for 10 min.
2. Resuspend splenocytes and myelomas in 20 mL of plain DMEM medium.
3. Enumerate splenocytes and myeloma cells using a hemocytometer.
4. Combine splenocytes and myeloma cells in a 2:1 ratio in a single 50 mL conical tube.
5. Centrifuge mixed cells at 1000× *g* for 10 min and perform a single wash with PBS.
6. Discard the supernatant and loosen the pellet by gently finger-flicking the bottom of the tube.
7. Slowly add 1 mL of a 50% PEG 1450 solution in PBS per 2×10^8 cells, along the side of the tube.
8. Mix the cells by gently swirling the tube for 2 min at RT.
9. Add a similar volume of plain DMEM and gently swirl the cells for a further 2 min at RT.
10. Slowly add 30 mL of plain DMEM medium (5 mL in the first minute, 10 mL in the second minute, and 15 mL in the third minute).
11. Incubate the cells for a further 2 min at RT.
12. Centrifuge the cells at 500× *g* for 5 min and discard the supernatant.
13. Resuspend the cells in 200 mL of DMEM conditioned medium containing 20% FBS, 2× HAT, and 10% of macrophage conditioned medium containing growth factors.
14. Transfer the cell suspension to an Erlenmeyer flask. Gently swirl flask to keep a homogeneous cell suspension.
15. Plate the entire cell suspension in 96-well cell culture plates (200 μL/well). Label the plates starting from 1. This is a crucial step to keep track of the wells in the future (each well will be identified as the number of the plate, row letter, and column number).
16. Incubate the cells at 37 °C in a 5% CO_2 incubator.
17. Observe cells regularly under an inverted microscope for hybridoma growth.
18. After 7–10 days, clones might be visible by the naked eye, and medium might turn orange to yellowish.
19. Supernatant may be tested by ELISA for hybridoma screening.

3.9 Screening of B Lymphocyte-Myeloma Hybridomas

Screening strategies for hybridomas may be performed by immunoassays to detect the production of antibodies in the supernatant of each well. An indirect ELISA assay is usually the fastest and most used technique. The method should be optimized before the screening itself and usually requires the same antigen used in the immunization strategies. Polyclonal serum obtained from immunizations may be used as a positive control, since it contains a complex mixture of antibodies targeting the desired antigen. Mock controls, only in the presence of antigen and medium, are also required. A simple indirect ELISA for hybridoma screening is described as follows:

1. Coat the same number of ELISA 96-well microplates that were generated for plating the hybridomas. Usually 0.5–1 μg of antigen is used per well (in a 50 μL volume). Protein antigens are usually diluted in carbonate buffer, whereas polysaccharides are usually diluted in PBS.
2. Wash the plates three times using washing buffer to remove unbound antigens.
3. Block the plates with 200 μL per well of blocking buffer.
4. Incubate the plates for 1 h at 37 °C.
5. Wash the plates three times with washing buffer.
6. Label the plates as described for the hybridomas plating.
7. Transfer 50 μL of the hybridoma supernatant to each well of the antigen-containing plates, following the set up described.
8. Incubate the plates for 1 h at 37 °C.
9. Wash the plates three times with washing buffer.
10. Add 50 μL per well of an anti-mouse Ig alkaline phosphatase conjugate diluted to 1 μg/mL in blocking buffer.
11. Incubate the plates for 1 h at 37 °C.
12. Wash the plates three times with washing buffer.
13. Add 50 μL per well of pNPP diluted to 1 mg/mL in bicarbonate substrate buffer.
14. Allow plates to develop, which usually takes 10–30 min.
15. Read plates at A_{405} in a microplate reader.

3.10 Scaling Up Selected Wells and Freezing in Liquid Nitrogen

Wells containing antibody-producing hybridomas should have absorbances at least twice that of the mock control wells. Next, hybridomas in the selected positive wells, usually with the highest absorbance, are transferred and scaled up to a 24-well plate (using 1–2 mL of DMEM HAT medium), where cells are maintained for 48 h at 37 °C/5% CO_2, and further to 6-well plates (using 3.5 mL medium). At this stage, cells can be frozen in liquid nitrogen to prevent the risk of losing the selected positive clones. To freeze the hybridomas, the following protocol is used:

1. Homogenize cells in the 6-well plate by pipetting up and down.
2. Transfer half of the volume to a new well and add fresh DMEM HAT medium.
3. Transfer the remaining cells to a new 15 mL conical tube and centrifuge at 1000× *g* for 10 min.
4. Discard the supernatant and resuspend the cells in a similar volume of cold freezing medium.
5. Immediately divide the cell suspension into two freezing vials (cryotubes).
6. Transfer immediately to dry ice (−20 °C) and let sit for 1–2 h.
7. Transfer the cells to a −80 °C freezer overnight.
8. For permanent storage, transfer the vials to liquid nitrogen.

3.11 Methods for Cloning Hybridoma Cells

As mAbs have a single specificity and belong to a specific antibody isotype, generated by an isolated B cell clone, the isolation and cloning of hybridomas are critical. Several methods may be used for single hybridoma cloning, including limiting dilution, soft agar cloning, and cell sorting. Each method will be explained in separate sections bellow.

3.11.1 Limiting Dilution

In the limiting dilution cloning of hybridomas, cells are counted and plated at a concentration of one cell per well in 96-wells culture microplates. This ensure that cells growing in a specific well consist of a single clone and the antibody secreted will be monoclonal. The steps for the limiting dilution cloning technique are described as follows:

1. From the wells of the 6-well plate, homogenize the cells by pipetting up and down.
2. Count the hybridoma cells in a hemocytometer and dilute to a concentration of 10 cells/mL in DMEM HAT medium.
3. Plate 100 μL per well to each well of a 96-well microplate.
4. Add an additional 100 μL of DMEM HAT medium to each well.
5. Incubate the plates at 37 °C in a 5% CO_2 incubator.
6. Observe the wells regularly under an inverted microscope for the growth of single clones.
7. Select the wells with single clones and upon growth to approximately 25% of the well's surface and test the supernatants using the indirect ELISA protocol described in Subheading 3.9, **steps 1–15**.

3.11.2 Soft Agar Cloning

In the soft agar cloning method, cells are diluted in a mixture of cell culture medium and agar to make the soft agar and plated over a semi-solid mixture of agar/cell culture medium. In this method, due to the gelatinous pressure of the soft agar, single cells will grow as clusters and may be visible by eye. Many clones can be obtained and extracted from a single plate and individually transferred to the wells of a 96-well plate. The steps for the soft agar cloning technique are as follows:

1. From the wells of the 6-well plate, homogenize the cells by pipetting up and down.
2. Count the hybridomas in a hemocytometer and dilute to a concentration of 50 cells/mL in DMEM HAT medium.
3. For each cloning, mix 1.4 mL of 0.5% low-melt agarose + DMEM HAT medium with 0.7 mL of the cell suspension. This will result in a final concentration of 0.33% low melt agarose in DMEM HAT medium (*see* **Note 9**).
4. Overlay approximately 2 mL of 0.33% soft agar cell suspension onto 50 mm cell culture plates containing 0.5% low melt agarose + DMEM HAT medium (*see* **Note 10**).
5. Incubate the plates at 37 °C in a 5% CO_2 incubator.
6. Monitor the plates for cluster growth using an inverted microscope.
7. Individual clusters can be transferred using a Pasteur pipette coupled with a rubber bulb, to the wells of a 96-wells microplate containing DMEM HAT medium supplemented with 10% FBS. To facilitate colony picking, use a magnifying loupe.
8. Incubate the plates at 37 °C in a 5% CO_2 incubator.
9. Monitor hybridoma growth under an inverted microscope and assay the supernatant using the indirect ELISA protocol described in Subheading 3.9, **steps 1–15**.

3.12 Production of mAbs from Cloned Hybridomas

Upon cloning of hybridomas, as described in the previous sections, clones may be cultivated in DMEM supplemented with 10% FBS, frozen, and upon every 48 h passages, cell culture may reach volumes of large flasks, from where supernatants may be obtained for mAb purification. However, if the desired, purified mAb needs to be free of contaminations from immunoglobulins contained in the serum used, a wide variety of hybridoma serum-free medium are commercially available. Depending on the requirements for how pure the mAbs preparation is, upon obtaining the supernatants, they may simply be precipitated, prior to further purification.

3.12.1 MAb Precipitation Using Ammonium Sulfate

1. Supernatants are centrifuged at 3000× *g* for 10 min to eliminate cell contamination.

2. Clarified supernatants are transferred to a large Erlenmeyer flask containing a stir bar, which is placed on a magnetic stirrer.
3. Add an equal volume of a saturated solution of ammonium sulfate (to a final concentration of 50%) while stirring.
4. Place the system at 4 °C overnight.
5. Transfer the contents to 50 mL conical tubes and centrifuge at 3000× *g* for 30 min. Carefully discard the supernatant and invert the conical tubes to remove any excess of liquid.
6. Resuspend the pellet in 1/10 of the starting supernatant volume.
7. Place the solution in a 12,000–14,000 Da dialysis bag and dialyze the system with three changes of PBS for approximately 12 h.
8. Transfer the antibody solution from the tubing to a new conical tube and centrifuge at 3000× *g* for 30 min.
9. Carefully transfer the supernatant to a new conical tube and determine the protein concentration.
10. The mAb solution may be stored at −80 °C.

3.12.2 MAb Purification

1. Culture supernatants or precipitated mAbs may be purified by affinity chromatography with protein A or G. These are commercially available as columns/cartridges or may be obtained as a resin slurry to be manually packed in-house.
2. Supernatants or mAb solutions may be passed through these columns, and upon washes with PBS, retained antibodies may be eluted with an elution buffer under acidic conditions.
3. The purified material may be pooled and the resulting solution neutralized using 1 M Tris–HCl.
4. Protein concentrations may be determined accordingly. Purified antibodies may be aliquoted and frozen at −80 °C.

3.13 New Methodologies for mAb Production

With advances in laboratory technologies, such as flow cytometry and cell sorting, B lymphocytes from the spleen or blood of immunized or infected animals or even the peripheral blood of infected patients may be individually obtained. For this task, peripheral blood mononuclear cells (PBMCs) may be initially isolated from blood through sedimentation using density gradients, such as Ficoll-Hypaque [29]. Upon PBMC isolation, B lymphocytes may be obtained by cell sorting using a combination of fluorescent labeling to target a specific B cell population. Initially, to target responsive cells to the desired antigen of a polysaccharide or protein nature, cells may be biotinylated and upon capture by the target cells, are sorted based on the Streptavidin-Alexa Fluor signal. For the further isolation of memory B cells, the following labeling panel

Table 1
Markers used for sorting single B cells from spleen or peripheral blood

Target molecule	B cell phenotype
Anti-CD3	CD3−
Anti-CD16	CD16−
Anti-CD235a	CD235a−
Anti-CD38	CD38−
Anti-CD27	CD27+
Anti-CD19	CD19+
Anti-IgG (or specific)	IgG+
Antigen-biotin (captured by B cell)	Alexa fluor/Streptavidin+

(Table 1) may be used, each with a specific fluorescent label, as designed by the user [30]:

Sorted B cells may be used for fusions with hybridomas by following specific requirements depending on the host species (as previously explained and depicted in Fig. 2). Fused B lymphocyte-myeloma hybridomas may be then further cloned using the methods described above, for cloning and mAb production.

3.14 Single B-Cell Sequencing and Expression of mAbs

Using the same sorting methodology as described above, B memory cells may be sorted to individual wells of 96-well plates (1 cell/well). Next, mRNA may be extracted from single-sorted B cells and used to generate cDNAs using a reverse transcriptase reaction. Then, complete V(D)J segments of each clone may be amplified by PCR using specific primers for the constant regions of the light and heavy chains. Upon DNA sequencing, sequences may be evaluated and the higher frequency V(D)J pair identified [31]. Each of the amplified and selected V(D)J pairs may be inserted into commercially available immunoglobulin expression cassettes. Upon construction of immunoglobulin expression vectors, these vectors may be transfected to mammalian cells for mAb expression with the desired selected specificity and isotype determined by the type of vector used. This methodology may even be used to generate antibodies of the same specificity from different species [32]. This technology has been recently used for the rapid production and generation of neutralizing mAbs against the SARS-CoV2 virus during the pandemic [30, 33]. Further, the presence of mAbs in the supernatant and their binding capacity may be evaluated using the indirect ELISA methodology, following a similar protocol as that described in the hybridoma screening strategy (Subheading 3.9, **steps 1–15**).

4 Notes

1. Minimal media may also be filtered through a 0.22 μm filter in a biosafety cabinet for sterilization. Ensure that all bottles and glassware are sterile before the procedure.
2. Working with animals or human serum requires approval by the Institutional Ethics Committee, following specific guidelines for Good Practices established by each National Council.
3. If the HT does not dissolve completely, heat the solution gently up to 70 °C. Allow the solution to cool down at RT before sterilizing by filtration.
4. Titrate the aminopterin solution to neutral pH before sterilizing by filtration.
5. The macrophage conditioned medium may be obtained upon cultivation of a mouse macrophage lineage (usually J774.16 or RAW 264.7 cells) in DMEM supplemented with 10% FBS + amino acids + antibiotics + 10% NCTC 109 medium for 48 h at 37 °C/5% CO_2. Supernatant is harvested, centrifuged for 10 min at 1100× *g*, and filtered through a 0.22 μm filter.
6. To prepare a homogeneous solution, it is mandatory that the DMEM HAT medium should be also kept at 50 °C before the addition of the 5% low-melt agarose solution.
7. Allow agar to solidify in the biosafety cabinet. Solidified 0.5% low melt agarose + DMEM HAT medium plates might become a little more opaque than the liquid medium.
8. At this point, when the salt is not able to further dissolve, stop stirring the solution and visualize the precipitation of the salt.
9. The 0.5% low melt agarose + DMEM HAT medium should be kept at 50 °C during the addition of cells, with gentle stirring to make a homogeneous suspension.
10. It is essential during this step that the suspension is maintained at 50 °C. Otherwise, the overlay will not be homogeneous.

Acknowledgments

This work was supported by the Brazilian agencies National Council for Scientific and Technological Development (CNPq grant 309736/2021-8) and Fundação de Amparo à Pesquisa do Estado do Rio de Janeiro (FAPERJ grant E-26/200.997/2022).

References

1. Maziarz EK, Perfect JR (2016) Cryptococcosis. Infect Dis Clin N Am 30:179–206. https://doi.org/10.1016/j.idc.2015.10.006
2. Mansour MK, Reedy JL, Tam JM et al (2014) Macrophage–*Cryptococcus* interactions: an update. Curr Fungal Infect Rep 8:109–115. https://doi.org/10.1007/s12281-013-0165-7
3. Elsegeiny W, Marr KA, Williamson PR (2018) Immunology of cryptococcal infections: developing a rational approach to patient therapy. Front Immunol 9:651. https://doi.org/10.3389/fimmu.2018.00651
4. Gibson JF, Johnston SA (2015) Immunity to *Cryptococcus neoformans* and *C. gattii* during cryptococcosis. Fungal Genet Biol 78:76–86. https://doi.org/10.1016/j.fgb.2014.11.006
5. Rohatgi S, Pirofski L (2015) Host immunity to *Cryptococcus neoformans*. Future Microbiol 10: 565–581. https://doi.org/10.2217/fmb.14.132
6. Gao Y, Huang X, Zhu Y et al (2018) A brief review of monoclonal antibody technology and its representative applications in immunoassays. J Immunoass Immunochem 39:351–364. https://doi.org/10.1080/15321819.2018.1515775
7. Köhler G, Milstein C (1975) Continuous cultures of fused cells secreting antibody of predefined specificity. Nature 256:495–497. https://doi.org/10.1038/256495a0
8. Scheen AJ (2009) International classification of various types of monoclonal antibodies. Rev Med Liege 64:244–247
9. Buss NA, Henderson SJ, McFarlane M et al (2012) Monoclonal antibody therapeutics: history and future. Curr Opin Pharmacol 12:615–622. https://doi.org/10.1016/j.coph.2012.08.001
10. Morrison SL, Johnson MJ, Herzenberg LA et al (1984) Chimeric human antibody molecules: mouse antigen-binding domains with human constant region domains. Proc Natl Acad Sci U S A 81:6851–6855. https://doi.org/10.1073/pnas.81.21.6851
11. Jones PT, Dear PH, Foote J et al (1986) Replacing the complementarity-determining regions in a human antibody with those from a mouse. Nature 321:522–525. https://doi.org/10.1038/321522a0
12. Feldmesser M (1998) Mechanism of action of antibody to capsular polysaccharide in *Cryptococcus neoformans* infection. Front Biosci 3: d136–d151. https://doi.org/10.2741/A270
13. Casadevall A, Pirofski L (2012) Immunoglobulins in defense, pathogenesis, and therapy of fungal diseases. Cell Host Microbe 11:447–456. https://doi.org/10.1016/j.chom.2012.04.004
14. Maitta RW, Datta K, Chang Q et al (2004) Protective and nonprotective human immunoglobulin M monoclonal antibodies to *Cryptococcus neoformans* glucuronoxylomannan manifest different specificities and gene use profiles. Infect Immun 72:4810–4818. https://doi.org/10.1128/IAI.72.8.4810-4818.2004
15. Rachini A, Pietrella D, Lupo P et al (2007) An anti-beta-glucan monoclonal antibody inhibits growth and capsule formation of *Cryptococcus neoformans in vitro* and exerts therapeutic, anticryptococcal activity *in vivo*. Infect Immun 75:5085–5094. https://doi.org/10.1128/IAI.00278-07
16. Rosas ÁL, Nosanchuk JD, Casadevall A (2001) Passive immunization with melanin-binding monoclonal antibodies prolongs survival of mice with lethal *Cryptococcus neoformans* infection. Infect Immun 69:3410–3412. https://doi.org/10.1128/IAI.69.5.3410-3412.2001
17. Rodrigues ML, Travassos LR, Miranda KR et al (2000) Human antibodies against a purified glucosylceramide from *Cryptococcus neoformans* inhibit cell budding and fungal growth. Infect Immun 68:7049–7060. https://doi.org/10.1128/IAI.68.12.7049-7060.2000
18. Figueiredo ABC, Fonseca FL, Kuczera D, Conte F de P, Arissawa M, Rodrigues ML (2021) Monoclonal antibodies against cell wall chitooligomers as accessory tools for the control of Cryptococcosis. Antimicrob Agents Chemother 65:e0118121. https://doi.org/10.1128/AAC.01181-21
19. Liedke SC, Miranda DZ, Gomes KX, Gonçalves JLS, Frases S, Nosanchuk JD, Rodrigues ML, Nimrichter L, Peralta JM, Guimarães AJ (2017) Characterization of the antifungal functions of a WGA-Fc (IgG2a) fusion protein binding to cell wall chitin oligomers. Sci Rep 7:12187. https://doi.org/10.1038/s41598-017-12540-y
20. Cordero RJB, Liedke SC, de S. Araújo GR, Martinez LR, Nimrichter L, Frases S, Peralta JM, Casadevall A, Rodrigues ML, Nosanchuk JD, Guimaraes AJ (2016) Enhanced virulence of *Histoplasma capsulatum* through transfer and surface incorporation of glycans from *Cryptococcus neoformans* during co-infection. Sci Rep 6:21765. https://doi.org/10.1038/srep21765

21. Casadevall A, Mukherjee J, Devi SJN, Schneerson R, Robbins JB, Scharff MD (1992) Antibodies elicited by a *Cryptococcus neoformans*-tetanus toxoid conjugate vaccine have the same specificity as those elicited in infection. J Infect Dis 165:1086–1093. https://doi.org/10.1093/infdis/165.6.1086
22. De Jesus M, Chow S-K, Cordero RJB, Frases S, Casadevall A (2010) Galactoxylomannans from *Cryptococcus neoformans* varieties *neoformans* and *grubii* are structurally and antigenically variable. Eukaryot Cell 9:1018–1028. https://doi.org/10.1128/EC.00268-09
23. Nosanchuk JD, Rosas AL, Lee SC, Casadevall A (2000) Melanisation of *Cryptococcus neoformans* in human brain tissue. Lancet 355:2049–2050. https://doi.org/10.1016/S0140-6736(00)02356-4
24. Rydahl MG, Kračun SK, Fangel JU, Michel G, Guillouzo A, Génicot S, Mravec J, Harholt J, Wilkens C, Motawia MS, Svensson B, Tranquet O, Ralet M-C, Jørgensen B, Domozych DS, Willats WGT (2017) Development of novel monoclonal antibodies against starch and ulvan – implications for antibody production against polysaccharides with limited immunogenicity. Sci Rep 7:9326. https://doi.org/10.1038/s41598-017-04307-2
25. Broecker F, Anish C, Seeberger PH (2015) Generation of monoclonal antibodies against defined oligosaccharide antigens. In: Lepenies B (ed) Carbohydrate-based vaccines. Springer, New York, pp 57–80
26. Chow S-K, Casadevall A (2011) Evaluation of *Cryptococcus neoformans* galactoxylomannan–protein conjugate as vaccine candidate against murine cryptococcosis. Vaccine 29:1891–1898. https://doi.org/10.1016/j.vaccine.2010.12.134
27. Feng Y, Han G, Chung T-S, Weber M, Widjojo N, Maletzko C (2017) Effects of polyethylene glycol on membrane formation and properties of hydrophilic sulfonated polyphenylenesulfone (sPPSU) membranes. J Membr Sci 531:27–35. https://doi.org/10.1016/j.memsci.2017.02.040
28. Greenfield EA (2018) Polyethylene glycol fusion for hybridoma production. Cold Spring Harb Protoc 2018:pdb.prot103176. https://doi.org/10.1101/pdb.prot103176
29. Fuss IJ, Kanof ME, Smith PD, Zola H (2009) Isolation of whole mononuclear cells from peripheral blood and cord blood. Curr Protoc Immunol 85. https://doi.org/10.1002/0471142735.im0701s85
30. Cao Y, Su B, Guo X, Sun W, Deng Y, Bao L, Zhu Q, Zhang X, Zheng Y, Geng C, Chai X, He R, Li X, Lv Q, Zhu H, Deng W, Xu Y, Wang Y, Qiao L, Tan Y, Song L, Wang G, Du X, Gao N, Liu J, Xiao J, Su X, Du Z, Feng Y, Qin C, Qin C, Jin R, Xie XS (2020) Potent neutralizing antibodies against SARS-CoV-2 identified by high-throughput single-cell sequencing of convalescent patients' B cells. Cell 182:73–84.e16. https://doi.org/10.1016/j.cell.2020.05.025
31. Tiller T, Meffre E, Yurasov S, Tsuiji M, Nussenzweig MC, Wardemann H (2008) Efficient generation of monoclonal antibodies from single human B cells by single cell RT-PCR and expression vector cloning. J Immunol Methods 329:112–124. https://doi.org/10.1016/j.jim.2007.09.017
32. Goldstein LD, Chen Y-JJ, Wu J, Chaudhuri S, Hsiao Y-C, Schneider K, Hoi KH, Lin Z, Guerrero S, Jaiswal BS, Stinson J, Antony A, Pahuja KB, Seshasayee D, Modrusan Z, Hötzel I, Seshagiri S (2019) Massively parallel single-cell B-cell receptor sequencing enables rapid discovery of diverse antigen-reactive antibodies. Commun Biol 2:304. https://doi.org/10.1038/s42003-019-0551-y
33. Horns F, Dekker CL, Quake SR (2020) Memory B cell activation, broad anti-influenza antibodies, and bystander activation revealed by single-cell transcriptomics. Cell Rep 30:905–913.e6. https://doi.org/10.1016/j.celrep.2019.12.063

Chapter 21

Fluorescence and Biochemical Assessment of the Chitin and Chitosan Content of *Cryptococcus*

Brian T. Maybruck, Rajendra Upadhya, Woei C. Lam, Charles A. Specht, and Jennifer K. Lodge

Abstract

The cell wall of the fungal pathogens *Cryptococcus neoformans* and *C. gattii* is critical for cell wall integrity and signaling external threats to the cell, allowing it to adapt and grow in a variety of changing environments. Chitin is a polysaccharide found in the cell walls of fungi that is considered to be essential for fungal survival. Chitosan is a polysaccharide derived from chitin via deacetylation that is also essential for cryptococcal cell wall integrity, fungal pathogenicity, and virulence. *Cryptococcus* has evolved mechanisms to regulate the amount of chitin and chitosan during growth under laboratory conditions or during mammalian infection. Therefore, levels of chitin and chitosan have been useful phenotypes to define mutant *Cryptococcus* strains. As a result, we have developed and/or refined various qualitative and quantitative methods for measuring chitin and chitosan. These techniques include those that use fluorescent probes that are known to bind to chitin (e.g., calcofluor white and wheat germ agglutinin), as well as those that preferentially bind to chitosan (e.g., eosin Y and cibacron brilliant red 3B-A). Techniques that enhance the localization and quantification of chitin and chitosan in the cell wall include (i) fluorescence microscopy, (ii) flow cytometry, (iii) and spectrofluorometry. We have also modified two highly selective biochemical methods to measure cellular chitin and chitosan content: the Morgan-Elson and the 3-methyl-2-benzothiazolone hydrazine hydrochloride (MBTH) assays, respectively.

Key words *Cryptococcus*, Chitin, Chitosan, Epifluorescence, Spectrofluorometer, Flow cytometry, Morgan-Elson, MBTH

1 Introduction

1.1 Chitin and Chitosan in Cryptococcus

Chitin is an insoluble and neutrally charged polysaccharide composed of N-acetylglucosamine (GlcNAc) joined by β-(1 → 4) linkages that form crystalline-like microfibers and is an essential component of fungal cell walls [1, 2]. It is synthesized by chitin synthases, which are integral membrane proteins that use UDP-GlcNAc from the cytoplasm to polymerize and extrude chitin to the extracellular side of the plasma membrane [3]. Chitin synthases are found in the plasma membrane, but have also been

Erin E. McClelland (ed.), *Cryptococcus neoformans: Methods and Protocols*, Methods in Molecular Biology, vol. 2775, https://doi.org/10.1007/978-1-0716-3722-7_21,

localized to small vesicles within the cell called chitosomes [4, 5]. The chitosomes likely act as transporters to deliver chitin synthases and proteins that regulate their activity to specific cellular locations. Chitin is typically cross-linked to β-glucans and is thought to play an important role in septa formation [6]. Chitin provides strength and structure to the fungal walls.

Chitosan is the deacetylated form of chitin. The polymer becomes positively charged in acidic solutions, as well as soluble and flexible by the conversion of N-acetylglucosamine to glucosamine [7]. Enzymatic deacetylation of chitin is catalyzed by chitin deacetylases and occurs after the extrusion of chitin through the plasma membrane [8]. These deacetylases remove the acetyl group from GlcNAc to generate glucosamine (GlcN) whose amino group has a pKa of approximately 6.3 and confers a positive charge, particularly in acidic medium [8]. Chitosan, by definition, may have varying degrees of deacetylation (DA), which can be determined by its solubility in dilute acids [9, 10]. In *Cryptococcus* and some other fungi (e.g., *Mucor rouxii*), chitosan is present during vegetative growth [11]. However, in *Saccharomyces cerevisiae*, chitin deacetylase expression is restricted to the generation of the cell wall of spores and chitosan is not found in vegetatively growing cells [8].

The biochemical differences between chitin and chitosan may be exploited to measure each polymer. Depolymerization of chitosan to monosaccharides results predominantly in the release of GlcN, whereas chitin releases GlcNAc [3]. The charge difference between chitin and chitosan, due to the presence of the acetyl or amino group on the monosaccharide, means that fluorescent dyes and lectins bind with different specificities to the two polymers.

Chitin and chitosan are both vital to *C. neoformans* for cell wall strength and stability [8]. *Cryptococcus* has eight putative chitin synthase (CHS) genes, only some of which are expressed during vegetative growth in rich medium [3]. Some chitin synthases also require a chitin synthase regulator (CSR) for function; three CSR genes are encoded in the *C. neoformans* genome. There are four chitin deacetylase (CDA) genes.

For *Cryptococcus* to generate chitosan during vegetative growth, the three deacetylases (Cda1, Cda2, Cda3) specifically target chitin produced by chitin synthase 3 (Chs3), which requires Csr2 for its function [12]. Chitin deacetylases are secreted proteins, but are sequestered in different extracellular locations. Experimental evidence suggests that they can be attached to the plasma membrane through their glycosylphosphatidylinositol (GPI) anchor and become inactive when located in the cell wall [3, 8, 13].

The lack of chitosan in *Cryptococcus* results in cell wall integrity, growth, budding, and melanization defects [8]. In addition to these chitosan deficiency phenotypes, we have also identified its importance for virulence and found that chitosan-deficient strains are avirulent [12]. Strains that lack chitosan were generated by the deletion of three CDA genes (*CDA1, CDA2,* and *CDA3*). These chitosan-deficient strains (*cda1Δcda2Δcda3Δ*) generate protective immune responses when used as live or heat-killed vaccines [12, 14]. Finally, in consideration that current antifungals have limited efficacy in the treatment of an AIDS-defining *C. neoformans* infection, novel targets and new therapies are needed. One of the strategies is to screen small molecules that target the chitosan biosynthetic pathway of *C. neoformans* and, in turn, recapitulate the phenotypes we observe with chitosan-deficient strains, including rapid clearance from an infected host. These studies have led to the development and/or refinement of fluorescent techniques and biochemical methods for the measurement of chitin and chitosan of *C. neoformans* and *C. gattii* (*see* **Note 1**).

1.2 Fluorescence Techniques for Measuring Chitin and Chitosan from C. neoformans

Fluorescence techniques allow one to identify a target signal from background using a fluorophore that targets a specific biopolymer, allowing one to understand the localization and abundance of that biopolymer in the cell under different experimental settings, such as using specific mutants or growing wild-type cells under diverse conditions. Furthermore, if the fluorophores have different excitation and emission spectra, it becomes possible to detect multiple targets simultaneously (*see* **Note 2**). These fluorescence techniques require a source of light (laser, mercury lamp, or light emitting diode) of particular wavelengths through coupling with an excitation filter, a fluorophore to absorb the light, and a filter to capture the emitted light of the fluorophore (emission filter), which is converted into an electric signal by a detector. Three commonly used fluorescence detection systems that utilize these concepts are as follows: (i) spectrofluorometers that will measure the average fluorescence from the sample, (ii) fluorescence microscopy, which will generate an image of where a fluorophore binds, and (iii) flow cytometry, which will quantitate the unique binding of multiple fluorophores simultaneously on individual cells as they stream through interrogating lasers and filters [15].

Using these three systems, we have developed and refined methods to measure chitin and chitosan using different fluorophores. For chitin measurements, calcofluor white (CFW; also called Fluorescent Brightener 28) and wheat germ agglutinin (WGA) conjugated to a fluorescent dye (e.g., Alexa) were used. CFW is a 916 Da fluorescent dye that targets the β-(1 → 4) glycosidic bond linking GlcNAc monomers of chitin [16]. WGA is a 36,000 Da protein (lectin) that binds to GlcNAc oligomers of chitin, a subset of total chitin present in the cell wall [17]. For chitosan measurements, the fluorescent anionic stains, eosin Y (molar mass of 648 Da) and cibacron brilliant red 3B-A (molar mass of 907 Da), are used because they are known to preferentially bind to chitosan. Staining occurs by an ionic bond forming between the anionic dye and the protonated amino groups of GlcN that make up chitosan [18, 19]. The cationic attribute of chitosan allows for selectivity because other polysaccharides are negatively or neutrally charged [20, 21].

The three fluorescent dyes and the WGA fluorescent conjugate have enabled us to observe the location and assess the amount of chitin and chitosan associated with *C. neoformans* via fluorescence microscopy [3, 8]. When used in flow cytometry, we were able to quantitate the production level of chitin and chitosan biopolymers and the GlcNAc oligomers of chitin. We recently developed a medium-throughput assay using cibacron to quantify chitosan in the cell wall and used flow cytometry to identify small-molecule inhibitors of chitosan biosynthesis [22].

1.3 Biochemical Techniques for Measuring Chitin and Chitosan from C. neoformans

To selectively and quantitatively assess chitin and chitosan, we modified the Morgan-Elson and the 3-methyl-2-benzothiazolone hydrazine hydrochloride (MBTH) methods, respectively [23–26]. The Morgan-Elson method is based on the formation of a chromogenic product by combining p-dimethylaminobenzaldehyde (DMAB) with N-acetyl hexosamine (e.g., GlcNAc) in a dilute alkali solution after chitin is hydrolyzed to monosaccharides with exogenous chitinases [23, 25]. This method is a direct measure of chitin and specifically measures GlcNAc monosaccharides. The MBTH method is specific for GlcN and develops a chromogen under dilute acidic conditions, from a 2,5-anhydrohexose product of hexosamine (e.g., GlcN) following deamination with sodium nitrite. This method is a direct measure of chitosan. By changing how samples of alkaline-extracted cell walls are processed, both methods may also be used to quantitate the other polymer (i.e., chitosan using the Morgan-Elson and chitin using the MBTH).

2 Materials

2.1 Fluorescence Microscopy: CFW Staining of Chitin

1. Calcofluor (Fluorescent Brightner 28).
2. Ultrapure water.
3. 1× Phosphate-buffered saline (PBS).
4. Epifluorescence microscope capable of ultraviolet (UV) excitation at 387 nm (e.g., 4,6-diamidino-2-phenylindole [DAPI]).
5. CoverGrip™ Coverslip Sealant.

2.2 Fluorescence Microscopy: WGA-AF488 Staining of Chitin

1. Wheat germ agglutinin, Alexa Fluor 488 (WGA-AF488).
2. Ultrapure water.
3. 1× PBS.
4. 1× Hanks' Balanced Salt Solution (HBSS).
5. Epifluorescence microscope capable of blue excitation at 472 nm (e.g., green fluorescent protein [GFP]).

2.3 Fluorescence Microscopy: Eosin Y Staining of Chitosan

1. Eosin Y disodium salt.
2. McIlvaine's buffer: 0.2 M Na_2HPO_4, 0.1 M citric acid ($pH = 6.0$).
3. Epifluorescence microscope capable of either green excitation at 543 nm (e.g., tetramethylrhodamine [TRITC]) or excitation at 472 nm (e.g., fluorescein isothiocyanate; FITC/GFP]).

2.4 Fluorescence Microscopy: Cibacron Staining of Chitosan

1. Cibacron (Cibacron Brilliant Red 3B-A).
2. Ultrapure water.
3. 1× PBS.
4. Epifluorescence microscope capable of green excitation at 543 nm (e.g., TRITC).

2.5 Flow Cytometry: CFW Staining of Chitin

1. Calcofluor white (Fluorescent Brightner 28).
2. Ultrapure water.
3. 1× PBS.
4. 4% Paraformaldehyde (PFA).
5. Flow cytometry buffer (FCB): PBS, 0.5% bovine serum albumin (BSA), 2 mM ethylenediaminetetraacetic acid (EDTA).
6. Flow cytometer with at least one channel for detecting emission from violet (405 nm) or UV (355 nm) laser excitation and dichroic and bandpass filters for collection of emission at 420LP and 440/50, respectively.
7. FlowJo (LLC) or other flow cytometry software for data analysis.

2.6 Flow Cytometry: WGA-AF488 Staining of Chitin

1. Wheat germ agglutinin, Alexa Fluor 488 (WGA-AF488).
2. Ultrapure water.
3. 1× PBS.
4. 1× HBSS.
5. 4% PFA.
6. FCB: PBS, 0.5% BSA, 2 mM EDTA.

7. Flow cytometer with at least one channel for detecting emission from blue (488 nm) laser excitation and dichroic and bandpass filters for collection of emission at 505LP and 530/30, respectively.
8. FlowJo (LLC) or other flow cytometry software for data analysis.

2.7 Flow Cytometry: Cibacron Staining of Chitosan

1. Cibacron (Cibacron Brilliant Red 3B-A).
2. Ultrapure water.
3. 1× PBS.
4. 4% PFA.
5. FCB: PBS, 0.5% BSA, 2 mM EDTA.
6. Flow cytometer with at least one channel for detecting emission from yellow-green (561 nm) laser excitation and emission collection filter from 586/15 nm.
7. FlowJo (LLC) or other flow cytometry software for data analysis.

2.8 Quantification of Chitosan Production from C. neoformans Using Spectrofluorometry

1. Cibacron (Cibacron Brilliant Red 3B-A).
2. Ultrapure water.
3. 1× PBS.
4. Clear-bottom black 96-well microplates.
5. Spectrofluorometer that will bottom-read with excitation at 550 nm and emission at 600 nm.
6. Spectrophotometer capable of measuring absorbance at 650 nm.

2.9 Biochemical Quantification of Chitin and Chitosan: Morgan-Elson Analysis of C. neoformans Chitin

1. Sterile culture tubes.
2. Sterile yeast peptone dextrose (YPD) broth.
3. 2 mL screw capped microcentrifuge tubes.
4. Shaking incubator.
5. Microcentrifuge.
6. Freeze dryer.
7. 6% Potassium hydroxide solution.
8. 1× PBS.
9. Acetic anhydride.
10. 1.0 M Sodium bicarbonate.
11. 0.27 M Sodium borate.

12. McIlvaine's buffer: 0.2 M Na_2HPO_4, 0.1 M citric acid (pH = 6.0).
13. *Trichoderma viride* chitinase.
14. Thermocycler for programmable heating.
15. Acetic acid.
16. Hydrochloric acid.
17. *p*-Dimethylaminobenzaldehyde (DMAB). To make a 10× DMAB stock, dissolve 5 g DMAB in 43.75 mL of glacial acetic acid and combine with 6.25 mL of concentrated HCl (store in the dark at 4 °C; stable for at least 1 month). Dilute DMAB stock 1:10 with glacial acetic acid just prior to use in the assay.
18. Flat-bottom 96-well half area microtiter plate.
19. N-Acetyl-D-glucosamine (GlcNAc).
20. Microplate reader capable of measuring absorbance at 585 nm.

2.10 Biochemical Quantification of Chitin and Chitosan: MBTH Analysis of *C. neoformans* Chitosan

1. Sterile culture flasks.
2. Sterile YPD broth.
3. Microcentrifuge tubes (2 mL).
4. Shaking incubator.
5. Microcentrifuge.
6. 15 mL conical centrifuge tubes.
7. Freeze dryer.
8. 6% KOH solution.
9. 1× PBS.
10. Ultrapure water.
11. HCl.
12. Sodium nitrite.
13. Ammonium sulfamate.
14. MBTH.
15. Iron (III) chloride hexahydrate.
16. D-(+)-Glucosamine hydrochloride (GlcN).
17. Fisher sonic dismembrator model 300 attached to GraLab model 451 high-accuracy digital electronic timer or any other type of probe sonicator.
18. Flat-bottom 96-well microtiter plate.
19. Spectrophotometer capable of measuring absorbance at 650 nm.

3 Methods

3.1 Fluorescence Microscopy: CFW Staining of Chitin (Fig. 1a)

1. Dissolve 1 mg/mL CFW in ultrapure water by gentle heating at 42 °C for 10 min.
2. Wash *C. neoformans* cells twice in PBS buffer (centrifuge at 3330× *g* for 10 min) and dilute cell concentration to 1×10^7 cells/mL in PBS.
3. Add 1 mg/mL of CFW to washed and diluted cells for a final concentration of 5 μg/mL and incubate at room temperature for 15 min in the dark.
4. Wash and resuspend cells once in PBS.
5. To deposit cells onto a slide with good separation and in a monolayer, add $2–4 \times 10^5$ cells in 5–10 μL to a slide.

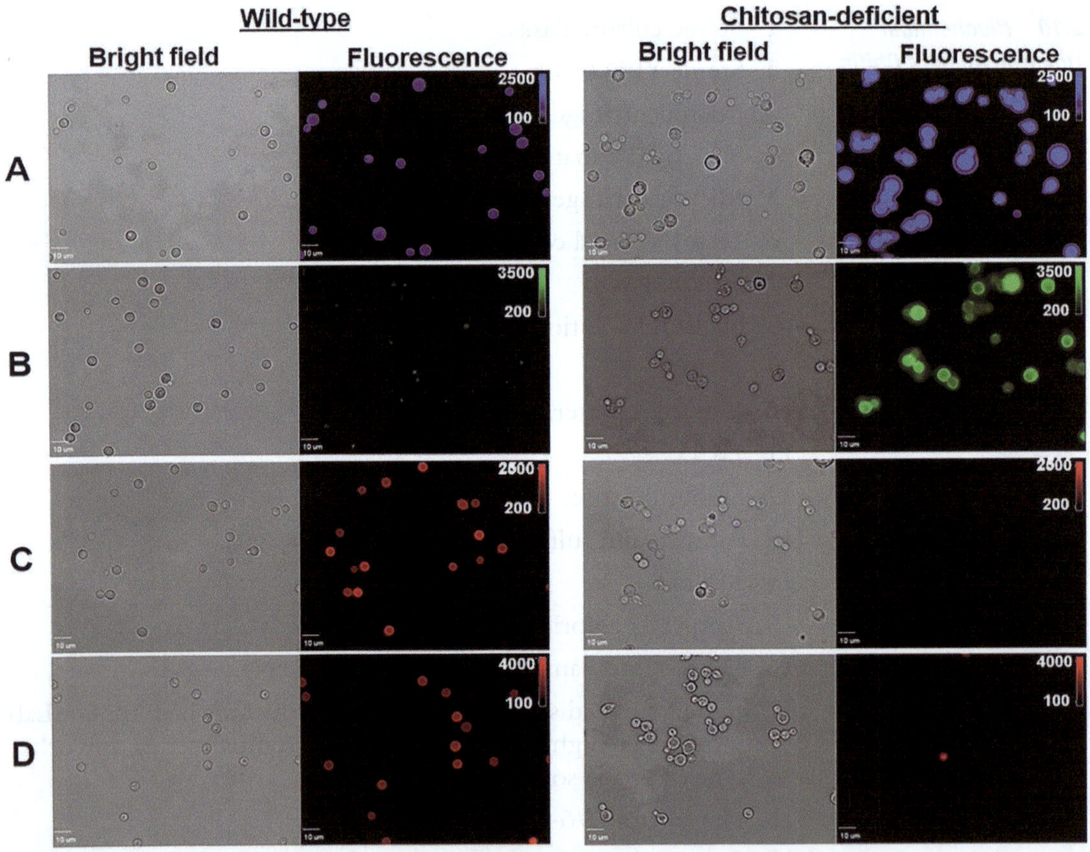

Fig. 1 Fluorescence microscopy of *C. neoformans* strains that differentially express chitin and chitosan. (**a**) Calcofluor white (CFW) and (**b**) wheat germ agglutinin (WGA-AF488) that preferentially stain chitin show that the chitosan-deficient strain (*cda1Δ cda2Δ cda3Δ*) fluoresces brighter for chitin compared to that of the wild-type strain (KN99α). Alternately, (**c**) eosin Y and (**d**) cibacron that preferentially stain for chitosan show minimal fluorescence in a chitosan-deficient strain (*cda1Δ cda2Δ cda3Δ*)

6. To prevent your sample from drying out and to increase the rate of cell settling, a cover slip is sealed over the sample using either coverslip sealant or nail polish.
7. Examine the slide by epifluorescence using filter sets for DAPI (excitation 376–398 nm; emission 387–507 nm).

3.2 Fluorescence Microscopy: WGA-AF488 Staining of Chitin (Fig. 1b)

1. Dissolve 1 mg/mL of WGA-AF488 conjugate in 4 °C ultrapure water.
2. Wash *C. neoformans* cells once in 4 °C HBSS buffer (centrifuge at 3330× *g* for 10 min) and dilute cell concentration to 1×10^7 cells/mL in HBSS.
3. Add 10 μg/mL of WGA-AF488 to washed and diluted cells and incubate at 37 °C for 10 min in the dark.
4. Wash cells once in HBSS.
5. To deposit cells onto a slide, follow Subheading 3.1, **steps 5–6**.
6. Examine the slide by epifluorescence using filter sets for GFP (excitation 442–502 nm; emission 485–555 nm).

3.3 Fluorescence Microscopy: Eosin Y Staining of Chitosan (Figs. 1c and 2)

1. Dissolve eosin Y in ultrapure water at 5 mg/mL.
2. Wash *C. neoformans* cells twice in McIlvaine's buffer, pH 6.0 (3330× *g* for 10 min), and dilute cell concentration to 1×10^7 cells/mL in McIlvaine's buffer.
3. Add 5 mg/mL of eosin Y to washed cells for a final concentration of 300 μg/mL and incubate at room temperature for 10 min in the dark.
4. Wash and resuspend cells once in McIlvaine's buffer.
5. To deposit cells onto a slide, follow Subheading 3.1, **steps 5–6**.
6. Examine the slide by epifluorescence using filter sets for either TRITC (excitation 521–565 nm; emission 553–633 nm) or FITC (excitation 472 nm; emission 520 nm) (Fig. 2).

3.4 Fluorescence Microscopy: Cibacron Staining of Chitosan (Fig. 1d)

1. Dissolve cibacron in ultrapure water at 10 mg/mL.
2. Wash *C. neoformans* cells once in PBS buffer (3300× *g* for 10 min) and dilute cell concentration to 1×10^7 cells/mL in PBS.
3. Wash cells once in PBS, pH 2.0.
4. Add 10 mg/mL of cibacron to the cells washed with PBS, pH 2, to a final concentration of 5 μg/mL and incubate at room temperature for 20 min in the dark.
5. Wash and resuspend cells once in PBS.
6. To deposit cells onto a slide, follow Subheading 3.1, **steps 5–6**.
7. Examine the slide by epifluorescence using filter sets for TRITC (excitation 521–565 nm; emission 553–633 nm).

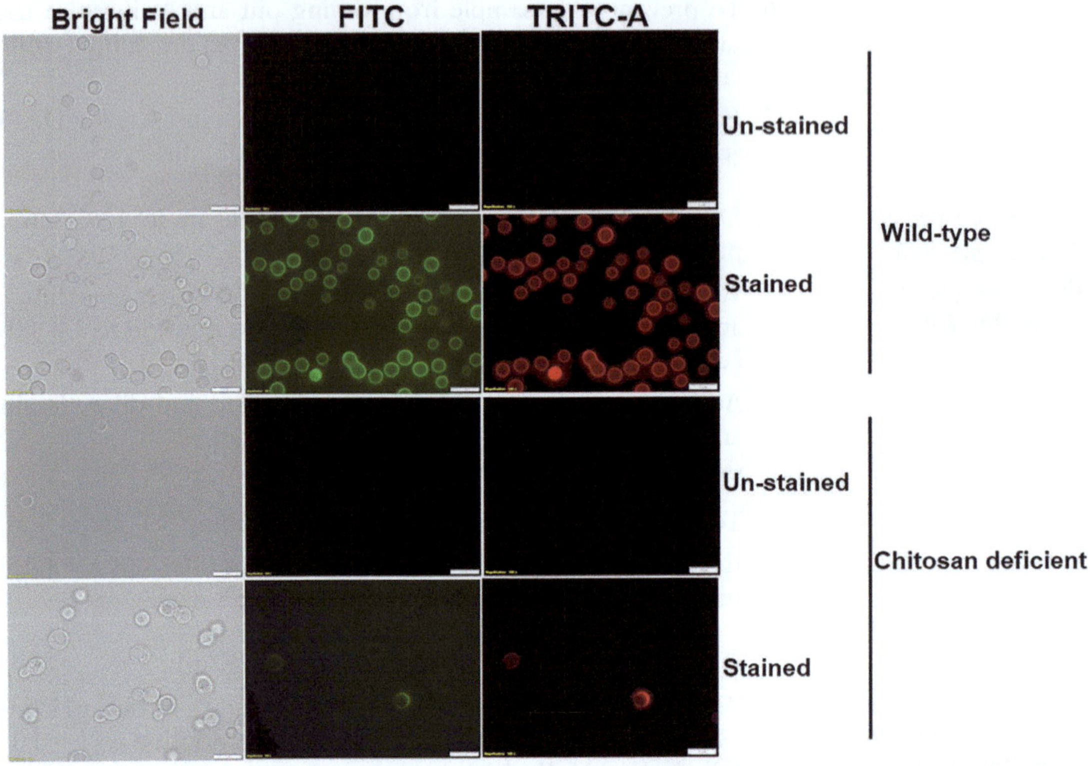

Fig. 2 Eosin-Y-stained *C. neoformans* cells may be imaged either with FITC/GFP (excitation at 472 ± 30 nm and emission at 520 ± 35 nm) or using TRITC-A (excitation at 543 ± 22 nm and emission at 593 ± 40 nm) filters. The wild-type (KN99) and the chitosan-deficient strains (*cda1Δ cda2Δ cda3Δ*) were stained with eosin-Y and the images were taken with an Olympus BX-61 fluorescent microscope. (All photos were taken at identical exposure times). Unstained control samples show no fluorescence. Scale bar 10 μm

3.5 Flow Cytometry (see Note 4): CFW Staining of Chitin (Fig. 3a)

1. Dissolve CFW in ultrapure water at 1 mg/mL by gentle heating.
2. Wash *C. neoformans* cells twice in PBS buffer (3330× *g* for 10 min) and dilute cell concentration to 1×10^7 cells/mL in PBS.
3. Staining. To one set of samples, add 5 μg/mL of CFW to the cells in PBS and incubate at room temperature for 15 min in the dark.
4. For the negative control, wash and dilute cells, and incubate at room temperature for 15 min in the dark in PBS without CFW.
5. Wash cells once in PBS.
6. Fix cells for 30 min in 1% PFA in FCB.
7. Pellet cells by centrifugation and resuspend in an appropriate volume of FCB for flow cytometry.

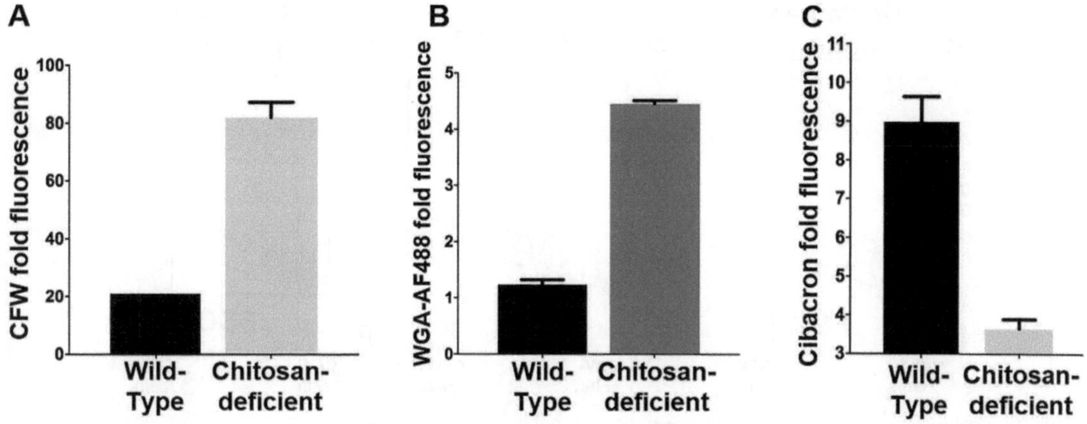

Fig. 3 (**a**) Calcofluor white (CFW), (**b**) wheat germ agglutinin (WGA-AF488), and (**c**) cibacron staining of *C. neoformans* strains that differentially express chitosan, as measured by flow cytometry. Fold fluorescence = geometric mean fluorescence intensity (gMFI) of cibacron stained strain/gMFI of unstained strain. Cell events (>1000) are equivalent among all strains within each experiment. Cells grown in complete RPMI (10% HI-FBS) at 37 °C and 5% CO_2 for 3 days, $N = 3$. *C. neoformans* wild-type (KN99α) and chitosan-deficient (*cda1Δ cda2Δ cda3Δ*) strains were analyzed

8. Flow cytometry acquisition considerations (Fig. 4): Use the UV laser (355 nm) for excitation with a dichroic and bandpass filter for collection of emission at 420LP and 440/50, respectively. Set photomultiplier tube (PMT) voltages for FSC vs SSC to exclude smaller cell events, as these small-sized events are normally considered cell debris or dead cells. Set PMT voltages to ensure that the negative control (unstained) and the CFW stained cells are on-scale. Use the negative control to set the gate for CFW^+ cells. Be sure that when comparing the amount of chitin produced (i.e., signal fluorescence of CFW) among different experimental groups, you are acquiring the same number of cell events for CFW^+ cells (>1000) and calculating the geometric mean fluorescence intensity (gMFI) for those populations.

3.6 Flow Cytometry: WGA Staining of Chitin (Fig. 3b)

1. Dissolve WGA-AF488 at 1 mg/mL in 4 °C ultrapure water.
2. Wash *C. neoformans* cells once in 4 °C HBSS buffer (3330× *g* for 10 min) and dilute cell concentration to 1×10^7 cells/mL in HBSS.
3. Staining. To one set of samples add 10 μg/mL WGA-AF488 (in HBSS) to washed and diluted cells and incubate at 37 °C for 10 min in the dark.
4. For the negative control, wash and dilute cells, and then incubate at 37 °C for 10 min in the dark in HBSS without WGA-AF488.
5. Wash cells once in HBSS.

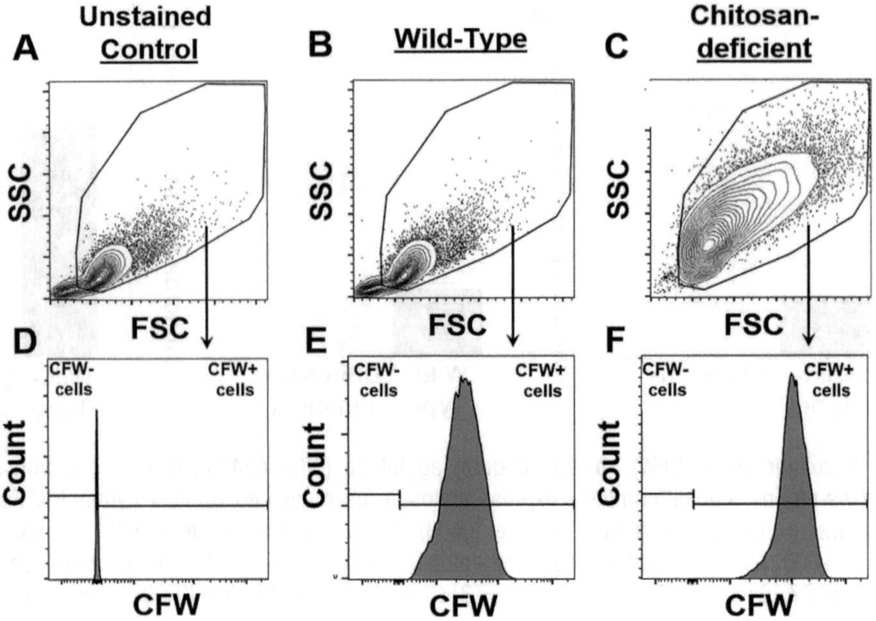

Fig. 4 Flow cytometric profiling of chitin staining of *C. neoformans* with calcofluor white (CFW). (**a–c**) *C. neoformans* gating excludes small cell events. (**d–f**) Gated regions from (**a–c**) are examined for CFW fluorescence intensity, respectively. Gating for CFW$^+$ cells is determined by applying a gate for fluorescence of unstained wild-type cells (**d**) to (**e**) wild-type and (**f**) chitosan-deficient experimental groups. All CFW$^+$ experimental groups were collected at 5000 cell events and calculated for geometric mean fluorescence intensity (gMFI). *C. neoformans* wild-type (KN99α) and chitosan-deficient (*cda1Δ cda2Δ cda3Δ*) strains were profiled

6. Fix cells for 30 min in 1% PFA in FCB.
7. Pellet cells by centrifugation and resuspend in an appropriate volume of FCB for flow cytometry.
8. Flow cytometry acquisition considerations: use the blue laser (488 nm) for excitation with a dichroic and bandpass filter for collection of emission at 505LP and 530/30, respectively. Follow the same instructions described in Sbheading 3.5, **step 8**. *See* **Note 2** for simultaneous analysis of chitin using WGA-AF488 and chitosan using cibacron (*see* **Note 3**).

3.7 Flow Cytometry: Cibacron Staining of Chitosan (Fig. 3c)

1. Dissolve cibacron in ultrapure water at 10 mg/mL.
2. Wash *C. neoformans* cells once in PBS buffer (3330× *g* for 10 min) and dilute cell concentration to 1 × 10^7 cells/μL in PBS.
3. Staining. To one set of samples: Add cibacron to cells washed with PBS, pH 2, dilute the cells for a final concentration of 5 μg/mL, and incubate at room temperature for 20 min in the dark.

4. For the negative control, wash and dilute cells, and then incubate in PBS (pH = 2) at room temperature for 20 min in the dark without cibacron.
5. Pellet cells by centrifugation (3300× *g* for 10 min).
6. Fix cells for 30 min in 1% PFA in FCB.
7. Pellet cells by centrifugation and resuspend in an appropriate volume of FCB for flow cytometry.
8. Flow cytometry acquisition considerations: use the yellow/green laser (561 nm) for excitation with bandpass filter for collection of emission at 586/15. Follow the same instructions described in Subheading 3.5, **step 8**.

3.8 Quantification of Chitosan Production from C. neoformans *Using Spectrofluorometry*

1. Dissolve cibacron in ultrapure water at 10 mg/mL.
2. Resuspend *C. neoformans* cells grown in media and determine the absorbance at OD_{650} to be later used for normalization.
3. Pellet cells by centrifugation (3330× *g* for 10 min) and then aspirate growth media supernatant.
4. Resuspend the cell pellet in 200 μL of PBS, pH 2, containing cibacron at a concentration of 25 μg/mL and incubate at room temperature for 30 min in the dark.
5. Transfer samples to clear-bottom black 96-well microplates.
6. Read plate within 30 min with a bottom-read spectrofluorometer with excitation at 550 nm and emission at 600 nm.
7. Normalize samples by dividing cibacron fluorescence by OD_{650} to compensate for differences in cell growth between samples (*see* **Notes 4**, **5**, and **6**).

3.9 Biochemical Quantification of Chitin and Chitosan: Morgan-Elson Analysis of C. neoformans *Chitin (Direct; Fig. 5a) and Chitosan (Indirect)*

1. *C. neoformans* cultures are initially grown in 3–4 mL YPD broth (15 mL culture tubes) for 24 h at 30 °C and 225 rpm.
2. Dilute culture into fresh YPD to an OD_{600} of 0.05 and shake as above for up to 3 days. Note that the amount and ratio of chitin and chitosan in *Cryptococcus* is highly dependent on growth conditions, including medium, pH, and cell density.
3. Cells (approximately 2–4 × 10^8 cells) are pelleted in preweighed 2.0 mL microfuge tubes by centrifugation at 14,000× *g* for 5 min. The supernatant is aspirated and this spin is repeated for 1 min to remove residual media.
4. Pelleted cells are then freeze-dried to normalize measurements to dry weights.
5. To include the indirect measurement of chitosan, a second identical sample of culture is pelleted.

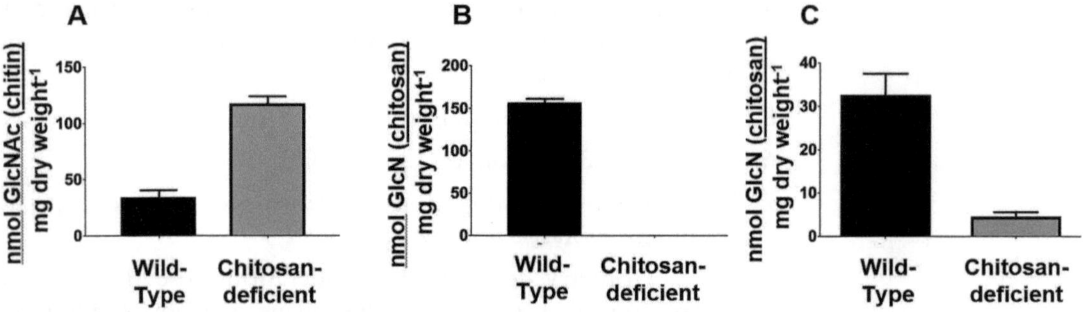

Fig. 5 Biochemical methods for measuring chitin and chitosan in *C. neoformans* strains: wild-type (KN99α) and chitosan-deficient (*cda1Δ cda2Δ cda3Δ*). (**a**) Morgan-Elson assay to measure chitin and (**b**) 3-methyl-2-benzothiazolone hydrazine hydrochloride (MBTH) assay to measure chitosan. Strains are grown in YPD at 30 °C and 300 RPM for 3 days, $N = 3$. (**c**) MBTH assay to measure chitosan. Cells are grown in complete RPMI (10% HI-FBS) at 37 °C and 5% CO_2 for 3 days, $N = 3$

6. This duplicate sample is treated with acetic anhydride to convert chitosan into chitin. Suspend the duplicate sample in 0.7 mL of 1 M sodium bicarbonate. Then, in a fume hood, add 50 μL acetic anhydride. *Caution:* The caps need to be loose to allow for venting of CO_2 that rapidly forms. Samples are periodically mixed over 20 min at room temperature and then incubated at 100 °C for 5 min. Acetylated cells are collected by centrifugation at 14,000× *g* for 10 min and the supernatant discarded.
7. Both the acetylated and untreated samples are alkaline extracted by the addition of 0.5 mL of 6% KOH to the cell pellets, followed by incubation at 80 °C for 90 min. Vortex tubes to suspend every 30 min. Deacetylation of chitin by KOH under these conditions has not been detected.
8. Samples are then centrifuged at 14,000× *g* for 10 min and the supernatant is discarded.
9. Cell pellets are washed with 1 mL of PBS, repelleted, and the PBS carefully aspirated from the pellet.
10. Suspend cell pellets by adding 100–200 μL McIlvaine's buffer, pH 6.0.
11. Samples are then frozen at −20 °C.
12. Once pellets are thawed, add 5–10 μL of *Trichoderma* chitinase (5 mg/mL in PBS; stocks stored at −80 °C) and incubate at 30 °C for 2 days; then, store at −20 °C. The *Trichoderma* chitinase contains chitinase and hexosaminidase activities to hydrolyze chitin to GlcNAc monosaccharides; GlcNAc oligosaccharides are not measureable using the Morgan-Elson assay.

3.10 Morgan-Elson Assay

1. Chitinase-treated samples are centrifuged at 14,000× *g* for 1 min.
2. In 0.2 mL PCR tubes, 10 μL of sample supernatant is combined with 10 μL of 0.27 M sodium borate.
3. Samples are then heated to 99.9 °C in a thermocycler for 90 s, mixed, and incubated for an additional 10 min at 99.9 °C.
4. Once cooled to room temperature, 100 μL of freshly made DMAB solution is added and samples are incubated at 37 °C for 20 min for color development.
5. Transfer 100 μL of each sample to a flat-bottom 96-well half area microtiter plate and read absorbance at 585 nm. Readings need to be made promptly as the color fades.
6. To quantitate GlcNAc in each sample, standard curves are prepared using 0.2–2.0 mM GlcNAc in water.
7. The amount of chitosan is calculated by subtracting the amount of chitin in the sample that was not treated with acetic anhydride (chitin only) from the sample that was treated with acetic anhydride (chitin + chitosan).

3.11 Biochemical Quantification of Chitin and Chitosan: MBTH Analysis of C. neoformans *Chitosan (Direct; Fig. 5b, c) and Chitin (Indirect)*

1. *C. neoformans* cultures are grown in 45 mL of YPD broth (5:1; flask/culture volume) for 72 h at 30 °C and 300 rpm. Note that the amount and ratio of chitin and chitosan in *Cryptococcus* are highly dependent on growth conditions, including medium, pH, and cell density.
2. A total of 10 mL of the culture is centrifuged for 10 min at 3300× *g* using pre-weighed 15 mL conical centrifuge tubes. The supernatant is aspirated and then samples are resuspended with PBS and the cells pelleted.
3. Pelleted cells are then weighed after being freeze-dried to normalize chitosan measurements to dry weights.
4. To reduce background chromogens, proteins are extracted from the sample by the addition of 10 mL of 6% KOH to cell pellets, followed by incubation at 80 °C for 30 min. Vortex tubes at maximum speed every 10 min.
5. Wash cell twice with 10 mL of PBS.
6. Wash cells twice with 10 mL of ultrapure water.
7. To include an indirect measurement of chitin, split the sample into two identical aliquots. To each 100 μL aliquot, add 100 uL of 1 M HCl. Fully deacetylate one of the subsamples at 110 °C for 2 h (chitin + chitosan). Allow sample to cool to room temperature. The other aliquot is not heated (chitosan).
8. Normalize all samples by adding enough ultrapure water to make all of them the same concentration. For example, we make concentrations at 10 mg/mL dry weight and measure the chitosan in 100 μL or 1 mg dry weight samples.

3.12 MBTH Colorimetric Determination of GlcN

1. To create the dilute acidic conditions needed for this reaction, make a 1:1 dilution of the above samples in a microcentrifuge tube with 1 M HCl for a 200 μL total volume.
2. In a fume hood, add 400 μL of 0.36 M sodium nitrite, vortex, and incubate for 15 min at room temperature.
3. In a fume hood, to avoid effervescence of the solution, slowly add 200 μL of 1.1 M ammonium sulfamate, vortex, and incubate for 5 min at room temperature.
4. Add 200 μL of 0.011 M MBTH and vortex. Then, incubate tubes for 30 min at 37 °C.
5. To generate blue color formation, add 200 μL of 0.018 M iron (III) chloride hexahydrate and vortex. Then, incubate tubes for 5 min at 37 °C.
6. Allow tubes to cool to room temperature. Transfer 100 μL of each sample to a flat-bottom 96-well microtiter plate and measure absorbance at 650 nm.
7. To quantitate chitosan in each sample, a standard curve is freshly prepared from 1:1 serially diluted 2 mM GlcN down to 0.016 mM GlcN using water as the diluent. By applying the MBTH colorimetric determination protocol on 100 μL of each standard, GlcN may be determined from the standard curve (i.e., 0, 1.6, 3, 6.3, 12.5, 25, 50, 100, and 200 ηmol). A linear regression equation is generated for the standard curve to determine the ηmol of GlcN (chitosan) per 1 mg dry weight of each sample.
8. To indirectly quantify chitin, subtract the amount obtained from the unheated aliquot (chitosan only) from the amount measured for the aliquot heated at 110 °C in Subheading 3.9, **step** 7 (chitin plus chitosan; *see* **Notes** 7, **8**, and **9**).

4 Notes

1. These methods may also be applied to *C. gattii*.
2. Flow cytometric analysis may be used to quantitate chitosan and GlcNAc oligomers of chitin simultaneously within the same samples using the flow cytometry protocols above and staining sequentially using cibacron and WGA-AF488.
3. All flow cytometry methods are optimized for single-tube or 96-well format analysis. Microtiter, 96-well plate format, using the BD LSRFortessa X-20 flow cytometer integrated with a BD High Throughput Sampler that is capable of a 20-parameter analysis, allowed us to have a medium-throughput analysis (>1000 wells/working day).

4. Measuring cibacron staining was optimized using a Biotek Synergy HT Multi-Mode Microplate reader in a 96-well format and for automated high-throughput analysis using a PerkinElmer EnVision Multimode plate reader to analyze >10,000 wells/day.
5. Please note that unbound cibacron is not fluorescent; as a result, washing is not necessary. This characteristic makes it ideal for high-throughput analysis of chitosan content from *Cryptococcus.*
6. Because cibacron strongly stains dead cells, it will interfere with fluorescence measurements using a spectrofluorometer. In such cases, the protocol for the plate reader must be configured so that fluorescence intensities above a threshold value are not considered for measurement.
7. The Morgan Elson and MBTH biochemical measurements are based upon the target monosaccharide, GlcNAc for the Morgan Elson and GlcN for the MBTH, so that each directly measures the quantity of chitin and chitosan, respectively. However, they both may be modified to indirectly measure the content of the other polysaccharide. To measure chitosan by the Morgan Elson assay, one splits the sample and chemically acetylates all the chitosan to chitin. The measurement represents the sum of chitin plus chitosan. Subtracting out the chitin measured from the untreated aliquot provides an indirect measurement of chitosan. The same is done in reverse for the MBTH assay. The sample is split and one aliquot is boiled in acid (1 M HCl) to deacetylate chitin to chitosan; the difference between the measurements of the boiled, deacetylated, and the untreated aliquot provides the indirect measurement of chitin.
8. For any of the above methods, if chitin or chitosan measurements from *C. neoformans* are desired when cultured in complete RPMI (cRPMI; ≤10% fetal bovine serum [FBS]), then cells are grown first in YPD, washed twice in PBS, and added to cRPMI at 5×10^5 cells/mL with 72 h incubation at 37 °C.
9. Depending on the type of growth media used, KOH-treated samples should be sonicated using a probe sonicator in Subheading 3.4, **step** 7 to generate a uniform cell suspension. We used a Fisher sonic dismembrator model 300 attached to a GraLab model 451 high-accuracy digital electronic timer to sonicate for 5 cycles of 60 s each to disrupt the clumps of cellular material.

Acknowledgments

We thank Joseph Vasselli for helpful comments on the manuscript. This work was supported by NIH grants AI072195 to JKL and AI125045 to JKL and CS.

References

1. Free SJ (2013) Fungal cell wall organization and biosynthesis. Adv Genet 81:33–82. https://doi.org/10.1016/B978-0-12-407677-8.00002-6
2. Gow NAR, Latge JP, Munro CA (2017) The fungal cell wall: structure, biosynthesis, and function. Microbiol Spectr 5(3). https://doi.org/10.1128/microbiolspec.FUNK-0035-2016
3. Banks IR, Specht CA, Donlin MJ et al (2005) A chitin synthase and its regulator protein are critical for chitosan production and growth of the fungal pathogen *Cryptococcus neoformans*. Eukaryot Cell 4:1902–1912. https://doi.org/10.1128/ec.4.11.1902-1912.2005
4. Rogg LE, Fortwendel JR, Juvvadi PR et al (2012) Regulation of expression, activity and localization of fungal chitin synthases. Med Mycol 50(1):2–17. https://doi.org/10.3109/13693786.2011.577104
5. Leal-Morales CA, Bracker CE, Bartnicki-Garcia S (1988) Localization of chitin synthetase in cell-free homogenates of *Saccharomyces cerevisiae*: chitosomes and plasma membrane. Proc Natl Acad Sci U S A 85(22):8516–8520. https://doi.org/10.1073/pnas.85.22.8516
6. Kollar R, Petrakova E, Ashwell G et al (1995) Architecture of the yeast cell wall. The linkage between chitin and beta(1-->3)-glucan. J Biol Chem 270(3):1170–1178. https://doi.org/10.1074/jbc.270.3.1170
7. Bulik DA, Olczak M, Lucero HA et al (2003) Chitin synthesis in *Saccharomyces cerevisiae* in response to supplementation of growth medium with glucosamine and cell wall stress. Eukaryot Cell 2(5):886–900. https://doi.org/10.1128/EC.2.5.886-900.2003
8. Baker LG, Specht CA, Donlin MJ et al (2007) Chitosan, the deacetylated form of chitin, is necessary for cell wall integrity in *Cryptococcus neoformans*. Eukaryot Cell 6(5):855–867. https://doi.org/10.1128/ec.00399-06
9. Sorlier P, Hartmann DJ, Denuziere A et al (2003) Preparation and development of anti-chitosan antibodies. J Biomed Mater Res A 67: 766–774. https://doi.org/10.1002/jbm.a.10132
10. Raafat D, Sahl HG (2009) Chitosan and its antimicrobial potential – a critical literature survey. Microb Biotechnol 2(2):186–201. https://doi.org/10.1111/j.1751-7915.2008.00080.x
11. Davis LL, Bartnicki-Garcia S (1984) The co-ordination of chitosan and chitin synthesis in *Mucor rouxii*. J Gen Microbiol 130(8): 2095–2102. https://doi.org/10.1099/00221287-130-8-2095
12. Baker LG, Specht CA, Lodge JK (2011) Cell wall chitosan is necessary for virulence in the opportunistic pathogen *Cryptococcus neoformans*. Eukaryot Cell 10(9):1264–1268. https://doi.org/10.1128/ec.05138-11
13. Gilbert NM, Baker LG, Specht CA et al (2012) A glycosylphosphatidylinositol anchor is required for membrane localization but dispensable for cell wall association of chitin deacetylase 2 in *Cryptococcus neoformans*. MBio 3 (1):e00007-00012. https://doi.org/10.1128/mBio.00007-12
14. Upadhya R, Lam WC, Maybruck B et al (2016) Induction of protective immunity to cryptococcal infection in mice by a heat-killed, chitosan-deficient strain of *Cryptococcus neoformans*. MBio 7(3):e00547-00516. https://doi.org/10.1128/mBio.00547-16
15. Anonymous (2018) Fluorescence fundamentals. In: the molecular probes handbook. ThermoFisher Scientific. https://www.thermofisher.com/us/en/home/references/molecular-probes-the-handbook/introduction-to-fluorescence-techniques.html. Accessed 1 May 2023
16. Gray LD, Roberts GD (1988) Laboratory diagnosis of systemic fungal diseases. Infect Dis Clin N Am 2(4):779–803
17. Rodrigues ML, Alvarez M, Fonseca FL et al (2008) Binding of the wheat germ lectin to *Cryptococcus neoformans* suggests an association of chitinlike structures with yeast budding and capsular glucuronoxylomannan. Eukaryot Cell 7(4):602–609. https://doi.org/10.1128/EC.00307-07
18. Mendelovits A, Prat T, Gonen Y et al (2012) Improved colorimetric determination of chitosan concentrations by dye binding. Appl

Spectrosc 66(8):979–982. https://doi.org/10.1366/12-06591a

19. Muzzarelli RA (1998) Colorimetric determination of chitosan. Anal Biochem 260(2):255–257. https://doi.org/10.1006/abio.1998.2705
20. Bueter CL, Specht CA, Levitz SM (2013) Innate sensing of chitin and chitosan. PLoS Pathog 9(1):e1003080. https://doi.org/10.1371/journal.ppat.1003080
21. Cheung RC, Ng TB, Wong JH et al (2015) Chitosan: an update on potential biomedical and pharmaceutical applications. Mar Drugs 13:5156–5186. https://doi.org/10.3390/md13085156
22. Maybruck BT, Lam WC, Specht CA et al (2019) The aminoalkylindole BML-190 negatively regulates chitosan synthesis via the cyclic AMP/protein kinase A1 pathway in *Cryptococcus neoformans*. MBio 10(6):e02264-02219. https://doi.org/10.1128/mBio.02264-19
23. Morgan WT, Elson LA (1934) A colorimetric method for the determination of N-acetylglucosamine and N-acetylchrondrosamine. Biochem J 28(3):988–995. https://doi.org/10.1042/bj0280988
24. Reissig JL, Storminger JL, Leloir LF (1955) A modified colorimetric method for the estimation of N-acetylamino sugars. J Biol Chem 217(2):959–966
25. Tsuji A, Kinoshita T, Hoshino M (1969) Analytical chemical studies on amino sugars. II. Determination of hexosamines using 3-methyl-2-benzothiazolone hydrazone hydrochloride. Chem Pharm Bull (Tokyo) 17(7):1505–1510. https://doi.org/10.1248/cpb.17.1505
26. Smith RL, Gilkerson E (1979) Quantitation of glycosaminoglycan hexosamine using 3-methyl-2-benzothiazolone hydrazone hydrochloride. Anal Biochem 98(2):478–480

Chapter 22

Induction of Dormancy in *Cryptococcus neoformans* In Vitro: The HypNOS Protocol

Ruchi Agrawal, Aude Sturny-Leclère, Raffael J. A. de Castro, and Alexandre Alanio

Abstract

Cryptococcus neoformans is the second major cause of death in patients with HIV. During a latent infection, this pathogenic fungus survives in the host for years without causing symptoms of active disease. Upon favorable conditions, such as immunosuppression due to HIV infection, or other conditions (steroid use or organ transplantation), the yeast may reactivate and cause active cryptococcosis. Hence, dormancy is an important phase in the pathogenesis of *C. neoformans*. Additionally, *C. neoformans* also persists during antifungal treatment and causes disease recurrence, which is a major medical problem, especially in low- and middle-income countries. To survive in the host, yeast cells must react to the stresses they are exposed to and generate a cellular response that is favorable for yeast survival. A prominent strategy used by *C. neoformans* to combat challenging surroundings is dormancy, which may translate into a viable, but nonculturable phenotype (VBNC). This chapter describes an in vitro protocol to generate and characterize dormant *Cryptococci*.

Key words *Cryptococcus neoformans*, In vitro dormancy protocol, Hypoxia, Normoxia, Dormant cells, Reactivation, Viable but nonculturable phenotype, Resuscitation

1 Introduction

Dormancy is an important life-phase of many pathogens, including *Cryptococcus neoformans* [1]. The dormant yeast form is able to survive difficult environmental conditions, such as hypoxia, within a granuloma, nutrient deprivation, and antibiotic treatment. The yeast reactivates and causes active cryptococcosis once the surroundings are favorable, sometimes decades after the primary infection. Recurrence of clinical infection, even after successful antifungal treatment, poses additional risks in cryptococcal meningitis and other cryptococcal infections [2, 3]. *C. neoformans* has recently been placed in the "critical priority pathogens" category in the World Health Organization's fungal priority pathogens list

Erin E. McClelland (ed.), *Cryptococcus neoformans: Methods and Protocols*, Methods in Molecular Biology, vol. 2775, https://doi.org/10.1007/978-1-0716-3722-7_22,

[4]. After in vivo and in vitro studies confirm the presence of dormant yeasts, it is important to characterize these dormant cells in detail to tackle the potential problems associated with clinical relapse. Since dormant yeast recovered from murine infection models are low in number (< 30/lungs) [5], multiple processing steps are required to obtain a homogenous population, which could affect the phenotype. Thus, it is essential to establish an in vitro protocol that generates a homogenous population of dormant cells in large enough amounts to allow in-depth characterization.

In this chapter, we describe in detail an in vitro protocol to generate viable, but nonculturable cells (VBNC) of *C. neoformans*, using low nutrients and low oxygen conditions [6]. We named this protocol HypNOS (HypoNutrition and hypOxia for Seven days), which generates yeast that exhibit high viability (> 90%), increased latency, and low culturability (< 3%) at day 7, which are hallmarks of dormant cells. It is important to note that cell/population phenotypes obtained via the in vitro HypNOS protocol may not completely resemble the subpopulation of dormant yeasts isolated from the lungs of infected mice from an in vivo model, due to obvious reasons. However, HypNOS protocol does allow the generation of VBNC cells for in-depth characterization of dormant phenotypes.

2 Materials

1. *C. neoformans* strain H99O (*see* **Note 1**) stored in 20% glycerol at −80 °C (*see* **Note 2**).
2. Yeast Peptone Dextrose (YPD) broth and solid media: 10 g yeast extract, 20 g peptone, 20 g D-glucose, 20 g agar (for solid medium) in 1 L of Milli-Q water, pH 6.5. Autoclave prior to use (*see* **Note 3**). Store at 4 °C.
3. Sabouraud Dextrose Agar (SDA): 10 g peptone, 20 g D-glucose, 20 g agar in 1 L of Milli-Q water, pH 6.5. Autoclave prior to use. Store at 4 °C.
4. T25 cm^2 flask with vented filter cap.
5. 96-well tissue culture plates with lid.
6. Incubator with the following settings: Temperature—30 °C. Shaking—lateral shaking at 150 rpm, with an artificial light bulb for illumination.
7. Minimal medium (MM): 15 mM D-glucose, 10 mM $MgSO_4$, 29.4 mM KH_2PO_4, 13 mM glycine, 3.0 μM thiamine, pH 5.5. Filter using 0.22 μm membrane filter and store at 4 °C (*see* **Note 4**).

8. Pantothenic acid (PA, D-Pantothenic acid hemicalcium salt): Dissolve in Milli-Q water to a concentration of 25 mM. Filter using 0.22 μm membrane filter and store at −20 °C for up to 6 months. Add to MM or MM10% to a final concentration of 125 μM before use. Alternatively, PA may be prepared fresh and added to the media while in use.
9. Viability stain (LIVE/DEAD™ Fixable Violet Dead Cell Stain).
10. Hypoxia generator: Airtight bags, chemical generator sachets, and clips to seal.
11. Sterile Dulbecco's phosphate-buffered saline.
12. Bioscreen apparatus and honeycomb 100-well plates.

3 Methods

3.1 Generation of Dormant *C. neoformans* Cells

In this protocol, stationary phase *Cryptococcus* cells are exposed to hypoxia in low-nutrient conditions for 7 days (*see* **Note 5**) to induce dormancy. This generates a population of cells that are mostly VBNC cells (*see* **Note 6**).

1. Streak *C. neoformans* strain H99O from a glycerol stock (stored in 20% glycerol at −80 °C) on a SDA or YPD agar plate and incubate at room temperature on the bench top for 2–5 days (*see* **Note** 7 and **8**).
2. Using a calibrated 1 μL loop, inoculate 1×10^7 yeast from the agar plate into 10 mL of YPD in a T25 cm^2 flask with vented cap (*see* **Note 9**). Incubate the flask at 30 °C with lateral shaking at 150 rpm and the light on, for 22 h to reach stationary phase (*see* **Note 10**). This is termed "Pre-culture-1." At this point, the cell concentration should be ~2×10^8 cells/mL.
3. Inoculate 100 μL (*see* **Note 11**) of the stationary phase culture (Pre-culture-1) into four T25 cm^2 flasks with vented caps containing 10 mL of YPD and incubate in the same conditions (30 °C with lateral shaking at 150 rpm and the light on for 22 h) (*see* **Note 12**). This is termed "Pre-culture-2."
4. After 22 h, the medium reaches stationary phase and is depleted of nutrients. Place two Pre-culture-2 flasks in an airtight bag with an anaerobic generator and seal to generate hypoxia (< 0.1% O_2) and incubate for 7 days in the dark at 30 °C in static conditions. These yeast cells are VBNC (*see* **Note 13**) (Fig. 1). Incubate the remaining two flasks in identical conditions, except they are not put in the anaerobic bag. These yeasts are termed NORMOx, since they are in normal oxygen conditions, and will be used as control cells (Fig. 1).

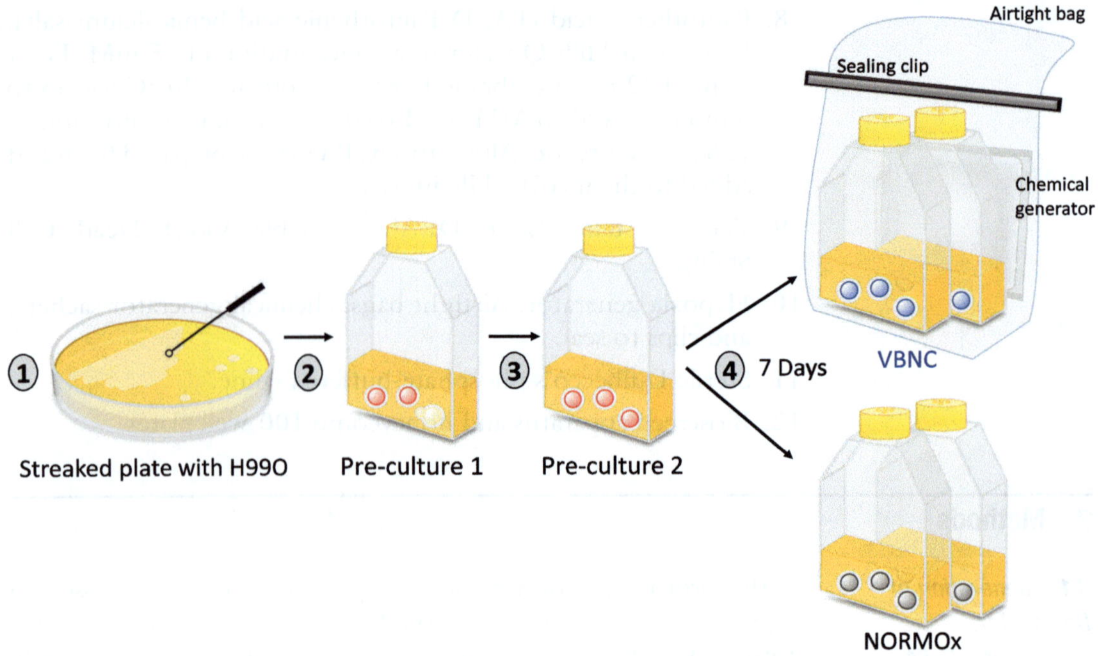

Fig. 1 Schematics to generate dormancy in *C. neoformans*: ***Step*** **1,** steak *C. neoformans* onto an SDA/YPD agar plate. ***Step*** **2,** inoculate 1×10^7 yeast into liquid YPD and incubate under agitation (150 rpm) at 30°C for 22 h in the presence of light to obtain Pre-culture-1 cells. ***Step*** **3,** inoculate 100 μL from Pre-culture-1 into 10 mL of fresh YPD, and incubate under agitation (150 rpm) at 30°C for 22 h in the presence of light to obtain Pre-culture-2 cells. ***Step*** **4,** after 22 h, the Pre-culture-2 flasks are incubated in a sealed anaerobic bag for 7 days, at 30°C, in the dark to obtain VBNC cells or in identical conditions, except the flasks are not placed in an airtight bag, to obtain NORMOx cells

5. At day 6, inoculate 1×10^7 yeast from the agar plate into 10 mL of YPD in a T25 cm^2 flask with vented cap and incubate as described in Subheading 3.1, **step 2**. These yeasts will also be used as control cells, consisting of yeast in stationary phase, and are termed STAT cells.

3.2 Characterization of Dormant C. neoformans Cells

VBNCs have certain characteristics, which categorize them as dormant cells. We use the following criteria: culturability, viability, and latency of growth to characterize VBNC cells generated by the HypNOS protocol. In summary, cells are exposed to different conditions (solid vs. liquid medium, rich vs. poor medium, or the presence or absence of cofactors) and their phenotypes are observed. It is important to note that these are basic criteria that we have outlined here; however, there are additional characteristics of VBNC cells that are found in the original research article [6].

The following two steps must be followed before any characterization method described below is performed.

1. After day 7, remove 1 mL of culture from each condition (VBNC, NORMOx, and STAT) and transfer to 1.5 mL Eppendorf tubes. Centrifuge at 1500 × *g*, at room temperature for 5 min, to obtain cell pellets. Wash the cell pellets twice with sterile phosphate-buffered saline (PBS) and resuspend in 1 mL sterile PBS (*see* **Note 14**).
2. Count the number of cells in each condition using a hemocytometer. Dilute the cells 1:100 or 1:200 in PBS to obtain a concentration that is within the range of the hemocytometer's recommendations for counting (*see* **Note 15**). Usually, the cell concentration at this point is approximately 3–4 × 10^8 cells/mL.

3.3 Assessment of Viability Using Live-Dead Staining

A successful stress exposure to generate dormant cells should produce enough viable cells to use in experiments. If the conditions are too harsh, most of the cells will die, and the protocol will not generate sufficient number of live cells for characterization. Thus, a viability stain is used to differentiate and quantify live and dead cells.

1. Add 1 × 10^7 yeast cells in 500 μL of PBS containing 0.5 μL of reconstituted LIVE/DEAD™ Fixable Violet Dead Cell Stain for each condition (*see* **Note 16**).
2. Use 1 × 10^7 heat-killed STAT cells as a positive control. Heat STAT cells in PBS in a heat block for 60–90 min at 70 °C to heat-kill the cells.
3. Incubate for 30 min at 30 °C in the dark.
4. Wash the cells with PBS three times and resuspend in PBS.
5. Quantify the number of dead and live cells using flow cytometry and microscopy with excitation and emission wavelengths of 416 nm and 451 nm, respectively.

The successful HypNOS protocol should result in more than 95% live cells. NORMOx and STAT cells should be 70%–90% and ~ 100% viable, respectively (Fig. 2a).

3.4 Assessment of Culturability on Solid Medium Using the Colony-Forming Unit (CFU) Method

After exposure to stressful conditions, when cells are allowed to grow in favorable conditions, such as on rich YPD agar, a subpopulation will resuscitate, divide, and give rise to colonies, which can be quantified to determine the culturability of the population.

1. Spread 3000–10,000 VBNC cells, 1000 NORMOx cells, and 100–200 STAT cells onto YPD agar plates in duplicate. Incubate plates at room temperature on the bench top for up to 5 days (*see* **Notes 8** and **17**).
2. After 5 days of incubation, record the number of colonies in each plate (*see* **Note 18**).

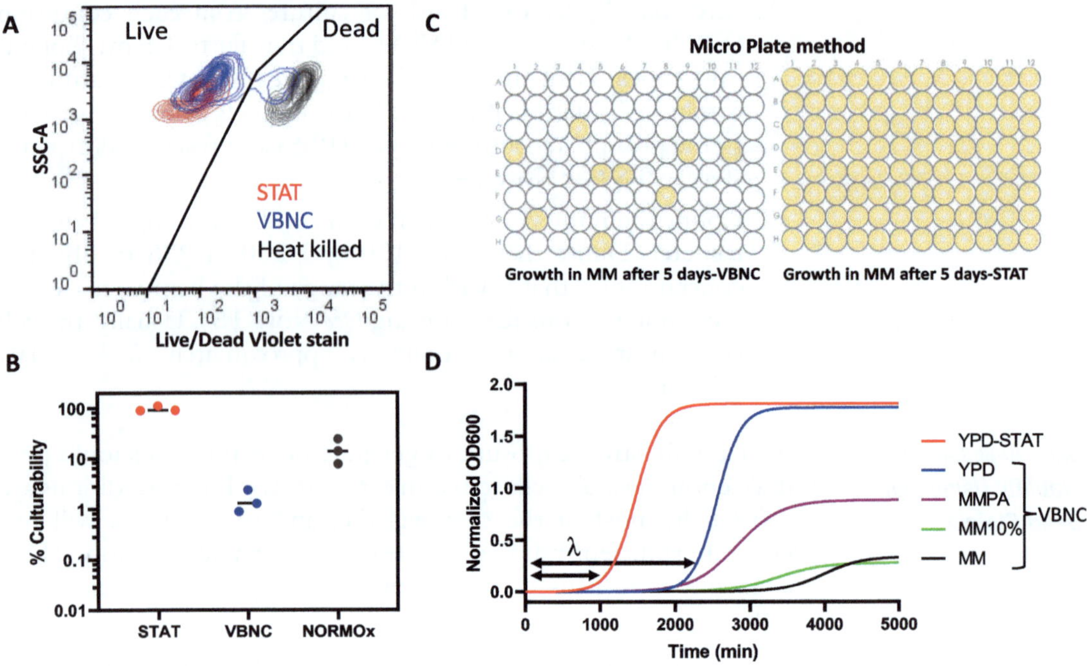

Fig. 2 Characterization of dormant cells: (**a**) Viability using the LIVE/DEAD staining method. Representative flow cytometry data after staining with LIVE/DEAD™ Fixable Violet Dead Cell Stain. (**b**) % Culturability using the colony-forming unit (CFU) method. Representative data from three independent experiments is shown. (**c**) Liquid culturability using the microplate method. Illustration of positive growth after 5 days of incubation in MM for STAT and VBNC cells. (**d**) Latency (λ) determination using growth curve analysis. Representative graph after nonlinear regression (curve fit) analysis using the "logistic model BH" in GraphPad Prism. For simplicity, the data is plotted for 3.5 days, instead of 5 days

3. Using the following formula, calculate the % Culturability for each condition.

$$\%\text{Culturability} = (\text{Average number of colonies/no.of cells plated}) \times 100.$$

The expected % culturability for VBNC, NORMOx, and STAT cells is usually in the range of 0.1%–3%, 5%–25%, and 95%–100%, respectively (Fig. 2b).

3.5 Assessment of Culturability in Liquid Medium Using the Microplate Method

This method is a modified version of the most probable number (MPN) technique commonly used to estimate microbial populations in water, soil, and food samples [7]. Here, this technique is used for two reasons: (i) a higher number of cells may be screened, and (ii) it will reduce the growth inhibition caused by incubation on solid media, if any. Two media conditions may be used to determine the probability of VBNC cell growth: (1) a well-defined MM; (2) MM containing the cofactor PA (MMPA). PA is used here because it is a growth hormone that triggers dormant *Cryptococci* to exit dormancy and allows resuscitation [6].

1. Dilute VBNC or STAT cells into MM and MMPA at a concentration of 5×10^2 cells/mL.
2. Using a multichannel pipette, distribute 200 μL of the above solutions in each well of three 96-well microplates per condition/medium, to obtain 100 cells per well (with 1%–2% culturability resulting in 1–2 culturable cells per well).
3. Incubate the plates for 5 days at 30 °C (*see* **Note 17**).
4. Count the number of positive wells at the end of the incubation time (*see* **Note 19**).
5. Calculate culturability, which is represented as the probability (p) of a cell to grow, using the following formula (based on the Poisson distribution), where X is the number of wells with growth, and Y is the total number of wells filled with yeast suspension:

$$X \sim B(100, p)\ p = 1 - \left(1 - \frac{X}{Y}\right)^{1/100}$$

Similar to culturability using the CFU method, p of a STAT cell to grow would be 100%, as all wells will show growth in both media (Fig. 2c). For VBNC cells, p varies from 0.1% to 3%, again similar to the CFU method. As expected, p (MMPA) > p (MM), as PA helps cells revive from dormancy.

3.6 Determination of Latency Using Growth Curve Measurements

A growth curve is classically composed of four phases: (1) Lag (Latent) phase. (2) Log (Exponential) phase. (3) Stationary phase. (4) Death phase. Dormant cells will take longer to reactivate since they have to reorganize their metabolic activity to the new nutrient conditions; hence, dormant cells show longer latency (λ). Another factor, which may affect their growth, is the type of medium used for resuscitation: rich medium (YPD), defined medium (MM), poor medium (MM diluted to 10% (MM10%)), and defined medium containing the cofactor PA (MMPA).

1. Dilute the cells resuspended in PBS into YPD, MM, MM10%, and MMPA at a concentration of 1×10^4 cells/mL.
2. Add 300 μL of each into honeycomb plate wells in triplicate (*see* **Note 20**).
3. As a control, use 300 μL of media alone in triplicate.
4. Incubate the plate/s for 120 h (5 days) at 30 °C with continuous, high amplitude and fast speed agitation using a Bioscreen apparatus. Record optical density (OD) at 600 nm every 20 min.
5. After 5 days, collect the data into an Excel file and subtract the ODs of the control media from the respective samples.

6. To calculate latency, plot the subtracted OD values on the Y-axis versus time on the X-axis using an XY table in GraphPad Prism. Then, perform an XY analysis using the nonlinear regression model "logistic model BH" (developed from logistic and Richards-modified models and written in GraphPad Prism) to determine latency and the slope of the curve. In the analyzed table, the best-fit values, L and m, represent latency and growth rate/slope of the curve, respectively. See Supplementary file (S1) for a sample calculation.

Overall, VBNC cells have a longer latency in all media tested compared to that of STAT cells. YPD is a very rich medium with many cofactors, including PA; hence, we expect the shortest latency in YPD among all media. MM10% shows shorter latency than MM, possibly due to nutrient shock. It is known that a change in nutrient conditions (sudden exposure of nutrient rich media) may cause the cells to go into shock, and they take longer to resuscitate [8, 9]. MMPA shows reduced latency compared to that of MM and MM10% because it contains the cofactor PA, which helps the cells revive from dormancy (Fig. 2d).

4 Notes

1. The HypNOS protocol has been established using the H99O strain of *C. neoformans* and has been demonstrated to be compatible with its engineered fluorescent derivatives, such as the GFP-tagged Poly(A) binding protein (PAB) transformant (PAB-GFP) [10]. The protocol may require some optimization for other species/strains of *Cryptococcus.*
2. *C. neoformans* is a Biosafety level (BSL) class-2 pathogen. Thus, it is important to follow BSL2 requirements, such as wearing appropriate PPE and working in a BSL2 hood.
3. YPD should be autoclaved at 105 °C for 20–30 min rather than the usual 121 °C, because at 121 °C, the sugar may be caramelized, and caramelized sugar affects the dormancy protocol. Alternatively, one can filter sterilize YPD using a 0.22 μm membrane filter.
4. Thiamine should be dissolved in Milli-Q water at a concentration of 3 mM and kept at −20 °C. It should be added to MM before use.
5. This protocol is designed to generate VBNC cells in 7 days, but 6–8 days can also be optimized.
6. VBNC cells, by definition, are viable, but nonculturable cells. However, in the presence of a cofactor, VBNC cells may resuscitate and revert to actively dividing cells.

7. This is a very critical step, as plates incubated for more than 5 days, do not generate cells with a good VBNC phenotype.
8. All agar plates in this protocol have been incubated on the benchtop at room temperature, exposed to ambient/day light. Alternatively, plates can be incubated at 30 °C, with the light on in an incubator.
9. For inoculation, inoculating with a single colony should be avoided. Instead inoculate using cells from the heavy confluent growth area of the plate.
10. Lateral shaking and illumination are critical for this protocol. Alternatively, if lateral shaking is not available, equal, or lower rpm should be tested on a rotational shaking.
11. This is a very critical amount. Inoculation with more or less than 100 μL does not result in similar outcomes.
12. It is highly critical that the culture achieves stationary phase.
13. The atmosphere in the GENbag becomes enriched in CO_2 by the biochemical process of oxygen depletion. It is important to keep the number of flasks consistent, sealed at similar height in a GENbag for every experiment so that cells are exposed to constant hypoxic conditions in each experiment. We recommend two flasks in one airtight GENbag.
14. Since yeast cells are prone to stick to each other, it is critical to pipette the cell suspension up and down several times before each use. VBNC cells are fragile, so vortexing should be avoided.
15. It is important to perform a tenfold serial dilution of cells. Direct 100-fold dilutions should be avoided to obtain consistent results.
16. Alternatively, LIVE/DEAD™ Fixable Dead Cell Stain (Invitrogen) available in other colors, such as green, may also be used.
17. It is important to keep a flask/beaker nearby that contains water, to maintain humidity and avoid evaporation.
18. It is important to incubate the plates for 5 days, as the colonies from VBNC and NORMOx cells appear later than STAT cells. One can count the STAT colonies once they are grown and then wait until day 5 for VBNC colonies to fully grow.
19. Positive growth is usually visible by eye. Alternatively, a magnifying inverted mirror may also be used to clearly distinguish yeast growth.
20. Utmost care should be taken when dealing with honeycomb plates, as the wells are difficult to track.

References

1. Alanio A (2020) Dormancy in *Cryptococcus neoformans*: 60 years of accumulating evidence. J Clin Investigation 130:3353–3360. https://doi.org/10.1172/jci136223
2. Kobayashi M, Murata K, Hiroshi HO et al (2004) Cryptococcosis: long-lasting presence of fungi after successful treatment. Acta Derm Venereol 84:320–321. https://doi.org/10.1080/00015550410025868
3. Spitzer ED, Spitzer SG, Freundlich LF et al (1993) Persistence of initial infection in recurrent *Cryptococcus neoformans* meningitis. Lancet 341:595–596. https://doi.org/10.1016/0140-6736(93)90354-j
4. World Health Organization (2022) WHO fungal priority pathogens list to guide research, development and public health action. https://www.who.int/publications/i/item/9789240060241. Oct 25th 2022
5. Alanio A, Vernel-Pauillac F, Sturny-Leclère A et al (2015) *Cryptococcus neoformans* host adaptation: toward biological evidence of dormancy. MBio 6:e02580–e02514
6. Hommel B, Sturny-Leclère A, Volant S et al (2019) *Cryptococcus neoformans* resists to drastic conditions by switching to viable but non-culturable cell phenotype. PLoS Pathog 15:e1007945. https://doi.org/10.1371/journal.ppat.1007945
7. Oblinger JL, Koburger JA (1975) Understanding and teaching the most probable number technique [1]. J Food Protect 38:540–545. https://doi.org/10.4315/0022-2747-38.9.540
8. Calcott PH, Postgate JR (1972) On substrate-accelerated death in *Klebsiella aerogenes*. J Gen Micro 70:115–122. https://doi.org/10.1099/00221287-70-1-115
9. Mukamolova GV, Kaprelyants AS, Kell DB et al (2003) Adoption of the transiently non-culturable state—a bacterial survival strategy? Adv Microb Physiol 47:65–129. https://doi.org/10.1016/s0065-2911(03)47002-1
10. de Castro RJA, Rêgo MTAM, Brandão FS et al (2022) Engineered fluorescent strains of *Cryptococcus neoformans*: a versatile toolbox for studies of host-pathogen interactions and fungal biology, including the viable but nonculturable state. Microbiol Spectr 10:e01504–e01522. https://doi.org/10.1128/spectrum.01504-22

Chapter 23

Analysis of *Cryptococcus* Extracellular Vesicles

Flavia C. G. Reis, Rafael F. Castelli, Diogo Kuczera, Haroldo C. Oliveira, and Marcio L. Rodrigues

Abstract

Extracellular vesicles (EVs) are produced by all domains of life. In fungal pathogens, they participate in virulence mechanisms and/or induce protective immunity, depending on the pathogenic species. EVs produced by pathogenic members of the *Cryptococcus* genus mediate virulence, antifungal resistance, as well as humoral and cell-mediated immunity. The isolation of cryptococcal EVs has been laborious and time-consuming for years. In this chapter, we detail a fast protocol for the isolation and analysis of EVs produced by members of the *Cryptococcus* genus.

Key words *Cryptococcus*, Extracellular vesicles, Nanoparticle tracking analysis, Solid media, Electron microscopy

1 Introduction

Fungal extracellular vesicles (EVs) were first reported in the basidiomycete *Cryptococcus neoformans* in 2007 [1]. Since then, EVs have been observed in dozens of fungal species, including model yeasts, filamentous fungi, phytopathogens, and human pathogens causing superficial, subcutaneous, and systemic diseases [2]. In *Cryptococcus* and *Candida*, EVs are vaccine candidates [3, 4] and mediators of drug resistance [5–7]. One major hurdle in the analysis of fungal EVs is the establishment of efficient and reproducible methods. As in most other organisms, fungal EVs can be isolated from liquid cultures through a combination of centrifugation steps [8]. This has been particularly challenging in the *Cryptococcus* genus, which is empirically known to be one of the least efficient models for EV isolation, most likely due to the presence of a polysaccharide capsule that, when surface-associated, decreases the efficacy of EV release and contaminates samples in the extracellular milieu [9]. In the search for easier methods of EV analysis in *Cryptococcus*, we developed a protocol in 2019 for vesicle release

Erin E. McClelland (ed.), *Cryptococcus neoformans: Methods and Protocols*, Methods in Molecular Biology, vol. 2775,
https://doi.org/10.1007/978-1-0716-3722-7_23,

and isolation from solid media [10]. Both protocols (liquid cultures and solid media) aim to collect cell-free, extracellular products of fungal cultures for final ultracentrifugation (usually 100,000 × *g*, 1 h) with consequent EV sedimentation. However, a major difference between the two approaches is the volume that will be ultracentrifuged. Isolation of EVs from the liquid medium requires at least 0.5 liters of cryptococcal culture supernatants [8]. In contrast, working volumes can be easily reduced when EVs are isolated from solid cultures. Based on the assumption that EVs will be entrapped in the extracellular matrices obtained in solid cultures, fungal cells are gently scraped from the plates and suspended in volumes of approximately 30 mL of any physiological buffer [10]. Multiple plates can be combined into a single tube, and this controlled volume is processed for EV isolation following the regular steps of removal of cells and debris, and final ultracentrifugation. Therefore, the volumes involved in EV isolation have dropped from a liter to a milliliter scale, changing a 2–3 week protocol to a 1–2 day protocol (*see* **Note 1**). It is important to highlight that the protocol is reproducible and applicable to different fungal species [9, 10].

2 Materials

1. Centrifuge and ultracentrifuge tubes of variable dimensions.
2. Centrifuges and ultracentrifuge.
3. Chemical hood.
4. Conical tubes (50 mL).
5. Filter paper.
6. Filtration system with membrane pores of 0.22, 0.45, and 0.8 μm.
7. Formvar-coated grids.
8. 2.5% Glutaraldehyde in 0.1 M sodium cacodylate buffer.
9. Incubators and shaker with temperature control and/or CO_2.
10. Inoculation loop (10 μL).
11. Insulin syringes (1 mL).
12. Microtubes.
13. Nanoparticle tracking analysis (NTA) system, including software.
14. Neubauer chamber or hemocytometer.
15. Parafilm.
16. Petri dishes (90 × 15 mm).

17. Phosphate-buffered saline (PBS): 137 mM NaCl, 2.7 mM KCl, 10 mM Na_2HPO_4, 1.8 mM KH_2PO_4, and pH 7.4. The solution is sterilized by filtration through a 0.22 μm pore filter.
18. Sabouraud agar and broth or yeast peptone dextrose (YPD) agar: solubilize the powdered reagent with deionized water according to the manufacturer's recommendations. Autoclave and store at room temperature.
19. Serological pipettes.
20. Sodium azide.
21. Sterile Erlenmeyer flasks (25 or 50 mL).
22. Titan cell solid medium (TCM): 5% Sabouraud broth, 5% inactivated fetal calf serum diluted in 50 mM MOPS, pH 7.3, 15 mM sodium azide, and 2% agar, distributed into 90 × 15 or 35 × 10 mm petri dishes.
23. Transmission electron microscope.
24. Tweezers.
25. Ultrapure water.
26. Uranyl acetate solution (5%).

3 Methods

3.1 Stock Cultures

1. According to the current literature, this protocol is known to be functional at least for 17 wild-type (WT) isolates of *C. neoformans, C. deuterogattii*, and *C. gattii* [9]. Based on literature reports, we expect it to be functional for any WT cryptococcal isolate, as well as for most of the genetically modified strains and even other species [10, 11].
2. For EV isolation, cryptococcal isolates can be maintained through successive passages in Sabouraud plates (90 mm × 15 mm petri dishes containing 25 mL of medium) as extensively reported [9, 10].
3. Stock cultures can be maintained at 4 °C. For fungal replication, cryptococcal isolates usually grow well on solid Sabouraud for 48 h at 25, 30, or 37 °C.

3.2 Preparation of Fungal Cultures

1. Colonies (enough for a half loop) of the solid Sabouraud's culture is inoculated into 5 mL of YPD medium for cultivation for 24 h at 30 °C with shaking (150 rpm).
2. Cell density is determined on a Neubauer chamber. The cell suspension is diluted to a density of 3.5×10^7 cells/mL in YPD.
3. Aliquots of 300 μL of the cell suspensions at 3.5×10^7 cells/mL in YPD are spread onto YPD agar plates (90 mm × 15 mm

petri dishes containing 25 mL of medium). Inoculated YPD plates are incubated for 24 h at 30 °C to reach confluence.

3.3 Preparation of Cell Suspensions for EV Isolation

1. Sterilize the buffer (0.01 M PBS) by filtration through 0.22 μm pore membranes for removal of nanoparticles that might contaminate fungal samples. Prepare at least 30 mL of filtered PBS for each three plates of fungal culture.
2. Fungal cells from confluent, solid cultures from three independent plates are gently removed from each of the three plates with an inoculation loop and transferred to a single 50 mL conical plastic tube containing 30 mL of PBS previously sterilized by filtration through 0.22 μm membranes (*see* **Note 2**). For highly encapsulated cultures, the limit of three plates per 30 mL of PBS must be maintained to avoid issues with cell pelleting in the further centrifugation steps. We highly recommend the determination of cell densities at this stage, making possible the further quantification of the EV/cell ratio, if necessary.
3. PBS-suspended cells are centrifuged at 5000 × *g* for 15 min at 4 °C. The supernatants are gently collected into a 50 mL centrifuge tube using plastic pipettes and the cells are discarded. The supernatants are then centrifuged at 15,000 × *g* for 15 min at 4 °C to remove debris.
4. The supernatants are again gently collected into a 50 mL conical centrifuge tube using plastic pipettes, and the pellet is discarded. Resulting supernatants are filtered through 0.8 or 0.45 μm pore syringe filters to remove residual large particles.

3.4 Obtaining EVs

1. Filtered supernatants free of cells and debris are ultracentrifuged at 100,000 × *g* for 1 h, at 4 °C. Supernatants are gently discarded, and pellets are suspended in 150–300 μL of filter-sterilized PBS. EV preparations may be maintained at 4 °C for week to months. The presence of EVs must be monitored by at least two independent nanoparticle-detecting approaches. In our experience, the fastest and most reliable approaches are NTA and transmission electron microscopy of negatively stained EVs as follows.

3.5 Transmission Electron Microscopy

1. For negative-stain electron microscopy of EVs, ultracentrifugation samples (20 μL) are placed over Formvar-coated grids with the use of a precision tweezer. Alternatively, the sample (30–50 μL) is placed over a parafilm-covered surface (a glass plate, for instance), and the grids are settled over the EV-containing sample drops, which are let to adhere to the grid for 60 min at room temperature. Excess liquid is removed by gently rubbing the bottom of the grid with filter paper. For

washing, this process is repeated three times using 0.22 μm-filtered PBS.

2. Sample-containing grids are placed over a droplet of 50 μL of 2.5% glutaraldehyde in 0.1 M sodium cacodylate buffer on a parafilm-covered surface. Samples are incubated at room temperature for 40 min, in a chemical hood. The grids are washed three times over droplets of 50 μL of 0.1 M sodium cacodylate buffer. Between the washes, the grids' two sides are dried by gentle touching with filter paper.
3. The samples are negatively stained for 2 min by placing the grids over droplets of 30–50 μL of 0.22 μm-filtered 5% uranyl acetate solution. To avoid eventual precipitates, pulse centrifugation of the uranyl acetate solution before use is recommended. This procedure must be carried out in a chemical hood. Excess uranyl is removed with filter paper. The samples are washed three times by placing the grids over 50 μL droplets of distilled water. The grids are dried with filter paper.
4. Observe the dried grids under a transmission electron microscope. Any model will be appropriate; a JEOL 1400Plus transmission electron microscope at 90 kV has been most frequently used by our group (*see* **Notes 3** and **4**).

3.6 NTA

1. NTA can be performed on multiple equipment from different manufacturers. Our most frequent experience has been with a Malvern Nanosight nanoparticle tracking analyzer coupled with a 488 nm laser, a sCMOS camera, and a syringe pump (Malvern Panalytical, Malvern, United Kingdom).
2. EV samples have their density adjusted to the equipment's optimal detection range, usually 1×10^8 to 1×10^9 particles/mL in filter-sterilized (0.22 μm) PBS.
3. The system must be cleaned using a syringe filled with distilled water. The syringe is placed onto the syringe pump, connected to the equipment's cannulas, and washed with 1 mL of distilled water three times. This step is repeated between the analyses of each sample.
4. EV samples are collected with a 1 mL syringe, which is connected to the syringe pump and the cannulas, where the samples are injected with a continuous flow (speed corresponding to 50).
5. The software is calibrated to capture three videos of 60 sec each; camera level is settled to 10–15; fluid viscosity is adjusted to that of water (0.654–0.655 cP).
6. When the script starts, the sample is injected through the equipment's chamber, and when particles pass through the laser, their concentration and diameter are determined by the

equipment's software (NTA 3.0 Software, Malvern Panalytical, in our case) using the Stokes-Einstein equation. The three captured videos are sequentially analyzed with the same software. At this step, it is possible to visually identify vesicle-like particles and set the detection threshold to 10–15 for a better view of nanoparticles (*see* **Note 5**).

7. After sample analysis, equipment cleaning must be performed to avoid particles contaminating the equipment's chamber or cannulas.

4 Notes

1. In contrast to what we have initially thought, EVs are not produced more abundantly in solid media. Instead, they seem to be more efficiently isolated because of the control of the culture volumes that will be ultracentrifuged.
2. The protocol can be adapted to different types of cells and growth conditions. It is known that *C. neoformans* undergoes morphological changes during infection, forming so-called Titan cells [12]. Recently, the EV isolation protocol was adapted to the use of Titan cells [9]. For that, fungal cells are first grown in liquid Sabouraud supplemented with 15 mM sodium azide overnight (30 °C). The cells are washed with PBS by centrifugation and adjusted to 5×10^7 cells/mL in PBS. This suspension (200 μL) is inoculated onto a TCM agar plate to reach a final density of 10^7 cells/plate. Importantly, the serum used for preparation of the TCM agar is depleted of EVs by ultracentrifugation at 100,000 × *g* for 16 h at 4 °C, followed by filtration through 0.22 μm membranes. The plates are incubated for 18 h at 37 °C under a 5% CO_2 atmosphere and processed as described for regular cultures. However, the yield of EVs using the TCM agar plates is approximately tenfold lower than that observed for regular cultures. We recommend at least 30 plates for NTA and TEM analysis.
3. Except for the steps of fixation and staining with uranyl, the entire protocol of sample preparation for TEM is carried out at room temperature and outside the chemical hood. Uranyl is a heavy metal and should be carefully handled with gloves in the chemical hood.
4. The formvar films are disrupted frequently during TEM analysis. Two alternative options can be used. First, samples can be carbon-coated in metal sputter coater, which increases the film's resistance, but also the sample conductivity. Second, electron beam tension can be modulated to avoid film disruption.

5. NTA parameters might change according to background noise and screen brightness. It is important that the captured video show the particles with adjusted focus, brightness, and contrast.

Acknowledgments

The authors are thankful to the Microscopy Platform of the Carlos Chagas Institute of Fiocruz for access to the electron microscope and technical advice.

References

1. Rodrigues ML, Nimrichter L, Oliveira DL et al (2007) Vesicular polysaccharide export in *Cryptococcus neoformans* is a eukaryotic solution to the problem of fungal trans-cell wall transport. Eukaryot Cell 6:48–59. https://doi.org/10.1128/EC.00318-06
2. Rodrigues ML, Nimrichter L (2022) From fundamental biology to the search for innovation: the story of fungal extracellular vesicles. Eur J Cell Biol 101:151205. https://doi.org/10.1016/j.ejcb.2022.151205
3. Rizzo J, Wong SSW, Gazi AD et al (2021) *Cryptococcus* extracellular vesicles properties and their use as vaccine platforms. J Extracell Vesicles 10:e12129. https://doi.org/10.1002/jev2.12129
4. Vargas G, Honorato L, Guimarães AJ et al (2020) Protective effect of fungal extracellular vesicles against murine candidiasis. Cell Microbiol 22:e13238. https://doi.org/10.1111/cmi.13238
5. Zarnowski R, Sanchez H, Covelli AS et al (2018) *Candida albicans* biofilm–induced vesicles confer drug resistance through matrix biogenesis. PLoS Biol 16:e2006872. https://doi.org/10.1371/journal.pbio.2006872
6. Chan W, Chow FW-N, Tsang C-C et al (2022) Induction of amphotericin B resistance in susceptible *Candida auris* by extracellular vesicles. Emerg Microbes Infect 1–40:1900. https://doi.org/10.1080/22221751.2022.2098058
7. Rizzo J, Trottier A, Moyrand F et al (2023) Coregulation of extracellular vesicle production and fluconazole susceptibility in Cryptococcus neoformans. mBio 14(4): e0087023. https://doi.org/10.1128/mbio.00870-23
8. Rodrigues ML, Oliveira DL, Vargas G et al (2016) Analysis of yeast extracellular vesicles. Methods Mol Biol 1459:175–190. https://doi.org/10.1007/978-1-4939-3804-9_12
9. Reis FCG, Gimenez B, Jozefowicz LJ et al (2021) Analysis of cryptococcal extracellular vesicles: experimental approaches for studying their diversity among multiple isolates, kinetics of production, methods of separation, and detection in cultures of titan cells. Microbiol Spectr 9:e00125. https://doi.org/10.1128/Spectrum.00125-21
10. Reis FCG, Borges BS, Jozefowicz LH et al (2019) A novel protocol for the isolation of fungal extracellular vesicles reveals the participation of a putative scramblase in polysaccharide export and capsule construction in *Cryptococcus gattii*. mSphere 4:e00080. https://doi.org/10.1128/mSphere.00080-19
11. Castelli RF, Pereira A, Honorato L et al (2022) Extracellular vesicle formation in *Cryptococcus deuterogattii* impacts fungal virulence and requires the *NOP16* gene. Infect Immun 90(8):e0023222. https://doi.org/10.1128/iai.00232-22
12. Zaragoza O, Nielsen K (2013) Titan cells in *Cryptococcus neoformans*: cells with a giant impact. Curr Opin Microbiol 16:409–413. https://doi.org/10.1016/j.mib.2013.03.006

Chapter 24

Preparation of Biologically Active Fractions Enriched with Glucuronoxylomannan, the Main Antigen of the Cryptococcal Capsule

Luisa J. Jozefowicz and Marcio L. Rodrigues

Abstract

Glucuronoxylomannan (GXM) is the principal capsular component in the *Cryptococcus* genus. This complex polysaccharide participates in numerous events related to the physiology and pathogenesis of *Cryptococcus*, which highlights the importance of establishing methods for its isolation and analysis. Conventional methods for GXM isolation have been extensively discussed in the literature. In this chapter, we describe two fast methods for obtaining extracellular fractions enriched with cryptococcal GXM.

Key words *Cryptococcus*, Capsule, Glucuronoxylomannan

1 Introduction

Glucuronoxylomannan (GXM) is the most abundant component of the cryptococcal capsule [1]. The huge impact of GXM on cryptococcal infection [2] and its potential as a target of therapeutic antibodies [3], as well as its ability to stimulate the immune system [4], justifies the establishment of methods for its isolation and characterization. GXM is a negatively charged heteropolysaccharide that is associated with the cell surface of *Cryptococcus*, but it is abundantly secreted into the extracellular space, both during in vitro growth and infection [1]. Therefore, GXM may be efficiently obtained from culture fluids based on the well-established ability of 75% ethanol to precipitate polysaccharides [5]. GXM may be separated from other polysaccharide components on the basis of its negative charge, which makes the polysaccharide prone to precipitate in the presence of cationic detergents. Alternatively, GXM may be purified using ionic exchange chromatography columns, also because of its negative charge [6].

Erin E. McClelland (ed.), *Cryptococcus neoformans: Methods and Protocols*, Methods in Molecular Biology, vol. 2775,
https://doi.org/10.1007/978-1-0716-3722-7_24,

Highly purified fractions of GXM are required for immunological studies. A previous issue of this book series presented a detailed procedure for the preparation of purified fractions of the capsular antigens of *Cryptococcus*, including GXM [5]. This methodology is well established and widely used by the mycology and immunology communities, and since the protocols have not changed, we recommend referring to Wozniak and Levitz for details of the classical methods for GXM purification [5]. However, faster protocols that provide inferior levels of purity are likely not useful for immunological studies, but highly helpful in other models, including capsule growth assays and cell wall coating models for phagocytosis determination, among others [7]. In this chapter, we present two fast methods for obtaining GXM-enriched fractions (*see* **Note 1**), including one for the preparation of GXM from large amounts of culture fluids and another for GXM enrichment at the milliliter/microliter levels.

2 Materials

1. Erlenmeyer flasks (1000 mL).
2. Incubator with temperature control and shaking.
3. Bench centrifuge.
4. Balance (analytical precision).
5. Stir plate.
6. Vacuum drier.
7. Tank for nitrogen (N_2) in its gas form.
8. CO_2 incubator.
9. Filter membranes (0.45 and 0.8 μm cutoffs).
10. Ultrafiltration cell (Amicon ultrafiltration cell, cutoff of 10 or 100 kDa and a total capacity of 200 mL, supplemented with 63.5 mm Biomax polyethersulfone ultrafiltration discs).
11. Centrifugation-based ultrafiltration system (Amicon® Ultra Centrifugal Filters, 0.5 mL volume, 100 kDa cutoff, regenerated cellulose membranes).
12. Plastic cell scrapers.
13. Cell culture flasks of 25 cm^2.
14. Conical tubes (15 mL).
15. Plastic Pasteur pipettes.
16. Micropipettes (20–1000 μL).
17. Neubauer chamber.
18. Plastic tubes (1.5 mL).

19. Inoculation loops.
20. Phosphate-buffered saline (PBS).
21. Glucose (15 mM).
22. $MgSO_4$.
23. KH_2PO_4.
24. Glycine.
25. Thiamine-HCl.
26. Roswell Park Memorial Institute (RPMI) medium.
27. Yeast peptone dextrose (YPD) medium.
28. Minimal medium: glucose (15 mM), MgSO4 (10 mM), KH2PO4 (29.4 mM), glycine (13 mM), and thiamine-HCl (3 μM), with a pH of 5.5.
29. Paraformaldehyde (4% solution).

3 Methods

3.1 GXM Preparation by Ultrafiltration (Suitable for Large Volumes of Culture Supernatants)

1. Any cryptococcal strain that produces GXM in large quantities is inoculated (10^9 cells) into 1000 mL Erlenmeyer flasks containing 400 mL of any media that induces capsule production. We recommend using minimal medium.
2. Fungal cells are cultivated for 3 days at 30 °C with shaking and separated from the culture supernatant by centrifugation at 4000× *g* for 15 min at 4 °C (*see* **Note 2**).
3. The supernatant is collected and centrifuged again at 15,000× *g* for 15 min at 4 °C to remove smaller debris. The pellets are discarded and the resulting supernatant is filtered through 0.8 μm membranes to remove any remaining cells (*see* **Note 3**).
4. The filtered supernatant is transferred to an ultrafiltration device. Our experience is limited to the Amicon (Millipore, Danvers, MA) ultrafiltration cell (cutoff of 100 kDa and a total capacity of 200 mL) with stirring and Biomax polyethersulfone ultrafiltration discs (63.5 mm), but any ultrafiltration system is expected to work similarly.
5. The supernatant is concentrated approximately 20-fold under a nitrogen (N_2) stream as the pressure gas. Due to its high molecular mass, GXM is mostly retained in the upper compartment of the ultrafiltration cell, becoming continuously more concentrated due to the loss of water during the ultrafiltration process. As mentioned earlier, GXM is negatively charged, resulting in the capture of divalent cations from the culture medium. The divalent cations form bridges between different GXM molecules, resulting in polysaccharide aggregation

[7]. Therefore, after the supernatant is concentrated, a thick, translucent GXM-containing film is observed in close association with the ultrafiltration disc, covered by a concentrated fluid phase.

6. The fluid phase is discarded and the viscous layer is gently collected with a cell scraper for storage at room temperature.
7. Fractions that pass through the 100 kDa filtration discs may be filtered again through 10 kDa membranes, resulting in film formation once again.
8. For polysaccharide quantification, we recommend a capture enzyme-linked immunosorbent assay (ELISA) [8], the carbazole reaction for hexuronic acid [9], and the method for hexose detection described by Dubois et al. [10]. For structural and compositional determinations, we recommend nuclear magnetic resonance (NMR) and gas chromatography coupled to mass spectrometry (GC-MS), respectively [11]. High concentrations of GXM (mg/mL) are expected.

3.2 GXM Preparation by Microfiltration Followed by Ethanol Precipitation (Suitable for Small Volumes of Culture Supernatants)

1. Fungal cultures are prepared under conditions that promote capsular enlargement and stimulate GXM secretion [12]. For microvolumes, we recommend incubating the cultures in Roswell Park Memorial Institute (RPMI) medium at 37 °C, under a CO_2 atmosphere, as detailed in the following sections.
2. Cryptococci are cultured for 48 h at 30 °C with shaking (200 rpm) in liquid YPD medium.
3. PBS-washed fungal cells (5×10^7) are inoculated into 25 cm^2 culture flasks containing 5 mL of RPMI. Flasks should be stood vertically during incubation, minimizing cell adhesion to the bottom of the flask.
4. The suspensions are incubated for 24 h at 37 °C with 5% CO_2.
5. The suspensions are transferred to 15 mL conical tubes and centrifuged for 10 min at 4000× *g*.
6. The supernatants are carefully recovered with a plastic Pasteur pipette and filtered through 0.45 μm pore membranes to ensure removal of cells.
7. A sample of 500 μL of each supernatant is transferred to a microvolume, centrifugation-based ultrafiltration system. Our experience is limited to the Amicon® Ultra Centrifugal Filters (0.5 mL volume, 100 kDa cutoff, regenerated cellulose membranes).
8. The ultrafiltration systems loaded with supernatant samples are centrifuged for 2 min at room temperature (4000× *g*), and the process of adding 500 μL samples of the supernatant is repeated as ultrafiltration progresses.

9. The concentrated fluids, retained in the upper part of the ultrafilters, are collected with a micropipette and transferred to previously weighed 1.5 mL plastic tubes. The tubes are weighed again for calculation of the volume recovered from the ultrafilters.
10. Three volumes of absolute, cold ethanol are added to the tubes to form 75% ethanol solutions, which enable the immediate precipitation of polysaccharides. The tubes are vortexed and centrifuged for 5 min at 4000× *g*.
11. The supernatants are discarded and the polysaccharide-containing pellets are vacuum-dried at 45 °C for 90 min.
12. The tubes are weighed again to determine the dry weight and PBS is added to form 1 mg/mL solutions. In our experience, recovered dry masses vary from 200 to 1100 μg.

3.3 Biological Validation of GXM Preparations: Coating of Acapsular Cells (see Note 4)

1. One colony from a solid culture of *C. neoformans* acapsular cells (our experience is limited to the *cap67Δ* and *cap59Δ* strains) is collected from any solid medium with an inoculation loop and transferred to 5 mL of liquid YPD in plastic conical tubes.
2. The suspension is incubated for 24 h at 30 °C with shaking (200 rpm).
3. The suspension is quantified in a Neubauer chamber, and 5×10^6 cells are transferred to a microcentrifuge tube and washed three times with PBS. The cells are then resuspended in 495 μL of PBS.
4. Five microliters of a GXM solution (1 mg/mL) are added to the cell suspension, resulting in a polysaccharide solution at 10 μg/mL.
5. The cell suspension is incubated at room temperature for 4 h, then washed three times with PBS and fixed with 100 μL of 4% paraformaldehyde at room temperature for 30 min.
6. Paraformaldehyde is removed by centrifugation and the cells are washed with PBS, also by centrifugation. The cells are processed for capsule observation by immunofluorescence, as extensively described in the literature [13, 14] (*see* **Note 5**). The typical patterns of GXM incorporation are illustrated in Fig. 1.

4 Notes

1. The methods described in this chapter allow for the preparation of relatively highly concentrated samples of GXM. To ensure purity, analytical methods, such as NMR and GC-MS,

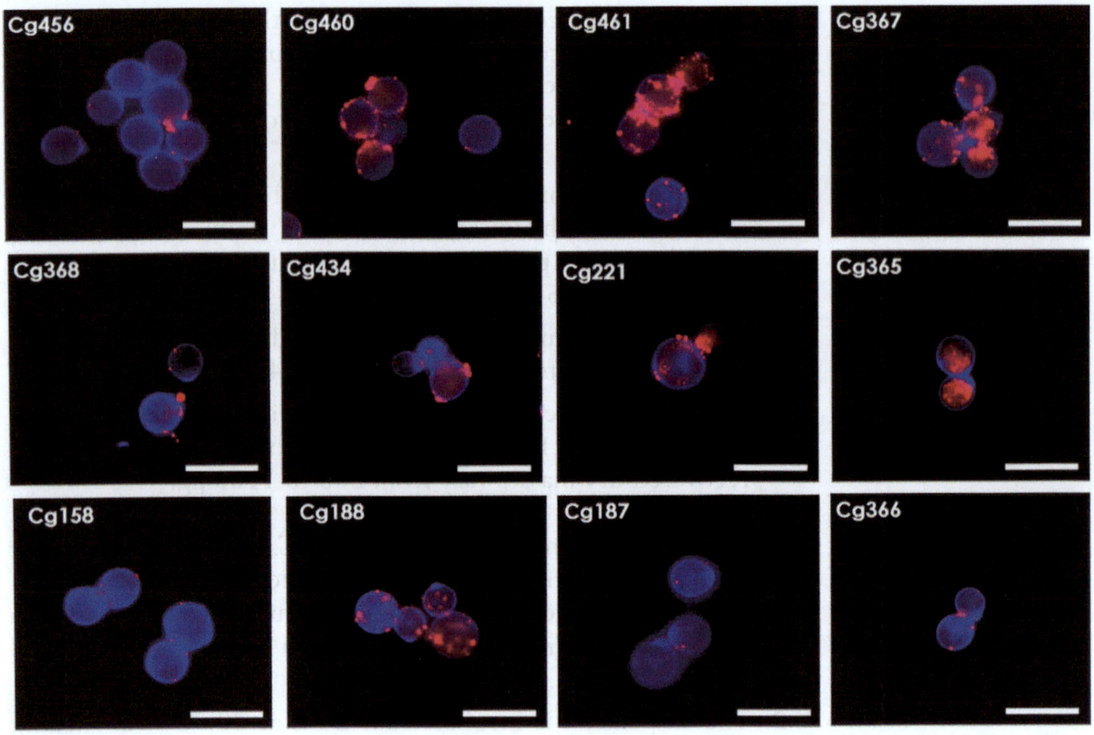

Fig. 1 Coating of an acapsular mutant of *C. neoformans* (*cap67Δ* strain) with GXM obtained from 12 different isolates of *Cryptococcus*. For details about each isolate, please see [15]. These images were prepared using GXM obtained according with the method detailed in Subheading 3.2 (small volumes), but any method is expected to work similarly. Blue fluorescence denotes staining of the cell wall with calcofluor white; red fluorescence denotes antibody binding to GXM. Scale bars, 5 mm

are required. Samples obtained through the methods we describe here may be further purified using the conventional, cationic detergent–based method of GXM purification. If only enriched samples are required for the desired biological model, the methods described in this chapter should suffice.

2. Under conditions of large capsule formation, inefficient pelleting is expected. Even if the supernatant is contaminated with cells, this issue will be resolved in the subsequent steps of centrifugation and filtration.
3. GXM is highly viscous, so filtration may take longer than usual for microbial supernatants in general.
4. This methodology is based on the ability of acapsular mutant cells of *C. neoformans* to incorporate externally added GXM [1], thereby validating the integrity of polysaccharide solutions obtained by any of the methods mentioned in this chapter.
5. Antibodies to GXM are required; our samples were a gift from Arturo Casadevall (Johns Hopkins University, Baltimore, USA) [3].

Acknowledgments

We thank our laboratory colleagues for helpful discussions about the methods for preparing GXM.

References

1. Zaragoza O, Rodrigues ML, De Jesus M et al (2009) The capsule of the fungal pathogen *Cryptococcus neoformans*. Adv Appl Microbiol 68:133–216
2. O'Meara TR, Andrew Alspaugh J (2012) The *Cryptococcus neoformans* capsule: a sword and a shield. Clin Microbiol Rev 25(3):387–408
3. Casadevall A, Cleare W, Feldmesser M et al (1998) Characterization of a murine monoclonal antibody to *Cryptococcus neoformans* polysaccharide that is a candidate for human therapeutic studies. Antimicrob Agents Chemother 42(6):1437–1446
4. Levitz SM (2002) Receptor-mediated recognition of *Cryptococcus neoformans*. Nippon Ishinkin Gakkai Zasshi 43:133–136. https://doi.org/10.3314/jjmm.43.133
5. Wozniak KL, Levitz SM (2009) Isolation and purification of antigenic components of *Cryptococcus*. Methods Mol Biol 470:71–83
6. Grijpstra J, Gerwig GJ, Wösten H et al (2009) Production of extracellular polysaccharides by *CAP* mutants of *Cryptococcus neoformans*. Eukaryot Cell 8:1165–1173. https://doi.org/10.1128/EC.00013-09
7. Nimrichter L, Frases S, Cinelli LP et al (2007) Self-aggregation of *Cryptococcus neoformans* capsular glucuronoxylomannan is dependent on divalent cations. Eukaryot Cell 6(8): 1400–1410. https://doi.org/10.1128/EC.00122-07
8. Casadevall A, Mukherjee J, Scharff MD (1992) Monoclonal antibody based ELISAs for cryptococcal polysaccharide. J Immunol Methods 154(1):27–35. https://doi.org/10.1016/0022-1759(92)90209-C
9. Bowness JM (1957) Application of the carbazole reaction to the estimation of glucuronic acid and glucose in some acidic polysaccharides and in urine. Biochem J 67:295–300. https://doi.org/10.1042/bj0670295
10. Dubois M, Gilles K, Hamilton JK et al (1951) A colorimetric method for the determination of sugars. Nature 168:167–167. https://doi.org/10.1038/168167a0
11. Chang YC, Cherniak R, Kozel TR et al (1997) Structure and biological activities of acapsular *Cryptococcus neoformans* 602 complemented with the CAP64 gene. Infect Immun 65: 1584–1592
12. Zaragoza O, Casadevall A (2004) Experimental modulation of capsule size in *Cryptococcus neoformans*. Biol Proced Online 6:10–15. https://doi.org/10.1251/bpo68
13. Rodrigues ML, Alvarez M, Fonseca FL et al (2008) Binding of the wheat germ lectin to *Cryptococcus neoformans* suggests an association of chitinlike structures with yeast budding and capsular glucuronoxylomannan. Eukaryot Cell 7(4):602–609. https://doi.org/10.1128/EC.00307-07
14. Fonseca FL, Nimrichter L, Cordero RJB et al (2009) Role for chitin and chitooligomers in the capsular architecture of *Cryptococcus neoformans*. Eukaryot Cell 8(10):1543–1553. https://doi.org/10.1128/EC.00142-09
15. Maufrais C, de Oliveira L, Bastos RW et al (2021) Population genomic analysis of *Cryptococcus* Brazilian isolates reveals an African type subclade distribution. G3 (Bethesda) 11(6): jkab107. https://doi.org/10.1093/g3journal/jkab107

Chapter 25

Measuring Replicative Lifespan in *Cryptococcus neoformans*

Vanessa K. A. Silva, Natalia K. Oliveira, and Bettina C. Fries

Abstract

Advances in understanding cellular aging research have been possible due to the analysis of the replicative lifespan of yeast cells. Studying longevity in the pathogenic yeast *Cryptococcus neoformans* is essential because old yeast cells with age-related phenotypes accumulate during infection and are associated with increased virulence and antifungal tolerance. Microdissection and microfluidic devices are valuable tools for continuously tracking cells at the single-cell level. In this chapter, we describe the features of these two platforms and outline technical limitations and information to study aging mechanisms while assessing the lifespan of yeast cells.

Key words Aging, Longevity, Microdissection, Microfluidic system, Yeast

1 Introduction

Replicative lifespan (RLS) depends on the number of cell cycle progressions and is defined as the number of daughter cells a mother cell produces before reaching senescence and death [1]. In the budding pathogenic yeast *Cryptococcus neoformans*, the asymmetric division ensures that the mother cell retains the classic aging-associated phenotypes, such as a decrease in vacuole acidity [2], increase in cell size [3], and cell wall thickness [2]. Investigating the impact of genetic, pharmacological, and metabolic modifications on controlling RLS [1, 4] constitutes a valuable tool for studying the aging process in yeast.

Currently, two methods are available to determine the RLS of *C. neoformans* cells: microdissection and using a microfluidic system [5, 6]. The first approach consists of manually separating the cells after each division and is the most commonly applied assay to determine the life span of a particular strain (Fig. 1) [7]. However, this traditional assay is labor-intensive and time-consuming. Other disadvantages include the reduced sample size that may not

Erin E. McClelland (ed.), *Cryptococcus neoformans: Methods and Protocols*, Methods in Molecular Biology, vol. 2775,
https://doi.org/10.1007/978-1-0716-3722-7_25,

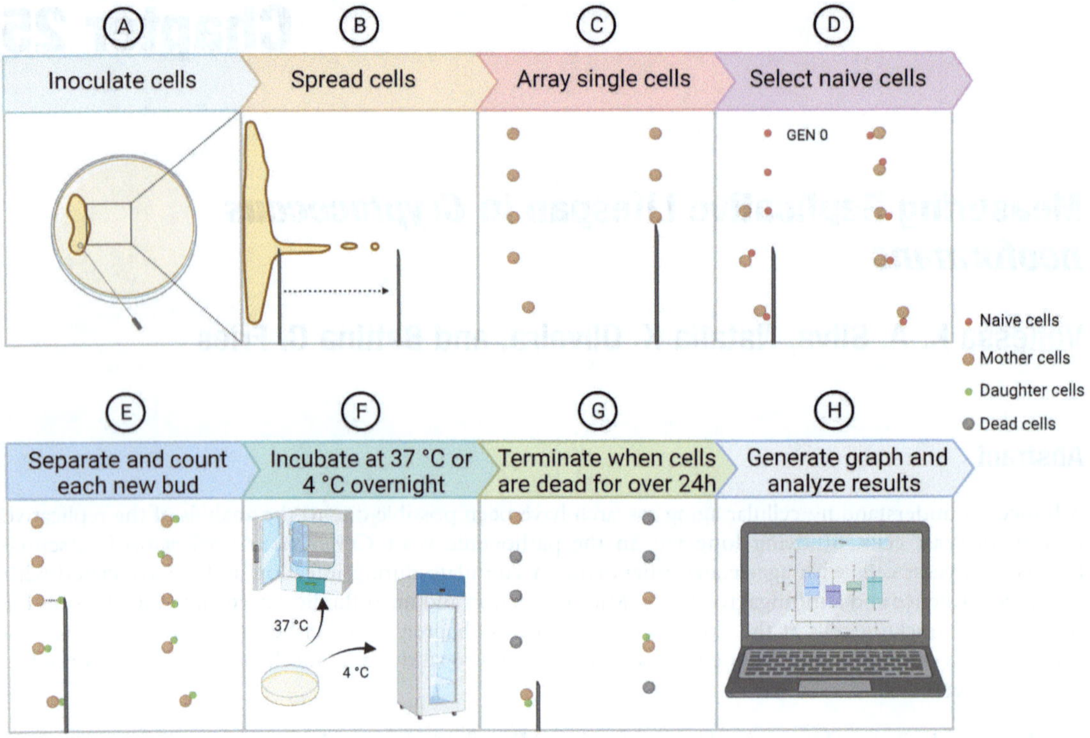

Fig. 1 Traditional microdissection workflow. (**a**) Inoculate *Cryptococcus neoformans* cells as a single thick line at the top of the desired agar plate using a sterile inoculation loop. (**b**) Once cells have grown, place the plate in the microscope and drag cells to the side using the fiber optic needle until single cells can be isolated. (**c**) Array 20–30 single cells in a line. (**d**) Once the mother cells generate a naïve cell, separate the cells, keeping the naïve cells in a line and dragging the mother cells away. The naïve cells will now be monitored as the mother cells. (**e**) Each time the mother cells bud, separate and count the daughter cells, dragging them away from the line. (**f**) Incubate the plate at 37 °C between budding events and store it at 4 °C overnight. (**g**) Terminate the study when cells have reached senescence and have not divided over 24 h. (**h**) Plot and analyze the data. The illustration was created using BioRender

accurately estimate the stochasticity of the RLS and the limited number of strains that can be analyzed simultaneously [8, 9]. Also, this system does not allow the characterization of the cells at the molecular level. Therefore, an automated microfluidic tracking system was developed that enables more cells to be evaluated at once in a reduced time frame. By using a microfluidic system combined with time-lapse microscopy for RLS characterization, it is possible to verify morphological and phenotypical changes, including protein expression and localization, at a single cell level during aging (Fig. 2) [4, 6].

The analysis of longevity in individual single yeast cells advances our understanding of the stochasticity of regulatory pathways that evolutionarily contribute to the mechanisms of aging [4]. Studying the RLS of pathogenic yeasts may shed light on important metabolic pathways and protein expression that directly promotes

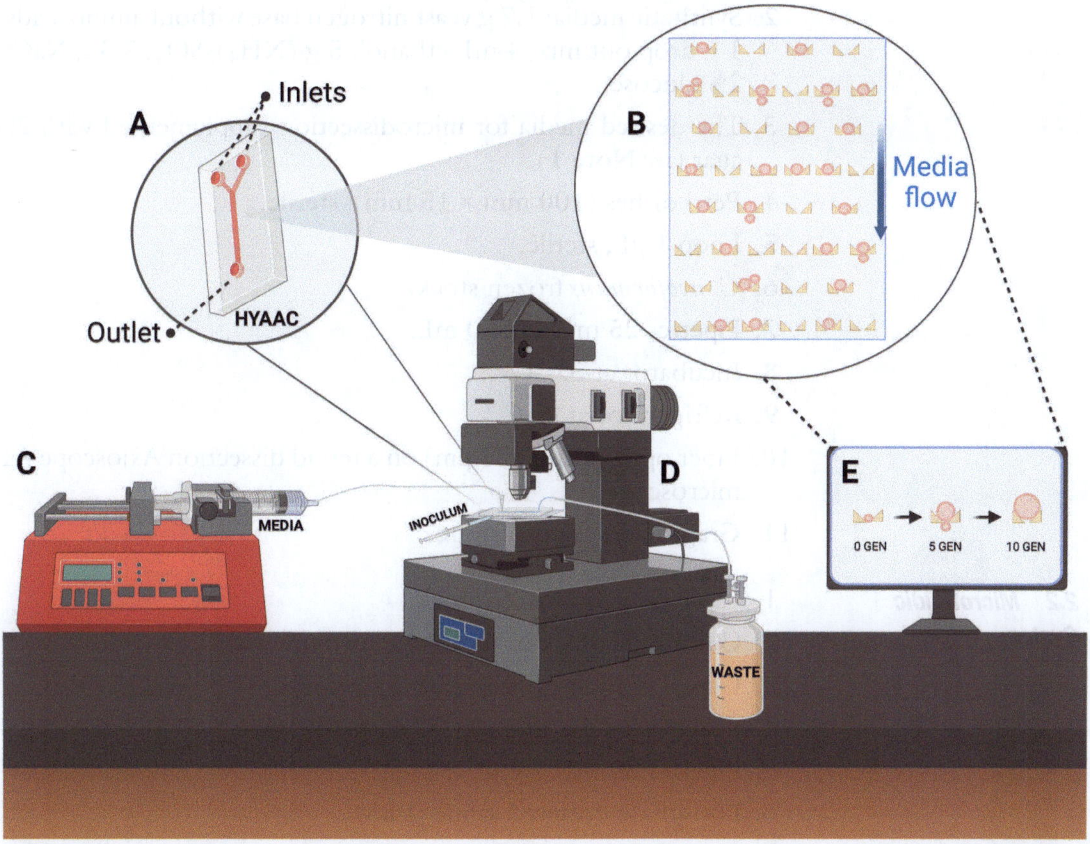

Fig. 2 Scheme of the microfluidic system. (**a**) High-throughput Yeast Aging Analysis for *Cryptococcus* (HYAAC) device. Channel has two upper inlets, one long channel, and a single outlet. (**b**) Microscopic image of yeast cells growing with age in buckets arrayed in channel while daughter cells are washed away. Blue arrow represents media flow direction. (**c**) Pump with larger syringe containing media flowing through tubing into one of the inlets of the device. (**d**) HYAAC device is set on inverted microscope stage. Inoculum is loaded using small syringe into the other upper inlet. Tubing is also connected to the outlet, which allows media to flow from the outlet into a waste container. (**e**) Images of *Cryptococcus* RLS are captured over time and displayed on the computer monitor

selective pressure leading to the accumulation of old cells, increasing tolerance to treatment, and the persistence of fungal infections [5, 6, 10]. In this chapter, we describe in detail these two procedures for quantifying the RLS of *C. neoformans* cells.

2 Materials

2.1 Traditional Microdissection

1. Yeast peptone dextrose (YPD) agar media: 1% yeast extract, 2% peptone, 2% glucose, 2% agar.

2. Synthetic media: 1.7 g yeast nitrogen base without amino acids, 1 g drop out mix, 4 mL ethanol, 5 g $(NH_4)_2SO_4$, 3.3 g NaCl, 2% glucose.
3. The desired media for microdissection supplemented with 2% agar (*see* **Note 1**).
4. Petri dishes (100 mm × 15 mm), sterile.
5. Loop 1 μL, sterile.
6. *C. neoformans* frozen stock.
7. Pipettes 25 mL and 50 mL.
8. Incubator at 37 °C.
9. Refrigerator at 4 °C.
10. Fiber optic needle (25 μm) on a tetrad dissection Axioscope A1 microscope.
11. GraphPad Prism software.

2.2 Microfluidic System

1. *C. neoformans* frozen stock.
2. Culture media (e.g., synthetic medium, SD).
3. Glass Erlenmeyer flask 250 mL, sterile.
4. Shaking incubator at 37 °C with 150 rpm.
5. Pipettes 50 mL, sterile.
6. Phosphate buffered saline (PBS).
7. Centrifuge.
8. Hemocytometer.
9. High-Throughput Yeast Aging Analysis for *Cryptococcus* (HYAAC) device: A polydimethylsiloxane (PDMS) chip bonded to a glass slide having a single channel (height 10–12 μm) with two inlets, one middle long channel (1 cm), and one single outlet (Fig. 2a). Each channel is comprised of 80 rows of staggered buckets that alternate between six or seven buckets per row (*see* **Note 2**). The buckets are modified triangles designed to be facing each other (Fig. 2B), where the bottom of the bucket is 3 μm wide and the top of the bucket is 9 μm wide (*see* **Note 3**).
10. Syringes (1 mL and 30 mL) with Luer-Loks to prevent leakage (*see* **Note 4**).
11. Tygon microbore tubing 0.020" × 0.060" outer diameter (OD).
12. Tubing adapter 23G.
13. Metal pin 0.025″ × 0.05″.
14. Metal tubes 0.025 OD × 0.017 inner diameter (ID) × 0.5″ (*see* **Note 5**).

15. Syringe pump, with multi-syringe capacity and speed.
16. Waste collector (Flask).
17. Inverted fluorescent microscope with slide adapter, 40× objective lens, automated stage with *x*, *y*, and *z* directions, temperature-controlled and attached camera to record images at regular intervals over time (*see* **Note 6**).
18. Fiji software to process the images.

3 Methods

3.1 Traditional Microdissection

1. To make the stock agar plates, aliquot approximately 15 mL of YPD agar media in each plate. After media solidification, recover and streak the desired yeast strain from a − 80 °C cryovial using a sterile loop onto the YPD agar plate.
2. Incubate for 48 h at 37 °C.
3. For making the microdissection plates, aliquot 35 mL of desired media with 2% of agar into each Petri dish (*see* **Note 7**).
4. Using a sterile loop, pick the desired yeast strain from the stock plate and streak it on the microdissection plate agar as a single thick line on the top part of the agar (Fig. 1a).
5. Incubate plates for 48 h at 37 °C, allowing growth of the cells.
6. On a tetrad dissection Axioscope A1 microscope (Zeiss) at 250× magnification, adjust the position of a 25 μm fiber optic needle so that it is located in the center of the microscope field of view.
7. Move the needle upward by lowering the joystick. The tip of the needle should touch the agar without perforating it.
8. To transfer the *C. neoformans* cells, very gently move the needle upward into the cell growth and drag it to the side while still in contact with the agar. Repeat the action until single cells can be separated (Fig. 1b).
9. Array 20–30 cells in a line in the center of the plate (Fig. 1c).
10. Allow at least 50 μm between each cell to ensure enough space to manipulate neighboring cells (*see* **Note 8**).
11. The first bud of each cell is identified as a naive cell. Keep the naive cell and drag the mother cell away from your line of arrayed cells (*see* **Note 9**). The bud cells can be separated by touching the needle to the bud and dragging it away from the mother cell. Alternatively, the needle can be positioned to touch both the mother and the bud cell and moved around slightly with gentle taping to the joystick to promote separation (Fig. 1d).

12. This naive daughter cell is now the mother cell that will be observed throughout the experiment.
13. New buds from the mother cell are separated at the end of each division (1–2 h). From this moment, each daughter cell separated from the mother should be counted and recorded as one generation (Fig. 1e) (*see* **Note 10**).
14. Between separations, the plate is returned to the incubator at 37 °C or maintained at 4 °C overnight to prevent excessive budding (Fig. 1f) (*see* **Note 11**). Seal the plates with parafilm to avoid contamination.
15. A cell is considered dead when it has not divided for a period of at least 24 h, with incubations at 37 °C (*see* **Note 12**).
16. Terminate the study when over 50% of arrayed cells have reached senescence. The RLS of each cell is the sum of the total budding events until the mother cell's death (Fig. 1g).
17. Plot and analyze the data using GraphPad Prism software (Fig. 1h). Data may be plotted as a survival graph (% survival vs. generations), where the Wilcoxon rank sum test is used to calculate p values. Alternatively, data can be plotted as a box or violin column graph (generations vs. strains), where a Student's t-test is used to calculate p values.

3.2 Microfluidic System: HYAAC

1. Start an overnight *Cryptococcus* culture at 37 °C and 150 rpm in desired liquid media (*see* **Note 13**). The next day, subculture cells 1:50 in a new flask using the same conditions. Grow *Cryptococcus* for 6 h to ensure a young, exponential culture (approximately 2 generations).
2. Centrifuge cells at 4000 × g for 5 min and wash yeast cells with PBS twice. Dilute *Cryptococcus* cells to a concentration of 10^5 cells/mL and load the suspension into a 1 mL syringe. Load a larger syringe with media.
3. Attach both syringes to a tubing adapter and connect each of them to Tygon microbore tubing. Then, insert a metal tube to the end of each Tygon microbore tubing.
4. Connect a metal tube to the outlet of the HYAAC system, attaching a Tygon microbore tubing to the end of the metal tube.
5. Load the media-filled syringe to the syringe pump and attach the end of the metal tube to one of the inlets of the HYAAC device. Pump media into the system at a constant rate of 10–15 μL/min until it fills the entire device and starts to come out of the outlet tubing into a waste container (Fig. 2c, d).
6. Pause the syringe pump to hold the pressure within the system and place the device atop an onstage incubator of an inverted

microscope that keeps the chip and its contents at 37 °C (Fig. 2d).

7. At the other upstream inlets of the HYAAC, place the metal tube end previously attached to the syringe containing the cell suspension and load the device. Check loading efficiency, loading more cells if necessary (*see* **Notes 14** and **15**).
8. Once the device is loaded, remove the metal tube and replace it with a metal pin.
9. Set the syringe pump to a speed of 10 μL/min and flow media through it (*see* **Note 16**). Monitor the experiment, checking on cell division (Fig. 2E), media levels, and clogging issues (*see* **Note 17**).
10. Capture images using a microscope with a 40× objective and equipped with an automated stage to allow for multiple *X* and *Y* positions to be imaged every 15 min until cells stop dividing (*see* **Note 18**).
11. The analysis is performed by comparing the sequential images acquired using Fiji software. Doubling time is measured by calculating the time it takes for the mother cell to complete each division. RLS is assessed by counting the total number of offspring produced by each single mother cell. Plot the percentage of viable cells/reproducing cells vs. the generation of daughter cells. Images can also be analyzed to determine cell size, morphological changes, and protein localization.

4 Notes

1. Agar media in plates must be originally made thicker; as the microdissection proceeds, the agar will desiccate. It is thus important to use thick-layered agar plates. YPD or synthetic media may be used for microdissection. Additionally, drugs/compounds may be added to the media to evaluate their effect on RLS.
2. Each channel enables the analysis of one strain; therefore, this device can quantify the RLS of up to ten different strains simultaneously.
3. This design of the microchip, an inverted triangle with angled walls, ensures that daughter cells can be washed off the system, not causing clumps, while allowing trapped mother cells to grow (4–10 μm) and bud in different directions.
4. According to the length of the assay, different syringe sizes can be used.
5. Tubing, pins, and adaptors must fit well into each other to prevent leakage.

6. Using a fluorescent microscope allows the analysis of aged-related phenotypical characteristics, including organelle morphology, gene expression, and protein localization.
7. Altering media types and nutrient content of the medium may affect the lifespan of *C. neoformans.*
8. Microdissections are performed at room temperature. It is useful to mark with a pen the area where the selected mother cells are outside the plate as a reference point on the plate. As a visual aid to keep count of the arrayed line, for every five cells, one cell can be kept slightly dislocated to the right of the line.
9. The discard site of cells must be at least two fields of view away from the arrayed line so that the cells will not impair the experiment as they keep budding.
10. Do not incubate the plate for more than 2 h so as not to generate multiple daughter cells, therefore making it difficult to differentiate the mother cell. As the experiment progresses, the agar medium will shrink, and the microdissection needle will have difficulty reaching the agar. If necessary, holes can be opened in the sides of the plates using hot pliers to facilitate needle movement.
11. Cells will replicate at 4 °C over a period of 1–2 days. They should not be kept at 4 °C for a period of over 2 days due to the risk of overgrowth.
12. Cells can have a delayed replication cycle; a minimum of 24 h is necessary before considering the cell under senescence. They should be watched for a couple of days to confirm death.
13. Using a chemically defined media is suggested to improve reproducibility.
14. It is noteworthy that the device is not able to selectively trap only naive cells at the beginning of the experiment. However, in an exponential culture, the majority (~ 80%) of yeast cells are unbudded [8].
15. The single-cell trapping efficiency rate should be approximately 60%, and there is a risk of channel clogging and loss of cells over time, especially within the first seven generations. Also, since this platform is in a stable, controlled temperature, it minimizes variations due to environmental temperature switching.
16. This constant flow of fresh medium provides continuous nutritional conditions to the cells. It removes daughter cells while the mother cell is trapped in the buckets and can be monitored for the entirety of the experiment. Due to the low flow resistance, only one cell tends to be trapped within an empty bucket. After the trap becomes occupied, there is an increase in the hydrodynamic resistance, ensuring that only single cells

are kept in their buckets. By diverting the fluid flow around trapped cells, oncoming cells bypass the occupied traps in preference for empty buckets while also preventing trapped cells from being pulled out of the array. This same system also aids the removal of daughter cells, reducing the physical stress at the cellular level caused by operators manipulating the daughter cells away from each mother cell. If the new bud is oriented upstream of the bucket, the diverted flow removes the daughter cell to the side; if the new bud is oriented downstream, it gets removed between the pillars of the bucket walls [10, 11].

17. If the media runs out during the experiment, the syringe pump should be paused to hold the pressure within the system. Load fresh media into a new syringe. Then, remove the metal pin from the inlet connected to the old media-filling syringe. Next, insert the tubing from the new syringe, and fill the new tubing with fresh media while the old syringe is still in place. This approach will hold the pressure within the system, preventing fluid pressure changes that can cause backflow and loss of cells within the device. Place the new syringe and restart the syringe pump. Once pressure has been reestablished, pause the pump to remove the old syringe tubing and reinsert the metal pin.
18. During the capture of images, care must be taken during prolonged assays, including overnight experiments, to avoid liquid media spills and leaks in the system that may cause damage to the microscope. Long experiments are more prone to clog and result in overgrowth of cells. Another limitation is providing enough media flowing through the system because the syringes only hold a small volume. Unfortunately, refilling the syringes may disturb the flow, and cells are flushed out of their buckets.

Acknowledgments

This work was supported by the National Institutes of Health (NIH 1R01- AI127704-01A1) and the Veterans Administration (VA) Merit (I01BX003741-01A2) to BCF. The contents of this article do not represent the views of the VA or the United States government.

References

1. Spivey EC, Jones SK Jr, Rybarski JR et al (2017) An aging-independent replicative lifespan in a symmetrically dividing eukaryote. eLife 6:e20340. https://doi.org/10.7554/eLife.20340
2. Silva VKA, Bhattacharya S, Oliveira NK et al (2022) Replicative aging remodels the cell wall and is associated with increased intracellular trafficking in human pathogenic yeasts. MBio

13(1):e00190–e00122. https://doi.org/10.1128/mbio.00190-22

3. Jain N, Cook E, Xess I et al (2009) Isolation and characterization of senescent *Cryptococcus neoformans* and implications for phenotypic switching and pathogenesis in chronic cryptococcosis. Eukaryot Cell 8(6):858–866. https://doi.org/10.1128/EC.00017-09
4. Coody TK, Hughes AL (2018) Advancing the aging biology toolkit. eLife 7:e42976. https://doi.org/10.7554/eLife.42976
5. Bouklas T, Alonso-Crisóstomo L, Székely T Jr et al (2017) Generational distribution of a *Candida glabrata* population: resilient old cells prevail, while younger cells dominate in the vulnerable host. PLoS Pathog 13(5): e1006355–e1006355. https://doi.org/10.1371/journal.ppat.1006355
6. Orner EP, Zhang P, Jo MC et al (2019) High-throughput yeast aging analysis for *Cryptococcus* (HYAAC) microfluidic device streamlines aging studies in *Cryptococcus neoformans.* Commun Biol 2(1):256. https://doi.org/10.1038/s42003-019-0504-5
7. Park PU, McVey M, Guarente L (2002) Separation of mother and daughter cells. In: Methods in enzymology, vol 351. Academic Press, pp 468–477. https://doi.org/10.1016/S0076-6879(02)51865-6
8. Lee SS, Vizcarra IA, Huberts DHEW et al (2012) Whole lifespan microscopic observation of budding yeast aging through a microfluidic dissection platform. Proc Natl Acad Sci 109(13):4916–4920. https://doi.org/10.1073/pnas.1113505109
9. Huberts DHEW, González J, Lee SS et al (2014) Calorie restriction does not elicit a robust extension of replicative lifespan in *Saccharomyces cerevisiae.* Proc Natl Acad Sci 111(32):11727–11731. https://doi.org/10.1073/pnas.1410024111
10. Jo MC, Liu W, Gu L et al (2015) High-throughput analysis of yeast replicative aging using a microfluidic system. Proc Natl Acad Sci 112(30):9364–9369. https://doi.org/10.1073/pnas.1510328112
11. Crane MM, Clark IBN, Bakker E et al (2014) A microfluidic system for studying ageing and dynamic single-cell responses in budding yeast. PLoS One 9(6):e100042. https://doi.org/10.1371/journal.pone.0100042

Chapter 26

In Vitro Titan Cell Generation in *Cryptococcus neoformans* and Automated Cell Size Measurements

Benjamin Hommel, Aude Sturny-Leclère, and Alexandre Alanio

Abstract

A special feature of the human fungal pathogen *Cryptococcus neoformans* is its morphological changes triggered by the interaction with the host. During infection, a specific increase in cell size is observed, particularly in lung tissue, from a typical cell size of 5–7 μm cells to cells larger than 10 μm, dubbed titan cells (TCs). However, the study of this specific cell subpopulation was, until now, only possible via recovery of TCs from lungs of mice during experimental infections where stable and reproducible generation of TCs occurs.

The protocol described here generates TCs using in vitro conditions and measures cell size using a rapid, automated method. TC generation in vitro is robust and reproducible, generating yeast cells harboring the same characteristics of TCs generated in vivo.

Key words *Cryptococcus neoformans*, In vitro, Titan cells, Cryptococcosis, Cell size counting

1 Introduction

Cryptococcus neoformans (*Cn*) is a ubiquitous environmental basidiomycetous yeast that causes 223,100 incident cases of meningoencephalitis and more than 181,100 deaths per year worldwide [1], mostly in immunocompromised individuals with acquired immunodeficiency syndrome [2]. From environment niches to its interaction with hosts, this yeast experiences such drastic changes that it requires remarkable capacities to rapidly adapt, survive, and proliferate in host tissues and ultimately cause disease [3–7]. In hosts, *Cn* is exposed to various stresses, such as high temperature, nutrient deprivation, drastic changes in pH, hypoxia, and increased concentration of free radicals or toxic metals [8].

To counteract host defenses, *Cn* initiates several phenotypic adaptations. Some of these include modification of capsule size and composition [9], increased melanization [10], changes in secretion patterns [11], phenotypic switching [12], aging [13], dormancy

Erin E. McClelland (ed.), *Cryptococcus neoformans: Methods and Protocols*, Methods in Molecular Biology, vol. 2775, https://doi.org/10.1007/978-1-0716-3722-7_26,

[14], and drastic cell enlargement [15]. Specifically, *Cn* produces enlarged cells referred to as giant or "titan cells" (TCs). This phenomenon has been observed in human lung and brain infections, animal infections, as well as insect models of cryptococcosis [16–19]. More recently, different teams reproduced TCs in vitro [20–22]. TCs are identified by increased cell body size (diameter > 10 μm) [15]. Studies exploring TC biology describe unique features, such as uninucleate polyploid cells harboring a large vacuole surrounded by a thick cell wall and a dense and highly cross-linked capsule [20–22]. TCs also exhibit increased resistance to phagocytosis [23], oxidative and nitrosative stress [23, 24], and resistance to the antifungal drug fluconazole [24]. Importantly, TC production also enhances dissemination, survival and, virulence in a mouse model of infection [25].

To date, TCs are generated in response to the lung environment in humans and mice or in *Galleria mellonella* models of infection and also in vitro in contact with macrophages, with minimal medium, or with serum [20–22]. Before 2018, in vitro protocols produced a low and unreproducible quantity of enlarged cells without all characteristics of TCs [17]. Additionally, there have been problems in easily and consistently implementing these protocols across laboratories [26]. We developed a robust in vitro protocol to generate TCs and validated an automated method for measuring cell size and counting using the Icy® free open source software [27].

2 Materials

2.1 Culture

1. Sabouraud agar plates: Suspend 50 g of powder in 1 L of purified water, mix thoroughly, and heat with agitation to completely dissolve the powder. Autoclave prior to use and store at 4 °C.
2. *Cryptococcus neoformans* isolate strain H99O stored in 20% glycerol at −80 °C (*see* **Note 1**).
3. Yeast Peptone Dextrose (YPD) liquid: 10 g yeast extract, 20 g peptone, 20 g D-glucose, and 1 L of Milli-Q water. Autoclave prior to use. Store at 4 °C.
4. T25 cm^2 flask with vented filter cap.
5. Incubator set at 30 °C with lateral shaking at 150 rpm.
6. Minimal medium (MM): 15 mM D-glucose, 10 mM $MgSO_4$, 29.4 mM KH_2PO_4, 13 mM glycine, 3.0 μM thiamine, pH 5.5. Filter using a 0.22 μm filter and store at 4 °C (*see* **Note 2**) [28].
7. Eppendorf Thermomixer® (Hamburg, Germany) set at 30 °C, 150 rpm.

2.2 Automatic Cell Size Measurements of Yeast After TC Generation

1. Eppendorf centrifuge.
2. Glass slide and 22 × 22 mm coverslip.
3. Interferential contrast microscope.
4. Camera.
5. Computer with windows 32/64 bit, Mac OS X 64 bit, or Linux platform.
6. Icy® free software downloaded from http://icy.bioimageanalysis.org/download, with plugins "HK" and "active contours".
7. *Cryptococcus* cell size batch protocol for automatic measurements (see Subheading 3.2).

3 Methods

3.1* Cryptococcus neoformans *Titan Cell Preparation (Fig. 1)

1. Seed *Cryptococcus neoformans* strain H99O from −80 °C stock cultures onto a Sabouraud agar plate for 48 h at 30 °C. Then, let the culture grow for a maximum of 5 days at room temperature.
2. Remove 1×10^7 H99O cells from the Sabouraud agar plate and inoculate 10 mL of YPD liquid medium in a 25 cm^2 flask (*see* **Note 3**).
3. Incubate the flask 24 h at 30 °C, with lateral shaking at 150 rpm, until the cells reach stationary phase (*see* **Note 4**) (final concentration ~ $2–3 \times 10^8$/mL).
4. Wash 1 mL of the YPD preculture twice with 1 mL of MM (pH 5.5) by centrifugation at 1150 × *g* for 5 min at room temperature. Count cells using a hemocytometer and adjust the cell concentration to 1×10^6 cells/mL in MM.
5. Remove 1 mL of the cell suspension and incubate in a 1.5 mL Eppendorf® tube with the cap closed, at 30 °C, shaking at 800 rpm for 24 h to 5 days in the Eppendorf Thermomixer® (*see* **Note 5**).

3.2 Cell Size Analysis

The Icy® protocol will use two plugins. The first is HK means [21], which uses the active contour model to automatically extract the cell boundaries (to deform an initial contour placed on the image until it fits the boundary of the desired entity). The second plugin is "active contours," which automatically segments the boundary of the cell starting from an approximate region of interest (ROI).

1. Centrifuge the 1.5 mL Eppendorf® tube at 1050 × *g* for 5 min.
2. Remove 900 μL of supernatant.
3. Gently homogenize the pellet using the remaining supernatant.

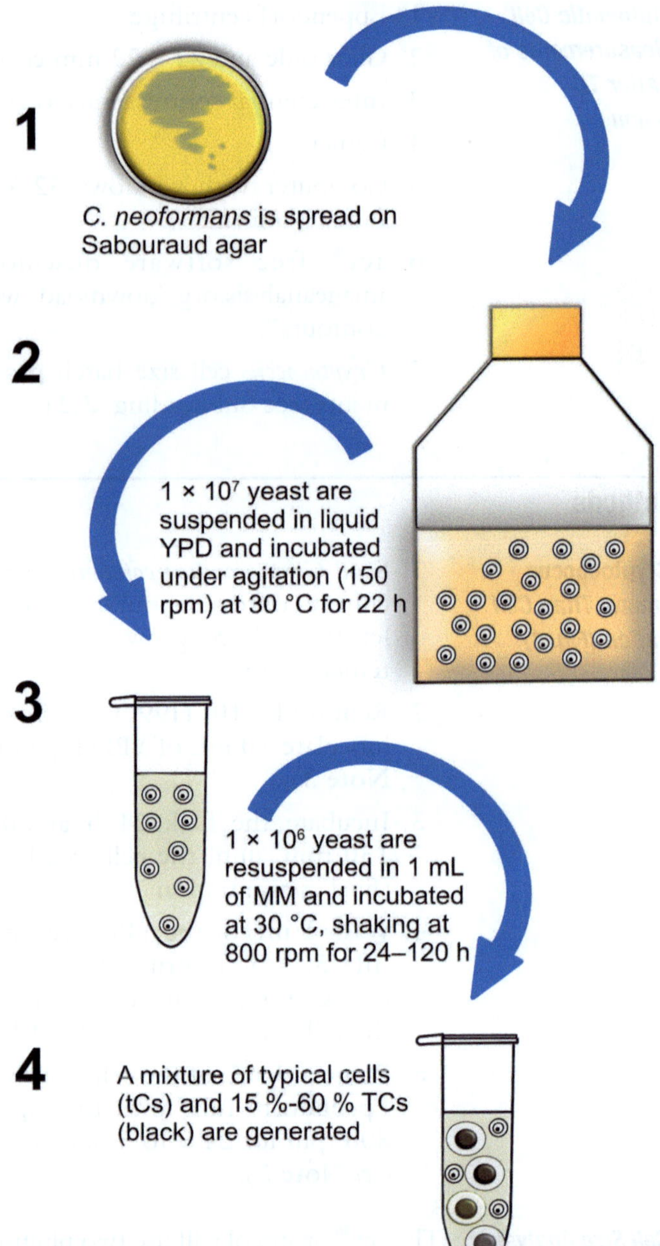

Fig. 1 Stepwise processes for the generation of titan cells (TCs) in vitro. (1) *C. neoformans* is spread on Sabouraud agar. (2) Approximately 1×10^7 yeast are suspended in liquid YPD and incubated under agitation (150 rpm) at 30 °C for 22 h. (3). The culture is washed twice in minimal medium (MM, pH = 5.5). Then, 1×10^6 yeast are resuspended in 1 mL of MM and incubated at 30 °C, shaking at 800 rpm for 24–120 h. (4) A mixture of typical cells (tCs) and generated TCs are ready for analysis. Depending on the experiments, strains, and the lab, between 15% and 60% TCs are generated using this protocol

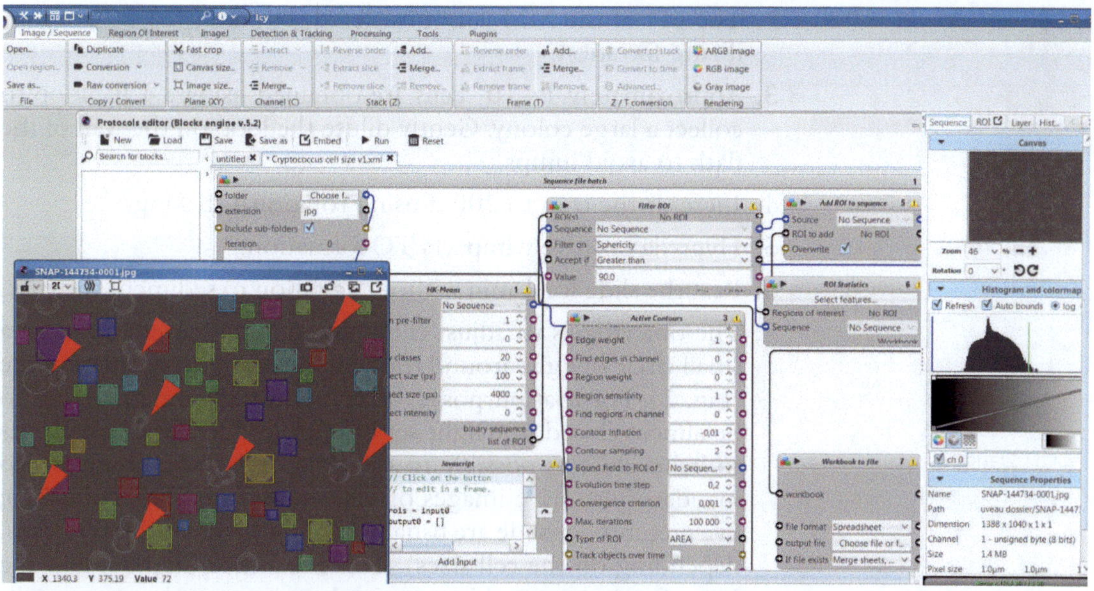

Fig. 2 Screenshot of the output after Icy software analysis of yeast cells obtained with this protocol (measured yeast cells are in color; the red arrow is pointing at excluded budding cells)

4. Drop 5 μL of the suspension onto a slide and cover it using a 22 × 22 mm cover slip (*see* **Note 6**).
5. Take pictures using the 40× objective on a microscope using the .tiff format at 1 × 1 binning (Fig. 2, *see* **Note 7**).
6. Run Icy® software. Search for "*Cryptococcus* cell size" in Icy search bar and open the protocol.
7. Enter your image extension (TIF for tiff or JPG for jpeg image) and your analysis folder.
8. In the box ROI statistics (Fig. 2, box 6), ensure that the following analysis parameters are ticked: full path, parent folder, dataset, and ellipse (column (a), *see* **Note 8**).
9. In the box workbook to file (Fig. 2, box 7), select the name, place, and the format of your file.
10. Press the arrow to run the current protocol.
11. The spreadsheet will provide the cell size measurements of the cells in the image under the ellipse (a) column, in pixels.

4 Notes

1. As there is variation in the amount of TCs generated among the H99 lineage [21], it is critical to use the H99O strain in the protocol.

2. The pH of the MM is not buffered, but adjusted before sterilizing filtration.
3. Approximately 1×10^7 cells are obtained using a 1 μL loop to collect a large colony. Gently dilute the loop on the wall of the flask to avoid lumps.
4. Increase the rpm to 200 if using rotational shaking.
5. Hypoxia positively impacts TC formation.
6. Rest the slide for 1 min without agitation to sediment the cells.
7. The objective is to adjust the light and condenser to obtain a small circle of light around the cells observed in the microscope (Fig. 2). For that purpose, the condenser diaphragm needs to be almost closed, the light set to intense, and the polarizer has to be as close as possible to the slide. A 1 × 1 binning is also mandatory to have images of high quality (at least 1000 Ko); five pictures per slide are usually enough (from 500–1000 cells, depending on the cell sizes). This is a critical step because the Icy software protocol is conceived to detect and measure cells with these adjustments.
8. The protocol is designed to exclude budding cells based on a sphericity criterion. Be careful to avoid cell-cell contact in the images, as cells in contact may be interpreted as non-spherical elements and excluded from the analysis. Therefore, cell density should not exceed 100–200 cells per image.

References

1. Rajasingham R, Govender NP, Jordan A et al (2022) The global burden of HIV-associated cryptococcal infection in adults in 2020: a modelling analysis. Lancet Infect Dis 17:873–881. https://doi.org/10.1016/s1473-3099(22)00499-6
2. Park BJ, Wannemuehler KA, Marston BJ et al (2009) Estimation of the current global burden of cryptococcal meningitis among persons living with HIV/AIDS. AIDS 23:525–530. https://doi.org/10.1097/qad.0b013e328322ffac
3. Alanio A, Vernel-Pauillac F, Sturny-Leclère A et al (2015) *Cryptococcus neoformans* host adaptation: toward biological evidence of dormancy. MBio 6:e02580–e02514. https://doi.org/10.1128/mbio.02580-14
4. Choi J, Vogl AW, Kronstad JW (2012) Regulated expression of cyclic AMP-dependent protein kinase a reveals an influence on cell size and the secretion of virulence factors in *Cryptococcus neoformans*. Mol Microbiol 85:700–715. https://doi.org/10.1111/j.1365-2958.2012.08134.x
5. Franzot SP, Mukherjee J, Cherniak R et al (1998) Microevolution of a standard strain of *Cryptococcus neoformans* resulting in differences in virulence and other phenotypes. Infect Immun 66:89–97
6. Hu G, Cheng P, Sham A et al (2008) Metabolic adaptation in *Cryptococcus neoformans* during early murine pulmonary infection. Mol Microbiol 69:1456–1475. https://doi.org/10.1111/j.1365-2958.2008.06374.x
7. Gerstein AC, Nielsen K (2016) It's not all about us: evolution and maintenance of *Cryptococcus* virulence requires selection outside the human host. Yeast 34:143–154. https://doi.org/10.1002/yea.3222
8. Brown SM, Campbell LT, Lodge JK (2007) *Cryptococcus neoformans*, a fungus under stress. Curr Opin Microbiol 10:320–325. https://doi.org/10.1016/j.mib.2007.05.014
9. Zaragoza O, Rodrigues ML, Jesus MD et al (2009) The capsule of the fungal pathogen *Cryptococcus neoformans*. Adv Appl Microbiol 68:133–216. https://doi.org/10.1016/s0065-2164(09)01204-0

10. Kronstad J, Saikia S, Nielson ED et al (2012) Adaptation of *Cryptococcus neoformans* to mammalian hosts: integrated regulation of metabolism and virulence. Eukaryot Cell 11: 109–118. https://doi.org/10.1128/ec.05273-11
11. Rodrigues ML, Nakayasu ES, Oliveira DL et al (2008) Extracellular vesicles produced by *Cryptococcus neoformans* contain protein components associated with virulence. Eukaryot Cell 7:58–67. https://doi.org/10.1128/ec.00370-07
12. Fries BC, Taborda CP, Serfass E et al (2001) Phenotypic switching of *Cryptococcus neoformans* occurs *in vivo* and influences the outcome of infection. J Clin Invest 108:1639–1648. https://doi.org/10.1172/jci13407
13. Bouklas T, Pechuan X, Goldman DL et al (2013) Old *Cryptococcus neoformans* cells contribute to virulence in chronic cryptococcosis. MBio 4:e00455–e00413. https://doi.org/10.1128/mbio.00455-13
14. Hommel B, Sturny-Leclère A, Volant S et al (2019) *Cryptococcus neoformans* resists to drastic conditions by switching to viable but non-culturable cell phenotype. PLoS Pathog 15:e1007945. https://doi.org/10.1371/journal.ppat.1007945
15. Zaragoza O, Nielsen K (2013) Titan cells in *Cryptococcus neoformans*: cells with a giant impact. Curr Opin Microbiol 16:409–413. https://doi.org/10.1016/j.mib.2013.03.006
16. Zaragoza O, García-Rodas R, Nosanchuk JD et al (2010) Fungal cell gigantism during mammalian infection. PLoS Pathog 6:e1000945. https://doi.org/10.1371/journal.ppat.1000945
17. Love GL, Boyd GD, Greer DL (1985) Large *Cryptococcus neoformans* isolated from brain abscess. J Clin Microbiol 22:1068–1070
18. Cruickshank JG, Cavill R, Jelbert M (1973) *Cryptococcus neoformans* of unusual morphology. Appl Microbiol 25:309–312
19. García-Rodas R, Casadevall A, Rodríguez-Tudela JL et al (2011) *Cryptococcus neoformans* capsular enlargement and cellular gigantism during *galleria mellonella* infection. PLoS One 6:e24485. https://doi.org/10.1371/journal.pone.0024485
20. Dambuza IM, Drake T, Chapuis A et al (2018) The *Cryptococcus neoformans* titan cell is an inducible and regulated morphotype underlying pathogenesis. PLoS Pathog 14:e1006978. https://doi.org/10.1371/journal.ppat.1006978
21. Hommel B, Mukaremera L, Cordero RJB et al (2018) Titan cells formation in *Cryptococcus neoformans* is finely tuned by environmental conditions and modulated by positive and negative genetic regulators. PLoS Pathog 14: e1006982. https://doi.org/10.1371/journal.ppat.1006982
22. Trevijano-Contador N, de Oliveira HC, García-Rodas R et al (2018) *Cryptococcus neoformans* can form titan-like cells *in vitro* in response to multiple signals. PLoS Pathog 14: e1007007. https://doi.org/10.1371/journal.ppat.1007007
23. Chrisman CJ, Albuquerque P, Guimarães AJ et al (2011) Phospholipids trigger *Cryptococcus neoformans* capsular enlargement during interactions with amoebae and macrophages. PLoS Pathog 7:e1002047. https://doi.org/10.1371/journal.ppat.1002047
24. Okagaki LH, Nielsen K (2012) Titan cells confer protection from phagocytosis in *Cryptococcus neoformans* infections. Eukaryot Cell 11: 820–826. https://doi.org/10.1128/ec.00121-12
25. Gerstein AC, Fu MS, Mukaremera L et al (2015) Polyploid titan cells produce haploid and aneuploid progeny to promote stress adaptation. MBio 6:e01340–e01315. https://doi.org/10.1128/mbio.01340-15
26. Okagaki LH, Strain AK, Nielsen JN et al (2010) Cryptococcal cell morphology affects host cell interactions and pathogenicity. PLoS Pathog 6:e1000953. https://doi.org/10.1371/journal.ppat.1000953
27. de Chaumont F, Dallongeville S, Chenouard N et al (2012) Icy: an open bioimage informatics platform for extended reproducible research. Nat Methods 9:690–696. https://doi.org/10.1038/nmeth.2075
28. Dufour A, Meas-Yedid V, Grassart A et al (2008) Automated quantification of cell endocytosis using active contours and wavelets. 2008 19th Int Conf Pattern Recognit 1–4. https://doi.org/10.1109/icpr.2008.4761748

Chapter 27

Chitosan-Deficient *Cryptococcus* as Whole-Cell Vaccines

Charles A. Specht, Woei C. Lam, Maureen M. Hester, Diana Lourenco, Stuart M. Levitz, Jennifer K. Lodge, and Rajendra Upadhya

Abstract

Creating a safe and effective vaccine against infection by the fungal pathogen *Cryptococcus neoformans* is an appealing option that complements the discovery of new small molecule antifungals. Recent animal studies have yielded promising results for a variety of vaccines that include live-attenuated and heat-killed whole-cell vaccines, as well as subunit vaccines formulated around recombinant proteins. Some of the recombinantly engineered cryptococcal mutants in the chitosan biosynthesis pathway are avirulent and very effective at conferring protective immunity. Mice vaccinated with these avirulent chitosan-deficient strains are protected from a lethal pulmonary infection with *C. neoformans* strain KN99. Heat-killed derivatives of the vaccination strains are likewise effective in a murine model of infection. The efficacy of these whole-cell vaccines, however, is dependent on a number of factors, including the inoculation dose, route of vaccination, frequency of vaccination, and the specific mouse strain used in the study. Here, we present detailed methods for identifying and optimizing various factors influencing vaccine potency and efficacy in various inbred mouse strains using a chitosan-deficient *cda1Δcda2Δcda3Δ* strain as a whole-cell vaccine candidate. This chapter describes the protocols for immunizing three different laboratory mouse strains with vaccination regimens that use intranasal, orotracheal, and subcutaneous vaccination routes after the animals were sedated using two different types of anesthesia.

Key words Cryptococcosis, Chitosan, Cell wall, Chitin deacetylase, Whole-cell vaccine, Attenuated vaccine, Vaccination dose, Vaccination schedule, Fungal virulence

1 Introduction

The human fungal pathogens *Cryptococcus neoformans* and *Cryptococcus gattii* are responsible for the infection known as cryptococcosis. While infections caused by *C. neoformans* are more common in immunocompromised hosts, infections by *C. neoformans* and *C. gattii* have also been reported in apparently healthy hosts [1–4]. This fungus is endemic worldwide and it is thought that the initial infection is caused by inhaling basidiospores or desiccated yeast cells. While the primary site of infection is the lung, yeast may spread to the brain and other organs, resulting in systemic infection

Erin E. McClelland (ed.), *Cryptococcus neoformans: Methods and Protocols*, Methods in Molecular Biology, vol. 2775, https://doi.org/10.1007/978-1-0716-3722-7_27,

and, if left untreated, the death of the infected host. According to estimates by Rajasingham et al. (2022), there are 152,000 cases of cryptococcal meningitis each year, and the illness results in 112,000 fatalities [5]. The World Health Organization has designated *Cryptococcus* as one of four fungal pathogens with unmet research needs, calling for increased research and development, as well as fresh public health intervention strategies [6]. The preferred anticryptococcal medication regimen is amphotericin B and 5-fluorocytosine. Unfortunately, this combination is extremely toxic and is not widely available in the developing world where the majority of reported cases occur [7, 8]. Fluconazole is frequently used in some areas as an alternative to amphotericin B. It is less effective and there have been reports of drug-resistant isolates emerging [9]. Echinocandins are effective against other fungal infections, but not against *C. neoformans* [10, 11]. As a result, it is critical to develop safe and effective treatments against cryptococcal infections. The development of vaccines against cryptococcosis would be an optimal strategy for preventing cryptococcal infections.

Various vaccine strategies have been developed for cryptococcal infections, including whole-cell vaccines, subunit vaccines, and glucan particles loaded with protein antigens [12–23]. In a murine infection model, both live and heat-killed forms of whole cells of either wild-type or avirulent recombinant cryptococcal strains have been effective. Multiple vaccination routes, including intradermal, intraperitoneal, subcutaneous, intravenous, intranasal, and intratracheal or orotracheal, were used in these studies. Multiple vaccination frequencies with varying doses of inoculum have been effective. Aside from the type of vaccine and its regimen for delivery, different inbred mouse lines demonstrated varying levels of vaccine effectiveness. A factor that also impacts testing vaccines in different mouse lines is that their inherent resistance to infection is not the same (Fig. 1). To reduce this effect, the number of cells of the virulent strain used for infection post-vaccination may be adjusted for the mouse line being used.

In 1960, the first whole-cell vaccines against *Cryptococcus* infection were tested, using various strains, such as those attenuated for virulence, hypocapsular strains, and heat-killed cells. When challenged with a virulent strain, all show varying degrees of protection [24]. Immunization with wild-type yeast cells whose capsules have been stripped by gamma irradiation or enzymatic treatments also elicit varying degrees of protective immunity [25, 26]. A robust immunity against infection with a virulent strain of *C. neoformans* was achieved when strain H99 was engineered to produce murine INF-γ (H99γ) and used for vaccination. In this case, a 5×10^4 colony-forming unit (CFU) dose of live H99γ administered intranasally is effective in inducing long-lasting protective immunity in BALB/c mice [18]. Following the success of this whole-cell vaccine, several strains of *Cryptococcus* were

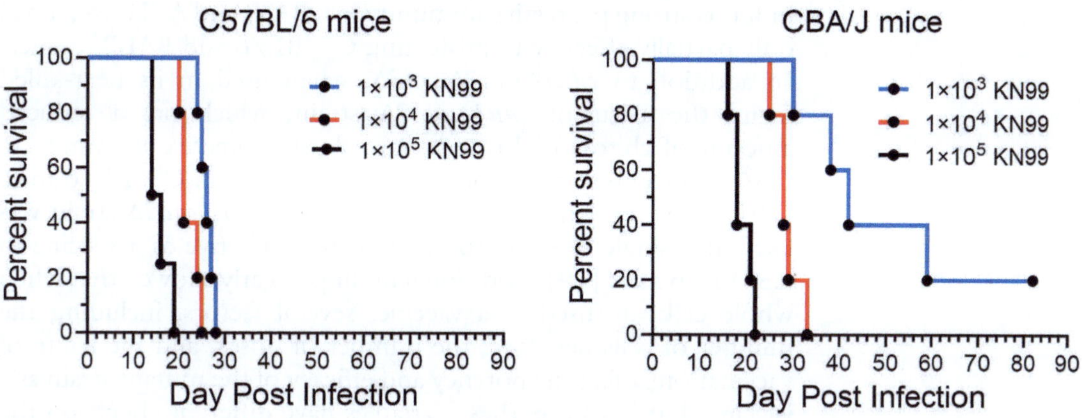

Fig. 1 Effect of the number of colony forming units (CFUs) of KN99 used for infection on the survival of unvaccinated C57BL/6 and CBA/J mice. Groups of five mice were infected intranasally with 1×10^3, 1×10^4, or 1×10^5 cells (CFUs) of KN99. Mice were monitored for survival for 84 days.

engineered to delete genes associated with virulence. These strains were avirulent in mice, and when tested as vaccines, they demonstrated effective protective immunity against challenge with virulent progenitor strains [17, 20–22]. In all of these cases, either the intranasal or orotracheal route was employed for primary vaccination. Subsequent booster doses of the vaccine improved protective immunity in various inbred mouse strains when administered intranasally or subcutaneously. A single dose of 10^7 cells of the live Znf2^{oe} strain is sufficient to induce protective immunity in A/J mice; however, a second intranasal booster dose is required for immunity when heat-killed cells are used for vaccination [22]. In CBA/J mice, intranasal vaccination with 5×10^5 live cells of a *Cryptococcus* mutant that lacks steryl glucosidase (*sgl1*Δ) elicits a robust protective immunity against a *C. neoformans* challenge [21]. However, in the case of another *Cryptococcus* mutant that lacks Fbp1, while two doses of vaccination with the *fbp1*Δ cells confer complete protection in BALB/c mice, it only confers partial protection in C57BL/6 J and A/J mouse strains [20]. Furthermore, it has been discovered that protective immunity is dose dependent: 5×10^7 cells provide complete protection, 2.5×10^7 and 1.0×10^7 cells provide partial protection, and 0.5×10^7 cells fail to induce protective immunity [27]. These tests have been carried out on BALB/c mice infected with the virulent H99 strain [27].

We have been characterizing the chitosan-deficient cryptococcal vaccine strain *cda1*Δ*cda2*Δ*cda3*Δ, which lacks three chitin deacetylases and produces no chitosan, and we have also observed dose-dependent protective immunity [17]. When delivered through intranasal inoculation, a minimum dose of 10^7 *cda1*Δ*cda2*Δ*cda3*Δ cells was required to induce protective immunity in CBA/J mice. While a single dose of 10^7 *cda1*Δ*cda2*Δ*cda3*Δ induced strong protective immunity in CBA/J and A/J mice, it was only partially effective in protecting C57BL/6 and BALB/c mice. In addition to *cda1Δcda2Δcda3Δ*, when used in its heat-killed form, the avirulent *cda1Δcda2Δ* strain, which has a reduced amount of chitosan, also induced robust protective immunity in CBA/J mice, but it was ineffective as a vaccine in C57BL/6 mice [19]. However, when a live culture of the *cda1Δcda2Δ* strain was used at a single dose of 10^7 cells, it was effective as a vaccine in C57BL/6 mice [19]. Additional findings clearly showed that when whole cells are used as a vaccine, several factors, including the number of cells per dose, the number of doses, and the route of vaccination, affect the potency and efficacy of the mutant strain as a vaccine. Furthermore, these variables have different effects on the nature of protective immunity in different inbred laboratory mouse strains. Therefore, using *cda1Δcda2Δcda3Δ* as an example, we will present a general method for determining the inoculum dose, mode of vaccination, and effect of these vaccination regimens on the protective immunity in different inbred laboratory mouse strains against a lethal pulmonary infection with the KN99 strain. KN99 is particularly virulent and was created by ten backcrosses to the extensively researched clinical strain H99 [28]. Over the course of several years at different institutions, we have developed two protocols that are equally effective at generating and testing the *cda1Δcda2Δcda3Δ* vaccine, but differ in some details. For example, we use different agents for anesthesia (ketamine/dexmedetomidine and isoflurane) and different routes to introduce *Cryptococcus* into the lungs (intranasal and orotracheal). We have been working principally with CBA/J mice at Washington University, St. Louis (WashU), and with BALB/c and C57BL/6 mice at University of Massachusetts Chan Medical School (UMASS). Working from a common set of materials, we note how the protocols differ by institution, although comparable vaccine efficacies are found.

2 Materials

2.1 Culturing of *Cryptococcus* Strains for Preparation of Live and Heat-Killed *cda1Δcda2Δcda3Δ* Vaccines and for Infection with KN99

Cryptococcus is a Biosafety Level 2 (BSL-2) organism. Personnel working in the BSL2 lab must have general training in handling pathogenic agents. It is necessary to have Institutional Biosafety Committee (IBC) approval to study *Cryptococcus*. As a BSL2 pathogen, work with *Cryptococcus* can be done on an open bench top as long as aerosols are not generated. But if the procedure involves the potential release of aerosols, working in a biosafety hood is preferable [29]. All culture-related materials are autoclaved prior to disposal. We recommend double distilled or Milli-Q water (18 megaohm) for making solutions and media.

1. YPD liquid medium: 10 g/L yeast extract, 20 g/L peptone, 20 g/L dextrose; for plates, add 20 g/L agar.
2. Sabouraud Dextrose Agar Emmons.
3. 30 °C incubator for incubating culture plates.
4. Incubator for shaking cultures capable of maintaining 300 rpm and 30 °C.
5. 14 mL polypropylene round-bottom tubes with cap for 4 mL cultures.
6. Wood applicators, 6 inches long, or inoculation loops to transfer cells.
7. 2 mL cryogenic vials.
8. 50% glycerol.
9. 250 mL polycarbonate Erlenmeyer flasks with vented cap.
10. 50 mL centrifuge tubes.
11. Centrifuge with a 1500 rpm (425 × *g*) minimum speed and rotor for 14 mL and 50 mL centrifuge tubes.
12. Dulbecco's phosphate buffered saline (PBS), endotoxin-free.
13. Hemocytometer: Standard glass, or disposable hemocytometer or a TC20 automated cell counter and counting slides.
14. Light microscope to view hemocytometer.
15. Water or dry bath incubator at 70 °C to heat-kill *Cryptococcus*.

2.2 Vaccination of Mice with cda1Δcda2Δcda3Δ or Infection with KN99 Cell Suspensions

1. Approved Institutional Animal Care and Use Committee (IACUC) protocols.
2. Ear tags or other means of identification to aid in recording the weight of individual mice during a study.
3. Digital scale accurate to 0.1 g to measure the weight of a mouse.
4. Anesthesia (WashU): ketamine (8 mg/mL)/dexmedetomidine (0.05 mg/mL) mixture.
5. Atipamezole (200 μL) (0.25 mg/mL) to reverse effects of anesthesia.
6. Syringe (1 mL with 27G 0.4 mm × 13 mm hypodermic needle) for injections.
7. Veterinary eye ointment to counter eye dryness during anesthesia.
8. Anesthesia (UMASS Dept. of Animal Medicine): isoflurane, USP.
9. Anesthesia chamber for small rodents equipped with F/air carbon canister to trap isoflurane. Alternatively, anesthesia with isoflurane may be done in a chemical fume hood without an anesthesia chamber.

10. Glass jar with screw-on lid (height 4.5–5″, diameter 4.0–4.5″).
11. Absorbent pad (Nestlet pad, 2″ × 2″).
12. Intubation stand to suspend the mouse for intranasal or orotracheal inoculation.
13. Dental tape to suspend mouse on intubation stand.
14. 100 μL pipettor and tips to administer 50 μL of cell suspension.
15. Vortex mixer for mixing the inoculum.
16. Heating pad for the postinoculation recovery of the animals.
17. 0.3 mL insulin syringe, 29G × 1/2″ for subcutaneous injection of vaccine.

2.3 Determining Lung CFUs

1. Pen/Strep: 10,000 U/mL penicillin, 10,000 μg/mL streptomycin.
2. PBS Pen/Strep: Adjust PBS to 200 U/mL penicillin and 200 μg/mL streptomycin to homogenize lungs and make serial dilutions of homogenates to plate for CFUs.
3. Homogenizer with appropriate probes: PRO200 homogenizer connected to a 7 mm diameter PRO quick connect probe with a saw-tooth bottom or Omni TH tissue homogenizer with tip adaptor for 7 mm × 110 mm hard tissue probe.
4. CFU counting pen.
5. GraphPad Prism software to analyze data.

3 Methods

Evaluate the Vaccine‘s Efficacy in CBA/J Mice (Wash U).

3.1 Make Stocks of *Cryptococcus* Strains for Long-Term Storage

Strains of *Cryptococcus* are stored at −80 °C and serve as the primary source for starting cultures used to inoculate mice. Under sterile conditions, cells from the frozen stock may be scraped using either a sterile wood applicator or a sterile loop, then transferred to an YPD agar plate and streaked.

1. Add 4 mL of liquid YPD medium to a 14 mL culture tube.
2. Use a sterile wood applicator to transfer cells from agar medium to the culture tube. Briefly vortex to suspend cells.
3. Shake culture tube at approximately a 70° angle at 220 rpm and 30 °C for 48 h.
4. Combine 0.9 mL of culture with 0.9 mL 50% sterile glycerol and mix.
5. Store at −80 °C.

3.2 Preparing the Inoculum

1. The required strains of *Cryptococcus* are always kept at −80 °C as stocks in YPD medium containing 25% glycerol. Strains from the glycerol stocks KN99 (Lodge Lab Stock # JLCN 434) and *cda1Δcda2Δcda3Δ* (Lodge lab Stock # LBCN 632) are each streaked on YPD agar plates and incubated for 72 h at 30 °C.
2. In a 250 mL Erlenmeyer flask with a vented cap, inoculate a loop of yeast cells into 50 mL of YPD medium, as in Fig. 2. Incubate the flask at 30 °C while shaking at 300 rpm for 48 h. These growth conditions are critical to the vaccine's efficacy.
3. Transfer the 50 mL culture to a 50 mL falcon tube and centrifuge for 10 min at 3300 × *g* and 22 °C. Carefully decant the medium without disturbing the cell pellet. Sterilize the collected medium with 10% bleach to avoid disposing of live yeast cells into the sink.
4. Wash the cell pellet twice by suspending the cell pellet each time with 50 mL endotoxin free-PBS followed by centrifugation. Ensure that the cell pellet is completely resuspended in PBS with no visible cell clumps. After the second wash, resuspend the cell pellet in 20 mL of PBS.
5. By serial dilution, make a 1:200 dilution of the *cda1Δcda2Δcda3Δ* mutant. Using a hemocytometer, load 10 μL of the diluted cell suspension. Count the yeast cells in the four corner squares, each is 1 mm^2 and has a volume of 0.1 mm^3, to determine concentration. When counting cells, consider cells with 2–3 attached buds as one cell and cell clumps of 4–8 as two cells, especially in the case of the *cda1Δcda2Δcda3Δ* mutant.
6. Figure 2 depicts the preparation of an inoculum of *cda1Δcda2Δcda3Δ* for intranasal vaccination. For example, we counted an average of 52 cells per corner square in the 1: 200 diluted sample. This results in a concentration of 1.04×10^8 cells/mL in the washed cell suspension of 20 mL. This starting yeast solution may be used to make various doses of inoculum. To make an inoculum with 2×10^6 cells/mL, add 100 μL of the cell suspension to 4.9 mL of PBS. In 50 μL of cell suspension, there will be 1×10^5 cells. Similarly, to make a suspension of 2×10^7 cells/mL, add 1 mL of the cell suspension to 4 mL of PBS. A 50 μL aliquot of this cell suspension will have 1×10^6 cells. To make a cell suspension with 2×10^8 cells/mL, centrifuge 10 mL of the initial cell suspension and resuspend the cell pellet in PBS to a final volume of 5 mL. In 50 μL of this cell suspension, there will be 1×10^7 cells. The cell count is confirmed by back plating the inoculum onto YPD. This is accomplished by spotting 50 μL of a 1:200,000 serially diluted cell suspension onto YPD. The plate was incubated at 30 °C for 3 days. There were 31 CFUs on average per spot in this case

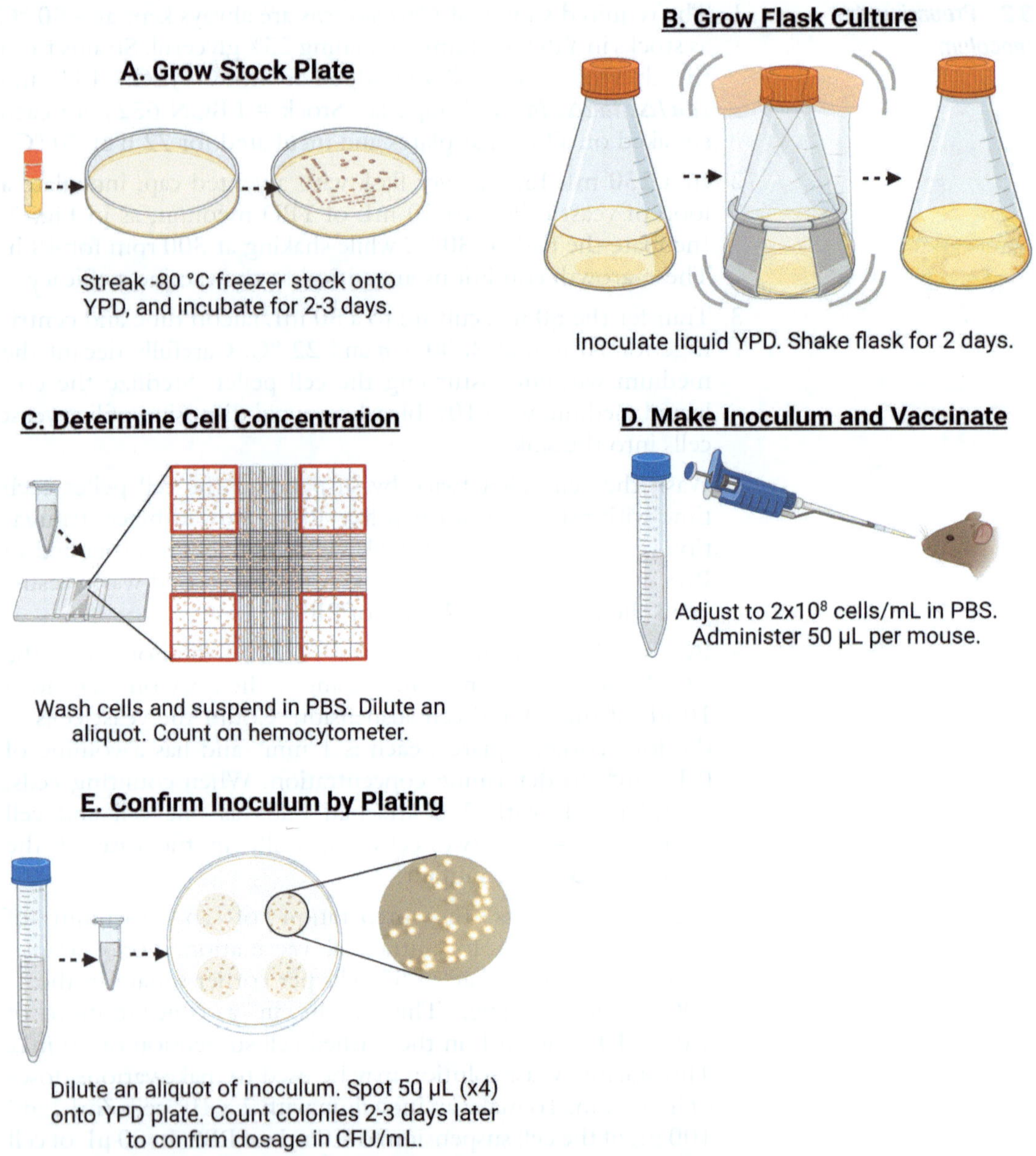

Fig. 2 Steps taken to prepare *cda1Δcda2Δcda3Δ* cells for vaccination of mice. (**a**) The −80 °C stock is streaked out on a YPD agar plate and after 2–3 days at 30 °C is used as the inoculum for a flask containing YPD medium. (**b**) The flask is shaken for 2 days at 30 °C. (**c**) Cells are harvested by centrifugation and washed twice with phosphate buffered saline (PBS). The washed cells are suspended in 20 mL of endotoxin free PBS. An aliquot is diluted 1:200 and counted using a hemocytometer. Cell counts of squares outlined in red are averaged. The calculated cell density of the suspension is adjusted to 2×10^8 cells/mL. (**d**) Mice are vaccinated intranasally or orotracheally with 50 μL. (**e**) The inoculum is diluted and plated to confirm that the mice were inoculated with 1×10^7 cells (colony forming units [CFUs]). These steps are also used to prepare the KN99 inoculum to challenge vaccinated mice. Adjustments are made to the concentration of the inoculum for the mouse strain being challenged. The illustration was created with Biorender.com.

(Fig. 2E), resulting in a concentration of the cell suspension as 1.2×10^8 cells/mL, which is comparable to the count determined by counting the cells on the hemocytometer.

3.3 Intranasal Vaccination with a Live Vaccine Strain

1. *Cryptococcus* strains are kept at −80 °C as glycerol stocks in YPD containing 25% glycerol.
2. Strains from the glycerol stocks of *cda1*Δ*cda2*Δ*cda3*Δ (Lodge Lab Stock LBCN 632) and KN99 (Lodge Lab Stock JLCN 434) are streaked onto YPD plates and incubated for 72 h at 30 °C.
3. Use CBA/J mice (*see* **Note 1**) that are 4–6 weeks old (sex matched). After receiving the mice, allow at least 10 days for the mice to acclimate in the animal facility before beginning the experiment.
4. Place five mice in each cage.
5. Provide the standard diet and water ad libitum during the entire length of the experiment.
6. Prepare the inoculum of the vaccine strains on the day of vaccination by following the steps in Subheading 3.1, **step 2**. Use an inoculum dose of 1×10^7 cells/50 μL of PBS.
7. For one experiment, a minimum of two groups of mice with five animals each should be used, and the experiment must be repeated with an independent preparation of the inoculum. As a control, the first group should only receive PBS as a vaccine. The vaccine strain should be administered to the second group of mice.
8. For vaccination, sedate the animals with an intraperitoneal injection (200 μL) of a ketamine (8 mg/mL)/dexmedetomidine (0.05 mg/mL) mixture. Apply veterinary eye ointment to the animals' eyes to keep them from drying out.
9. Tag each mouse with an ear tag and record its weight.
10. Suspend the animal by its top incisors at a height approximately twice the mouse's body length (e.g., 20 cm) above the surface of the biosafety hood on a strong thin string (e.g., dental tape) secured to a fixed object (e.g., tied to two titration stands; Fig. 3).
11. Vortex the inoculum, then pipette 50 μL followed by slowly dripping it into the mouse's nares (see Fig. 3). Allow the animal to hang for an additional 10 min.
12. Reverse the anesthesia by an intraperitoneal injection of atipamezole (200 μL) (0.25 mg/mL) to each mouse.
13. Transfer the animal to its cage with care and check on it until it is fully awake.

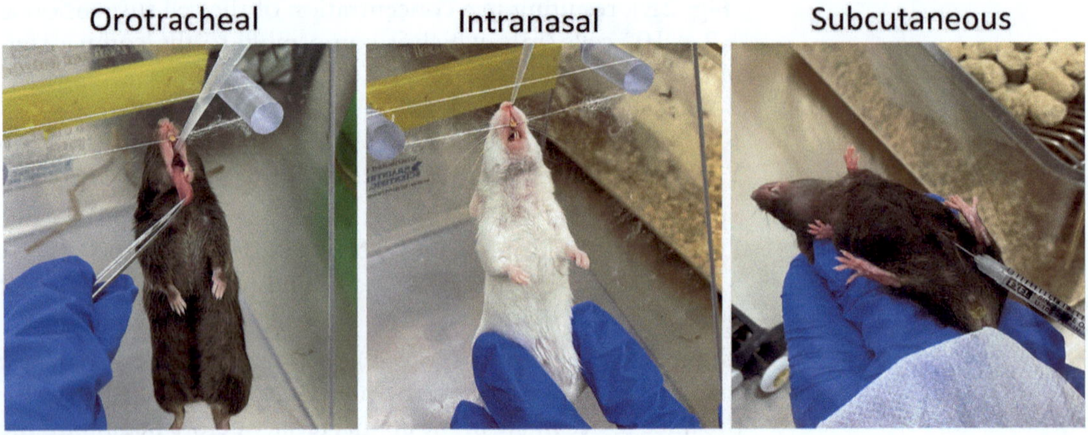

Fig. 3 Administration routes in mice using the *cda1Δcda2Δcda3Δ* vaccine

14. Monitor the mice and record their body weight twice daily until their weight has stabilized or begun to rise.
15. Allow the vaccinated animals to recover for 40 days (*see* **Notes 2** and **3**).
16. On the 41st day, challenge the animals intranasally with 5×10^4 cells (CFUs) of wild-type KN99 cells in 50 μL of PBS by repeating steps 9–13 above. Prepare the KN99 inoculum by following the steps outlined in Subheading 3.1, **step 2**. (*see* **Note 4**).
17. Monitor the mice twice daily and record their weight every day for 80 days after the post-challenge infection (DPI).
18. Euthanize the animals using CO_2 asphyxiation followed by cervical dislocation when their body weight falls below 80% of their pre-challenge infection weight. Determine the fungal burden in the lungs, brain, spleen, kidney, and liver.
19. Place a harvested organ in a preweighed 8 mL flat base screw cap tube containing 2 mL of PBS, and keep the organs on ice. Record the weight of the tube containing the organ. Calculate the weight of the organ.
20. Homogenize the organs in a biosafety hood with a PRO200 homogenizer connected to a 7 mm diameter PRO quick connect generator homogenizer probe with a saw-tooth bottom.
21. In a 24-well non-tissue culture treated plate, serially dilute the homogenized samples in PBS and spot 50 μL of the sample in triplicate onto an YPD plate with a wide-orifice pipet tip. Incubate the plates at 30 °C for 48 h and count the CFU to determine the fungal burden. Express the fungal burden as CFUs/organ or CFUs/g of tissue.

3.4 Intranasal Vaccination with a Heat-Killed Preparation of cda1Δcda2Δcda3Δ

1. When determining the survival of mice vaccinated with heat-killed *cda1Δcda2Δcda3Δ*, use heat-killed KN99 grown in YPD as a control vaccine since it is ineffective at developing protective immunity. As an unvaccinated control, use PBS for inoculation. Grow the necessary yeast strains and prepare the inoculum as described in Subheading 3.1, **step 2** to a final concentration of 2×10^8 cells (CFUs)/mL.
2. In a 70 °C dry bath, incubate the inoculum for 30 min.
3. Plate 100 μL of undiluted inoculum onto an YPD plate and incubate for 48 h at 30 °C to ensure that the incubation at 70 ° C has killed all the yeast cells.
4. Perform intranasal vaccination as in Subheading 3.1, **step 3**.
5. After 40 days, follow the steps in Subheading 3.1, **step 3**. to infect the vaccinated mice with wild-type KN99 cells, monitoring for survival and fungal burden analysis at the end of the experiment.

Evaluating the *cda1Δcda2Δcda3Δ* Vaccine in BALB/c and C57BL/6 Mice (UMASS).

3.5 Preparation of Cryptococcus Cultures of cda1Δcda2Δcda3Δ for Vaccination

Prior to use in mouse studies, yeast cells from a − 80 °C stock are first grown on YPD agar medium to make a working stock plate, which may be stored at 4 °C for one month when sealed with a strip of parafilm. The stock plate is used to inoculate liquid YPD medium to generate live and heat-killed *cda1Δcda2Δcda3Δ* vaccines. Each mouse is vaccinated three times: the first vaccination is administered into the lungs, and the second and third vaccinations are by subcutaneous injection in the abdominal region (Fig. 3). Vaccinations are spaced at two-week intervals. The timeline for vaccination(s) and challenge is shown in Fig. 4.

1. Without thawing the −80 °C glycerol stock, use a wood applicator to transfer the cells of the *cda1Δcda2Δcda3Δ* strain to an YPD agar plate. Spread cells over an area approximately 2.5 cm in diameter. Incubate the plate at 30 °C for 3 days. This stock plate is then stored at 4 °C using a strip of parafilm to seal the plate.
2. Add 25 mL of liquid YPD medium to a 250 mL culture flask.
3. To prepare the first vaccine, inoculate cells from the stock plate to the flask.
4. Shake the flask at 220 rpm and 30 °C for approximately 2 days.
5. Transfer the culture to a 50 mL centrifuge tube and spin at 425 × *g* for 5 min.
6. Suspend cells in 10 mL of PBS and centrifuge to wash the cells. Wash the cells twice.
7. Suspend cells in 10 mL of PBS.

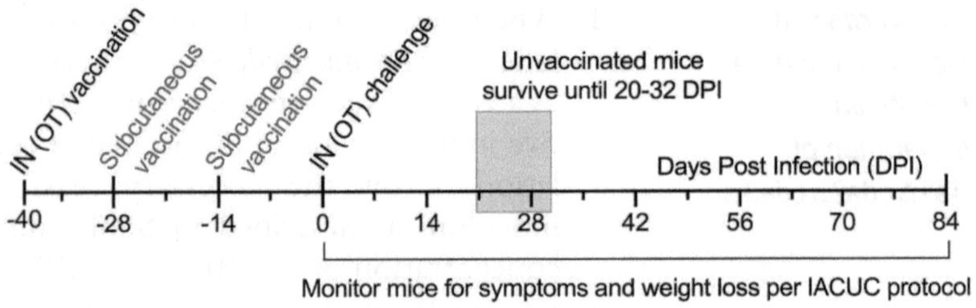

Fig. 4 Timeline to evaluate the efficacy of a vaccine in mice. The intranasal (IN) or orotracheal (OT) vaccination is done 40–42 days prior to challenge (Day 0). Two subsequent subcutaneous vaccinations of BALB/c and C57BL/6 mice are conducted at 28 days and 14 days before challenge. Following challenge, studies are typically ended 70–84 days (10–12 weeks) later. A study takes 16–18 weeks to complete

8. Dilute an aliquot of the suspension 1:100 with PBS.
9. Count cells with a hemocytometer or a TC20 automated cell counter.
10. Adjust the concentration of cells to 2×10^8 cells/mL using PBS to administer 1×10^7 cells into the lungs. Make aliquots of the vaccine.
11. Measure CFUs. Dilute an aliquot of the vaccine 1:100,000 with PBS and spot 10 μL four times onto a Sabouraud agar plate for drip plating.
12. Count CFUs after 3 days of incubation at 30 °C and average the four counts.
13. For the second and third vaccinations, culturing is done in 14 mL tubes and 4 mL of YPD (as above in Subheading 3.1, **step 2**) with shaking at 30 °C for 2 days.
14. Collect cells by centrifugation and suspend in 4 mL of PBS.
15. Dilute the cell suspension 1:100 and count the cells.
16. Adjust the cell suspension to 2×10^7 cells/mL with PBS to administer 2×10^6 cells in 100 μL by subcutaneous injection of the skin at the midline of the abdomen of each mouse (Fig. 3).

3.6 Vaccination of Mice

For the first vaccination, the mice are anesthetized with isoflurane to administer the vaccine into the lungs. Steps for the orotracheal vaccination are listed first (steps 1–8) followed by steps for administering the second and third vaccinations using subcutaneous injection (steps 9–11). Vaccination of mice by the orotracheal, intranasal, and subcutaneous methods is depicted in Fig. 3.

1. Attach dental tape between posts of the intubation stand.
2. Moisten an absorbent pad by placing the pad over the opening of the isoflurane bottle and invert the bottle 2–3 times. Tape

the pad to the lid of the glass jar. Place the mouse in the jar and hold the lid on the jar. Wait approximately 30–45 s for the mouse to stop moving and for the tail to drop.

3. Using the intubation stand, hang the mouse vertically on the dental tape by its incisors.
4. With a pair of tweezers pull the tongue out to one side.
5. Briefly vortex the 2×10^8 cells/mL suspension of *cda1Δcda2Δcda3Δ*.
6. Pipette 50 μL of suspended cells.
7. Insert the pipettor tip to the back of the throat and dispense the 50 μL. The mouse should gasp for air and inhale the vaccine.
8. Remove the mouse from the intubation stand and allow it to recover from the anesthesia. Place the mouse in a new cage.
9. Subcutaneous injections are done using a 0.3 mL insulin syringe. Briefly, vortex the 2×10^7 cells/mL suspension of *cda1Δcda2Δcda3Δ*.
10. Withdraw 100 μL of suspension into the syringe.
11. Hold the mouse securely with one hand and inject subcutaneously into the midline of the abdomen. Place the mouse in a new cage.

3.7 Preparation of *Cryptococcus* Cultures of KN99 for Challenge

Vaccinated mice are typically challenged with KN99 two weeks after the third vaccination. The amount of KN99 used is dependent on the mouse strain under study: C57BL/6 mice are challenged with 1×10^4 cells (CFUs), BALB/c mice with 2×10^4 cells (CFUs), and CBA/J mice with 5×10^4 cells (CFUs). The basis for this is empirical. The challenge doses have been adjusted so that infection with each unvaccinated strain of wild-type mouse is lethal within the third to fourth week of pulmonary infection.

1. Add 4 mL of liquid YPD medium to a 14 mL culture tube.
2. Use a sterile wood applicator to transfer cells from the stock plate to the culture tube. Briefly vortex to suspend cells.
3. Shake culture tube at approximately a 70° angle at 220 rpm and 30 °C for 18 h.
4. Collect cells by centrifugation for 5 min at $320 \times g$, suspend with 4 mL of PBS, and again centrifuge and suspend cells in 4 mL of PBS.
5. Dilute 1:100 and count cells (as in Subheading 3.1, **step 2**). Then dilute cells so that they are 20 times the amount (listed above for each mouse strain) per mL of PBS. For example, C57BL/6 mice are challenged with 50 μL of KN99 at a concentration of 2×10^5 cells/mL to deliver 1×10^4 cells into the lung.

6. Plate the inoculum used to infect the mice. Dilute 1:100, 1: 200, or 1:500 depending on which inoculum is administered and spot 10 μL four times onto a Sabouraud dextrose agar plate for drip plating. Incubate for 2 days at 30 °C, count the CFUs for each drip, and determine the average. The CFUs and cell counts should be equivalent.

3.8 Infection of Mice with KN99

As in Subheading 3.2, **step 2**, the mice are anesthetized with isoflurane for orotracheal inoculation of the highly virulent strain KN99. Follow steps 1–4 listed in Subheading 3.2, **step 2** to anesthetize and position the mouse on the intubation stand for inoculation.

1. Briefly vortex the suspension of KN99 prepared in Subheading 3.2, **step 3**.
2. Pipette 50 μL of suspended cells.
3. With the tongue drawn to one side, insert pipettor tip to the back of the throat and dispense the 50 μL. The mouse should gasp for air and inhale the KN99 cells.
4. Remove the mouse from the intubation stand and allow it to recover from the anesthesia.

3.9 Monitoring Mice Following Pulmonary Infection

Mice are monitored daily following infection, and vaccination studies are typically terminated 70–84 days postinfection. Unvaccinated wild-type mice succumb to infection during the third to fourth week postinfection. There are visual clues that indicate the health status of the infected mouse, including ruffled fur, hunched back, tip-toe walking, swollen cranium, and paralysis. Mice will begin to lose weight at approximately 2–3 weeks postinfection if unvaccinated. Weight loss of 70% from preinfection weight indicates a mouse is not likely to survive.

3.10 Euthanizing Mice

Protocols for euthanizing mice include asphyxiation with CO_2, followed by bilateral thoracotomy, or an anesthetic overdose of isoflurane followed by heart puncture to remove approximately 1 mL of blood.

1. CO_2 asphyxiation is performed according to institutional protocols.
2. Isoflurane on a small ball of absorbent material is placed in the bottom of a 50 mL centrifuge tube.
3. Place the head of the mouse in the opening of the tube. Breathing rate slows.
4. When the mouse does not respond to pinching of the hind foot, remove the mouse.
5. Use a 1 mL TB syringe with 25G × 5/8″ needle to remove blood from the heart. If collecting blood for plasma, coat the

syringe by drawing up and expelling a 10,000 U/mL heparin solution. If collecting blood for serum, then omit coating the syringe with heparin and allow the blood to coagulate.

6. Remove approximately 1 mL of blood by heart puncture using the syringe.
7. Complete euthanasia according to institutional protocols.

3.11 Determining CFUs of Mouse Lungs and Other Organs

Cryptococcal CFUs of the lungs are measured following homogenization of the lungs. Dilutions are plated on Sabouraud dextrose agar and plates are incubated for 2–3 days at 30 °C. The fungal load of the lung may accumulate to $>10^9$ CFU/lung for *C. neoformans* strain KN99. The weight of the lung increases as the infection progresses. Lung weight may increase > fivefold for lungs with the higher fungal burdens. The *cda1*Δ*cda2*Δ*cda3*Δ vaccinated mice that survive 70 days following challenge with KN99 often clear the infection. Those that do not clear the infection also appear healthy, but may retain a fungal burden of up to 10^7 CFU/lung.

1. Make PBS with Pen/Strep by combining 490 mL of PBS with 10 mL of Pen/Strep 100× stock for final concentrations of 200 U/mL penicillin and 200 μg/mL streptomycin. Store at 4 °C.
2. Add 4 mL of PBS plus Pen/Strep to a 14 mL culture tube.
3. Remove the lungs from the euthanized mouse and place into the tube.
4. Homogenize lungs using the OMNI homogenizer for approximately 5–10 s in a Biosafety hood.
5. Make tenfold serial dilutions of the homogenate into PBS plus Pen/Strep. Store homogenates at 4 °C for 3–7 days in case they need to be replated.
6. Plate the serial dilutions on Sabouraud dextrose agar plates.
7. Incubate plates for 2 days at 30 °C and count CFUs.

3.12 Analysis of Data

GraphPad Prism Software generates Kaplan-Meier plots of survival data and will do pair-wise comparisons of survival curves to determine whether they are statistically different. To ensure rigor and reproducibility, each experiment is repeated at least one time and the total number of mice/group is at least 10. The organ CFU data of mice that, for example, were vaccinated and survived the lethal challenge with KN99 may also be analyzed using standard parametric and non-parametric statistical methods provided in the software package. Two studies were done that informed us during the development of a vaccination protocol: (1) We determined the dose of the live and heat-killed *cda1*Δ*cda2*Δ*cda3*Δ strain that generates protection (Fig. 5) and (2) we determined the amount of KN99 needed for each mouse line so that unvaccinated mice of each line succumbed to infection in 3–4 weeks (Fig. 1).

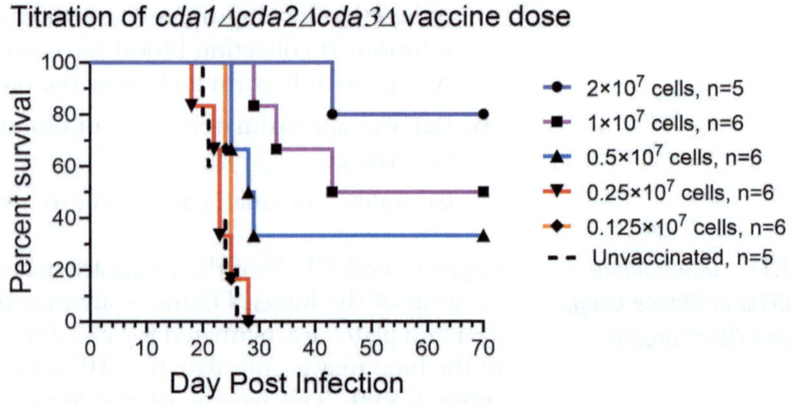

Fig. 5 Effect of the number of *cda1Δcda2Δcda3Δ* cells (colony forming units [CFUs]) used for orotracheal vaccination of BALB/c mice on survival. BALB/c mice were divided into six groups of mice with 5–6 mice per group. Five groups received a single vaccination at the indicated dosage and the sixth group was unvaccinated. Six weeks following vaccination, mice were infected with 2×10^4 cells (CFUs) of strain KN99. Mice were monitored for 70 days.

In Fig. 1, groups of unvaccinated C57BL/6 and CBA/J mice ($n = 5$ mice per group) were challenged with 1×10^3, 1×10^4, or 1×10^5 cells (CFUs) of KN99. The CBA/J mice were more resistant to the infection than C57BL/6 mice. From studies like these, which were also done for BALB/c mice (not shown), we deduced challenge doses for each mouse line that would normalize each line's susceptibility to KN99 infection. We decided on challenge doses of 1×10^4 cells (CFUs) for C57BL/6, 2×10^4 cells (CFUs) for BALB/c mice, and 5×10^4 cells (CFUs) for CBA/J mice so that mice succumb to infection in the same time frame (3–4 weeks) postinfection.

In Fig. 5, groups of BALB/c mice ($n = 5$ or 6 mice per group) were vaccinated with a series of twofold dilutions of *cda1Δcda2Δcda3Δ* cells (CFUs) that were cultured in YPD. Survival decreased with vaccine dose with the highest doses tested (2×10^7 CFU) generating the best protection. Vaccine doses of 0.25×10^7 CFU and 0.125×10^7 CFU did not protect. Heat-killed *cda1Δcda2Δcda3Δ* cells protect CBA/J mice given an intranasal vaccination of 1×10^7 cells, while 0.1×10^7 cells do not protect [17].

4 Notes

1. A single intranasal vaccination with 10^7 CFU of *cda1Δcda2Δcda3Δ* is effective in inducing robust protective immunity in 129 and A/J mice [17].
2. If the protective immunity induced by a single dose of vaccination is ineffective, more than one intranasal dose of vaccination

may be tested. For example, in CBA/J mice, a single intranasal dose of chitosan-deficient *cda1Δcda2Δ* is sufficient to induce strong protective immunity, but in C57BL/6 mice, two intranasal doses of 10^7 CFU of *cda1Δcda2Δ* are required to induce robust protective immunity [19].

3. Different postvaccination times may be used to test vaccine efficacy. After a 10-day recovery period, a single intranasal dose of 10^7 CFU of the chitosan-deficient *cda1Δcda2Δcda3Δ* mutant induces robust protective immunity in CBA/J mice. A recovery period of 20 and 30 days was equally protective as a recovery period of 40 days.
4. A 1:1000 serial dilution is usually required to count a reasonable number of cells in 10 μL of wild-type KN99 grown in YPD.

Acknowledgments

This work was supported by NIH grants AI125045 to JKL, CAS, and SML and AI172154 to SML. MMH was partially supported by NIH Training Grant T32 AI095213.

References

1. Chau TT, Mai NH, Phu NH et al (2010) A prospective descriptive study of cryptococcal meningitis in HIV uninfected patients in Vietnam - high prevalence of *Cryptococcus neoformans* var *grubii* in the absence of underlying disease. BMC Infect Dis 10:199. https://doi.org/10.1186/1471-2334-10-199
2. Chen J, Varma A, Diaz MR et al (2008) *Cryptococcus neoformans* strains and infection in apparently immunocompetent patients. China Emerg Infect Dis 14(5):755–762. https://doi.org/10.3201/eid1405.071312
3. Jarvis JN, Harrison TS (2008) Pulmonary cryptococcosis. Semin Respir Crit Care Med 29(2):141–150. https://doi.org/10.1055/s-2008-1063853
4. Byrnes EJ 3rd, Li W, Lewit Y et al (2010) Emergence and pathogenicity of highly virulent *Cryptococcus gattii* genotypes in the Northwest United States. PLoS Pathog 6(4): e1000850. https://doi.org/10.1371/journal.ppat.1000850
5. Rajasingham R, Govender NP, Jordan A et al (2022) The global burden of HIV-associated cryptococcal infection in adults in 2020: a modelling analysis. Lancet Infect Dis 22(12): 1748–1755. https://doi.org/10.1016/S1473-3099(22)00499-6
6. WHO (2022) WHO. Fungal priority pathogens list to guide research, development and public health action. World Health Organization, Geneva
7. Johnson MD, Perfect JR (2010) Use of antifungal combination therapy: agents, order, and timing. Curr Fungal Infect Rep 4(2):87–95. https://doi.org/10.1007/s12281-010-0018-6
8. Krysan DJ (2015) Toward improved anti-cryptococcal drugs: novel molecules and repurposed drugs. Fungal Genet Biol 78:93–98. https://doi.org/10.1016/j.fgb.2014.12.001
9. Smith KD, Achan B, Huppler Hullsiek K et al (2015) Increased antifungal drug resistance in Ugandan clinical isolates of *Cryptococcus neoformans*. Antimicrob Agents Chemother 59(12):7197–7204. https://doi.org/10.1128/AAC.01299-15
10. Bartizal K, Gill CJ, Abruzzo GK et al (1997) *In vitro* preclinical evaluation studies with the echinocandin antifungal MK-0991 (L-743, 872). Antimicrob Agents Chemother 41(11): 2326–2332. https://doi.org/10.1128/AAC.41.11.2326
11. Abruzzo GK, Flattery AM, Gill CJ et al (1997) Evaluation of the echinocandin antifungal MK-0991 (L-743, 872): efficacies in mouse

models of disseminated aspergillosis, candidiasis, and cryptococcosis. Antimicrob Agents Chemother 41(11):2333–2338. https://doi.org/10.1128/AAC.41.11.2333

12. Chaturvedi AK, Wormley FL Jr (2013) *Cryptococcus* antigens and immune responses: implications for a vaccine. Expert Rev Vaccines 12(11):1261–1272. https://doi.org/10.1586/14760584.2013.840094
13. Levitz SM, Huang H, Ostroff GR et al (2015) Exploiting fungal cell wall components in vaccines. Semin Immunopathol 37(2):199–207. https://doi.org/10.1007/s00281-014-0460-6
14. Mirza Z, Soto ER, Dikengil F et al (2017) Beta-glucan particles as vaccine adjuvant carriers. Methods Mol Biol 1625:143–157. https://doi.org/10.1007/978-1-4939-7104-6_11
15. Hester MM, Lee CK, Abraham A et al (2020) Protection of mice against experimental cryptococcosis using glucan particle-based vaccines containing novel recombinant antigens. Vaccine 38(3):620–626. https://doi.org/10.1016/j.vaccine.2019.10.051
16. Specht CA, Homan EJ, Lee CK et al (2021) Protection of mice against experimental cryptococcosis by synthesized peptides delivered in glucan particles. MBio 13(1):e0336721. https://doi.org/10.1128/mbio.03367-21
17. Upadhya R, Lam WC, Maybruck B et al (2016) Induction of protective immunity to cryptococcal infection in mice by a heat-killed, chitosan-deficient strain of *Cryptococcus neoformans*. mBio 7(3):e00547–e00516. https://doi.org/10.1128/mBio.00547-16
18. Wormley FL Jr, Perfect JR, Steele C et al (2007) Protection against cryptococcosis by using a murine gamma interferon-producing *Cryptococcus neoformans* strain. Infect Immun 75(3):1453–1462. https://doi.org/10.1128/IAI.00274-06
19. Upadhya R, Lam WC, Hole CR et al (2021) *Cryptococcus neoformans* Cda1 and Cda2 coordinate deacetylation of chitin during infection to control fungal virulence. Cell Surf 7:100066. https://doi.org/10.1016/j.tcsw.2021.100066
20. Masso-Silva J, Espinosa V, Liu TB et al (2018) The F-box protein Fbp1 shapes the immunogenic potential of *Cryptococcus neoformans*. mBio 9(1):e01828–e01817. https://doi.org/10.1128/mBio.01828-17
21. Rella A, Mor V, Farnoud AM et al (2015) Role of sterylglucosidase 1 (Sgl1) on the pathogenicity of *Cryptococcus neoformans*: potential applications for vaccine development. Front Microbiol 6:836. https://doi.org/10.3389/fmicb.2015.00836
22. Zhai B, Wozniak KL, Masso-Silva J et al (2015) Development of protective inflammation and cell-mediated immunity against *Cryptococcus neoformans* after exposure to hyphal mutants. mBio 6(5):e01433–e01415. https://doi.org/10.1128/mBio.01433-15
23. Caballero Van Dyke MC, Wormley FL Jr (2018) A call to arms: quest for a cryptococcal vaccine. Trends Microbiol 26(5):436–446. https://doi.org/10.1016/j.tim.2017.10.002
24. Louria DB (1960) Specific and non-specific immunity in experimental cryptococcosis in mice. J Exp Med 111(5):643–665. https://doi.org/10.1084/jem.111.5.643
25. Gadebusch HH (1960) Specific degradation of *Cryptococcus neoformans* 3723 capsular polysaccharide by a microbial enzyme. III. Antibody stimulation by partially decapsulated cells. J Infect Dis 107:406–409. https://doi.org/10.1093/infdis/107.3.406
26. Fromtling RA, Kaplan AM, Shadomy HJ (1983) Immunization of mice with stable, acapsular, yeast-like mutants of *Cryptococcus neoformans*. Sabouraudia 21(2):113–119
27. Wang Y, Wang K, Masso-Silva JA et al (2019) A heat-killed *Cryptococcus* mutant strain induces host protection against multiple invasive mycoses in a murine vaccine model. mBio 10(6):e02145–e02119. https://doi.org/10.1128/mBio.02145-19
28. Nielsen K, Cox GM, Wang P et al (2003) Sexual cycle of *Cryptococcus neoformans var. grubii* and virulence of congenic a and alpha isolates. Infect Immun 71(9):4831–4841
29. Matthew J, Arduino WDA, Bailin H et al (2020) Biosafety in Microbiological and Biomedical Laboratories, 6th edn. U.-S. Department of Health and Human Services Public Health Service, Centers for Disease Control and Prevention, National Institutes of Health, HHS Publication No. (CDC) 300859

Chapter 28

Vaccine Strategies for *Cryptococcus neoformans*

Veronica Soares Brauer, Samyr Machado Querobino, Gabriel Soares Matos, Deveney Dasilva, and Maurizio Del Poeta

Abstract

Cryptococcus neoformans infections are a major worldwide concern as current treatment strategies are becoming less effective in alleviating the infection. The most extreme and fatal cases are those of immunocompromised individuals. Clinical treatments for cryptococcosis are limited to a few classes of approved drugs, and due to a rise in drug resistance, these drugs are becoming less effective. Therefore, it is essential to develop innovative ways to control this infection. Vaccinations have emerged as a safe, viable, and cost-effective solution to treat a number of diseases over the years. Currently, there are no clinically available vaccines to treat cryptococcal infections, but a number of studies have shown promising results in animal models. Here, we present step-by-step experimental protocols using live-attenuated or heat-killed *C. neoformans* cells as a vaccination strategy in a preventive or in a therapeutic murine model of cryptococcosis.

Key words *Cryptococcus neoformans*, Vaccines, T cell depletion, Heat-killed inactivation, $CD4^+$ T cells, Preventive model, Therapeutic model, Intranasal administration, Intramuscular administration

1 Introduction

Vaccination is a key asset of primary health care to prevent the spread of harmful agents and boost the immune system, which also prevents further disease-associated complications. These features rely on their ability to trigger short- and long-term protections, activating both innate and adaptive immune cells, thus creating a memory system that protects against future infections [1, 2]. This prevention strategy has become increasingly important for fungal infections, such as cryptococcosis, which is responsible for most of the opportunistic infections and meningoencephalitis in immunocompromised groups, including those with HIV/AIDS-mediated lymphopenia, those under chemotherapy treatment, as well as recipients of organ transplants [3–5]. However, for these immunocompromised patients, there are limitations in maintaining

Erin E. McClelland (ed.), *Cryptococcus neoformans: Methods and Protocols*, Methods in Molecular Biology, vol. 2775, https://doi.org/10.1007/978-1-0716-3722-7_28,

a long-lasting memory response, which represents the actual challenge in cutting-edge research related to vaccine development [6].

Regarding this issue, the current treatments for general invasive fungi infections are dependent on drugs that, due to limited options, often lead to fungal resistance, which is currently a reality for *Candida auris* and *Candida glabrata* species [7]. Besides resistance, renal toxicity and a narrow range of therapeutic activity are other common problems related to current treatments [8, 9]. Therefore, the development of fungal vaccines that work both as a prevention and as a therapeutic strategy will help to improve the life expectancy of immunocompromised groups, since mortality rates may be nearly 70% when affected by cryptococcosis [10]. However, the cost of vaccine development targeting such a small high-risk group has pushed away the attention of pharmaceutical companies [9]. Coupling vaccination with antifungal therapy could improve efficacy and reduce resistance and treatment cost, allowing a more inclusive coverage for low-income and endemic countries [11]. Thus, investment in a preventive strategy is almost always the best approach for healthcare systems to decrease the costs related to the treatment of the disease [11].

Currently one *Candida* vaccine has reached a phase 2 clinical trial and different research groups have discovered the potential of *C. neoformans* mutant strains to promote protection against a secondary infection [12–15]. The first vaccination strategy designed for cryptococcosis in a mouse model, however, was based on the ability of the immune system to recognize different components of the fungi surface, especially the capsule molecule glucuronoxylomannan (GXM), which is able to induce a strong antibody response, independent of T cells [16]. However, when used as an immunization strategy, GXM was not able to induce a protective response in mice challenged with virulent cryptococcal cells [16–18]. Protective effects against cryptococcosis were observed by different research groups using either live-attenuated or heat-killed mutant cells, with promising results in the murine model [19–27]. Considering that some of these vaccines are also active in conditions of immunodeficiency, the perspective of this strategy is an exciting opportunity that should be seriously considered and promoted to improve the quality of life of individuals at risk of developing invasive fungal infections. Thus, in this chapter, we provide a step-by-step vaccination protocol that considers the intranasal and the intramuscular administration route in an immunocompetent mouse and in a mouse in which T cells are depleted, mimicking subjects mostly at risk of developing cryptococcosis.

2 Materials

2.1 Strains

1. *C. neoformans* var. *grubii* strain H99 serotype A.
2. Male and female CBA/J mice (4–5 weeks old, average weight 20 g).

2.2 Reagents and Media

1. Yeast Peptone Dextrose (YPD) broth.
2. Agar.
3. Sterile phosphate buffered saline (PBS) (pH 7.4).
4. In vivo MAb anti-mouse CD4 clone: GK1.5.
5. In vivo MAb rat IgG2b isotype control, anti-keyhole limpet hemocyanin, clone LTF-2.
6. Roswell Park Memorial Institute (RPMI)-1640 medium.
7. Fetal bovine serum (FBS), heat inactivated.
8. Penicillin-streptomycin (10,000 U/mL).
9. Collagenase IV.
10. DNase I.
11. Digestion buffer: RPMI, supplemented with 10% FBS, 1% penicillin-streptomycin, 150 U/mL of collagenase IV, and 10 U/mL of DNase I.
12. Fluorescence-activated cell sorter (FACS) buffer: PBS + 2% FBS.
13. Ammonium–chloride–potassium (ACK) lysis buffer: 150 mM NH_4Cl, 10 mM $KHCO_3$, 0.1 mM Na_2EDTA.
14. Anti-mouse CD16/CD32 antibody.
15. Fluorochrome-conjugated antibodies: Alexa Fluor™ 700 NHS Ester, CD45-BV711, CD8-BV785, and CD4-BV605.
16. Ketamine/xylazine: 95 mg of ketamine and 5 mg of xylazine/kg of mouse body weight, respectively, in sterile PBS.
17. 70% Ethanol (v/v).

2.3 Equipment

1. Incubator shaker.
2. Centrifuge.
3. Heat block.
4. Microscope.
5. LSR II flow cytometer.
6. Stomacher 80.

2.4 Miscellaneous

1. Disposable inoculating loop.
2. Reaction tube containing a 0.22 μm filter.
3. Petri dishes.
4. Hemocytometer.
5. Sterile surgical scissors and forceps.
6. Six-well plate.
7. Cell strainer with a 70 μm filter.
8. Syringes and needles: 1 mL, 25G × 5/8 [0.5 mm × 16 mm].
9. 1.5 mL/2 mL Eppendorf tubes.
10. Blender bag.
11. L-shaped inoculation loop.
12. 50 mL Falcon tubes.

3 Methods

3.1 Cryptococcus neoformans Growth and Maintenance

1. For long-time storage, *C. neoformans* cells are frozen at −80 °C or in liquid nitrogen. For this, a cell suspension of 1–5 × 10^7 cells/mL in 15% (v/v) glycerol in YPD broth is placed into a 2 mL sterile tube. The tube is vortexed for 30 sec and then immediately submerged in 100% ethanol containing dry ice for 2 min, so cells will instantly freeze. Tubes are then transferred to a − 80 °C freezer or to liquid nitrogen.
2. *C. neoformans* is recovered from a frozen 15% glycerol stock. For this, remove the stock tube from the −80 °C freezer or liquid nitrogen, and with a disposable inoculating loop, collect a sample by scraping the surface of frozen cells and then plate them onto a YPD agar plate. Incubate the plate for 48 h at 30 ° C.
3. After growth, pick a single colony with an inoculating loop and inoculate into 10 mL of YPD broth for 24 h at 30 °C under agitation (200 rpm) (*see* **Note 1**).

3.2 C. neoformans Inoculum Preparation

1. Pellet the cells by centrifugation (1734 × *g*) at room temperature for 10 min.
2. Discard the supernatant and wash the cells twice with 10 mL of sterile PBS (pH 7.4) by vortexing for 30 s. Centrifuge for 10 min at 1734 × *g* at room temperature. Remove the supernatant.
3. Discard supernatant and resuspend in 10 mL sterile PBS.
4. Count cells using a hemocytometer.
5. Dilute cells in sterile PBS to the desired concentration to be used for the inoculum.

6. Make sure to plate the same volume that is used to inoculate the mice on a YPD plate to double check that the desired cell concentration is accurate. Incubate plates at 30 °C for 2 days and count colonies. The colony count should be similar to that determined using the hematocytometer.

3.3 *C. neoformans* Inactivation by Heat Killing (HK)

1. For HK inactivation, grow the yeast and prepare the inoculum, as previously described in Subheading 3.2.
2. Preheat a heat block to 80 °C.
3. Heat the desired cell concentration (inoculum) at 80 °C for 75 min.
4. Confirm cell death by plating the same volume that is used to inoculate mice onto a YPD agar plate. Incubate plates at 30 °C for 2 days and count colonies. Colony count should be zero (*see* **Note 2**).

3.4 Mice and Ethical Statement

1. Male and female CBA/J mice, 4–5 weeks old, with an average weight of 20 g are used throughout the described vaccination model (*see* **Note 3**).
2. Mice are housed in pathogen-free conditions, 3–4 mice/cage with ad libitum access to food and water.
3. Animal procedures are performed following Biosafety level 2 (ABSL-2) guidelines.
4. Mice are acclimated for 1 week before beginning any experimental procedure.

3.5 In vivo CD4+ T-Cell Depletion

1. The $CD4^+$ T-cell depletion should start 48 h before administration of *C. neoformans.*
2. To deplete $CD4^+$ T cells, prepare an anti-CD4 antibody solution (clone GK1.5) at a concentration of 400 μg/100 μL, and inject 100 μL/mouse intraperitoneally (IP) (*see* **Note 4**).
3. The control group should receive 400 μg/100 μL of isotype-matched antibody, also via IP (e.g., clone LTF-2—Rat IgG2b isotype control or anti-keyhole limpet hemocyanin).
4. To maintain the $CD4^+$ T-cell depletion, administration of anti-CD4 antibody should be done every 4 days, until the end of the experiment (*see* **Note 5**). Control antibody administration should follow the same timeline.

3.6 Confirming CD4+ T-Cell Depletion Using Flow Cytometry

1. After at least 2 days of antibody administration, aseptically collect mouse blood or any organ (e.g., lungs). If an organ is collected, transfer the organ into a six-well plate with 5 mL of digestion buffer.

2. Cut the organ into small pieces using a surgical scalpel and incubate in digestion buffer for 90 min at 37 °C, 5% CO_2, under agitation (80 rpm).
3. Homogenize the tissue by passing it through a 70 μm filter.
4. Centrifuge the cells at 300 × *g* for 10 min at 4 °C and discard the supernatant.
5. Lyse the red blood cells by adding 5 mL of ACK lysis buffer. If blood is collected, skip the previous three steps, and add ACK lysis buffer to the blood sample to lyse red blood cells.
6. Centrifuge the solution at 300 × *g* for 5 min at 4 °C.
7. Discard the supernatant and wash the remaining cells twice with FACS buffer.
8. Count the cells using a hemocytometer and transfer 1×10^6 cells to a FACS tube.
9. Centrifuge the cells at 300 × *g* for 5 min at 4 °C and discard the supernatant.
10. For Fc block, prepare a solution of purified anti-mouse CD16/CD32 antibody (0.5 μg/50 μL) and pipette 50 μL/tube. Incubate the samples for 20 min on ice.
11. Prepare the cocktail antibody solution using the fluorochrome-conjugated antibodies: Alexa Fluor™ 700 NHS Ester (succinimidyl ester) (viability), CD45-BV711, CD4-BV605, and CD8-BV785. Then, pipette 50 μL/tube (*see* **Notes 6** and 7).
12. Incubate the cells for 30 min on ice, in the dark.
13. Wash the cells twice in FACS buffer, discard the supernatant, and resuspend the cells in 300 μL of FACS buffer.
14. Filter the cells using a 70 μm filter and analyze using an LSR II flow cytometer (*see* **Note 8**).

3.7 Vaccination as a Prevention Model (See Note 9)

1. Live or HK *C. neoformans* strains may be used here as vaccines.
2. Intranasal (IN) or intramuscular (IM) is the route of administration employed for the study, as a single dose or as multiple doses. See Subheadings 3.9 and 3.10.
3. Administration of a lethal *C. neoformans* wild-type (WT) strain (e.g., H99) should be done at least 30 days after the administration of the first dose.

3.8 Vaccination as a Therapeutic Model

1. Vaccination with live attenuated or HK *C. neoformans* vaccine is performed *after* challenge (e.g., 3–7 days) with the *C. neoformans* WT strain.
2. The route of administration of the vaccine employed for therapeutic vaccination could be IN or IM.

3.9 IN Route of Administration

1. Prepare the *C. neoformans* inoculum, as described in Subheading 3.2, at the desired concentration in sterile PBS.
2. Anesthetize the mice by injecting IP 100 μL solution of xylazine/ketamine (95 mg of ketamine and 5 mg of xylazine/kg of mouse body weight, respectively, in sterile PBS). Check that the mouse is completely anesthetized by a lack of footpad reflexes.
3. Hold the animal by the scruff, in a supine position, with the nose up to straighten the trachea and facilitate IN administration.
4. With a micropipette, carefully and slowly inoculate 20 μL of the yeast solution into the mouse nostrils. The release speed of the inoculum must be adjusted to avoid bubble formation in the nostrils while the mouse aspirates the inoculum.
5. The breathing rate increases during inoculation, so keep holding the mouse, in a hanging position, until breathing returns to normal.
6. Return mice to the cages and monitor them until they wake up from anesthesia.

3.10 IM Route of Administration

1. Prepare a *C. neoformans* vaccine inoculum (live attenuated or HK), as described in Subheadings 3.2 and 3.3 at the desired concentration in sterile PBS. For IM injection, a 40 μL volume is normally used.
2. Anesthetize the mouse, as described in Subheading 3.9, and wait until there is no footpad reflexes.
3. The site of injection is the thigh muscle of the hind limb. The area should be cleaned using 70% (v/v) ethanol before the IM injection.
4. Restrain the mouse by holding its back with one hand and then insert the needle (25–27G) into the thigh muscle through the skin using the other hand. Aspirate gently and briefly to make sure no blood comes back into the syringe. Then, proceed by slowly injecting the 40 μL of inoculum (*see* **Notes 10**, **11**, and **12**).

3.11 Challenge (see Notes 13 and 14)

1. Prepare the inoculum as described in Subheading 3.2.
2. Count the yeast using a hemocytometer and prepare a solution of 2.5×10^7 yeast cells/mL in sterile PBS.
3. As described in Subheading 3.9, anesthetize the mice and infect them IN with 20 μL of inoculum (5×10^5 cells/mouse).
4. It is essential that *C. neoformans* var. *grubii* strain H99 serotype A is virulent. This strain may lose virulence, particularly when

kept on an agar plate for a few weeks. To recover its virulence, the following steps may be used.

5. Prepare the *C. neoformans* var. *grubii* strain H99 serotype A (WT) (Subheading 3.2) at a concentration of 5×10^5 cells in 20 μL of sterile PBS. Inject this into the mouse IN, as described in Subheading 3.9.
6. After 7 days, euthanize the mouse and aseptically collect the brain.
7. Transfer the brain to a blender bag and homogenize the brain in 10 mL of sterile PBS for 2 min at high speed using a Stomacher 80.
8. Transfer the homogenate to a 50 mL tube and make serial dilutions of the homogenate using sterile PBS.
9. Plate 100 μL of each serial dilution onto a YPD plate and incubate the plates at 30 °C for 48 h.
10. Select a single colony and prepare a new inoculum as described in Subheading 3.2 and infect the mice using this inoculum, as described in Subheading 3.9.
11. The leftover inoculum may be cryopreserved, as described in Subheading 3.1.

3.12 Survival Studies

1. Mice are observed twice a day for signs of infection, such as weight loss, abnormal posture (hunched back), increased/decreased movement, trembling, ruffled coat due to reduced grooming, inability to walk, or to obtain food or water.
2. For survival studies, daily observation of the animal and weight monitoring is essential. In all experiments, animals are handled such that minimal pain and trauma are inflicted. If a significant weight loss (25% of body weight) is observed, or an inability to walk or to obtain food/water, and/or if mice are shaking and appear to be in pain, then animals are euthanized as described below. Survival or tissue burden culture or other analysis will be performed/calculated from this point.

3.13 Animal Euthanasia, Organ Harvest, and Burden Analyses

1. Euthanasia of mice is performed using CO_2 asphyxiation with 100% $FiCO_2$ (3 L/min) for 2 min followed by cervical dislocation. These methods are consistent with the Panel on Euthanasia of the American Veterinary Medical Association (JAVMA 188, 252–268, 1986). All animal protocols should be approved and reviewed yearly by the University Institutional Animal Care and Use Committee (IACUC).
2. Next, clean tools and the bench with 70% ethanol (v/v) and place surgical tools (scissor and forceps) in 100% ethanol.
3. Spray 70% ethanol on the euthanized mouse and pin it to a clean surface.

4. For brain dissection, keep the mouse face down, elevate the skin with sterilized forceps and cut the skin exposing the skull.
5. With flambeed scissors, remove the skull and meninges to access the brain.
6. Remove the brain with surgical forceps and place it in a previously identified tube containing 10 mL of PBS.
7. Flip the mouse over so the abdomen faces up.
8. Spray 70% ethanol (v/v) over the entire abdomen.
9. Elevate the skin with the forceps and cut along the ventral axis from the groin to the top of the sternum.
10. Spray ethanol 70% on the peritoneal membrane and wait until it is dry.
11. Cut the peritoneum. Locate the kidney in the dorsal wall of the abdominal cavity.
12. Remove both kidneys and place them in a PBS solution until processing.
13. Using clean and sterile forceps and scissors, first remove the spleen and then the liver, and store in two separate tubes each containing 10 mL of sterile PBS.
14. For lung dissection, cut the ventral diaphragm muscles with a scissor, releasing it from the rib cage. Cut the ribs on both sides. Once the rib cage is removed, the entire thoracic cavity is exposed. Cut the pulmonary artery and vein. Then, remove the lungs and place them in a tube containing 10 mL of sterile PBS.
15. Place the 10 mL of PBS containing the organ in a sterile blender bag. Tape the bag to the homogenizer and homogenize the tissue for 5 min using a Stomacher 80 blender.
16. Transfer the homogenized tissue into a sterile tube. Prepare 1: 10 serial dilutions of the homogenate in PBS and plate 100 μL of each dilution onto YPD plates. Incubate plates at 30 ° C for 48 h.
17. Count the number of colonies and calculate the colony forming units per total organ (10 mL).

4 Notes

1. Use a reaction tube containing a 0.22 μm filter to allow gas exchange and avoid contamination of the culture.
2. No growth should be observed after heat inactivation.
3. All in vivo experiments should be performed following the guidelines of the American Veterinary Medical Association

and approved by the University Institutional Animal Care and Use Committee (IACUC).

4. Based on the stock antibody concentration, prepare the 400 μg/100 μL solution of antibody in sterile PBS.
5. The defined antibody concentration for $CD4^+$ T cell depletion, as well as the interval between the doses was established using 3- to 4-week-old CBA/J mice. The effectivity of the cell depletion was confirmed using flow cytometry of cell populations in mouse lungs. It is strongly suggested that the antibody dose is standardize based on the strain, weight, and age of the mice.
6. The step before cell surface staining with fluorochrome-conjugated antibodies is the Fc block. As the cells are in 50 μL, the fluorochrome-conjugated antibody solution should be 2× concentrated to avoid diluting the antibody concentration.
7. For flow cytometry analysis, always prepare FACS tubes containing unstained cells to gate the negative population. Also, it is important to have single-stained cells for each antibody used for the analysis as compensation controls.
8. The gating strategy for T cells may be done following Normile et al. [28].
9. Always keep an unvaccinated group of mice as a control.
10. Proceed with the injection with caution to avoid damage to the sciatic nerve that is located near the mouse femur.
11. If blood comes back into the syringe during the aspiration process, the needle should be repositioned.
12. The IM technique may be done by two care givers; one can restrain the mouse and the other can locate the thigh muscle in the hind limb and perform the injection.
13. The maintenance of *C. neoformans* strains in the laboratory, either via cryopreservation or via subcultures in vitro, may result in changes to fungal metabolic fitness and virulence [29, 30]. Thus, it is important to keep *C. neoformans* strains freshly isolated from animal passages, before doing in vitro and in vivo experiments.
14. Use a *C. neoformans* strain freshly isolated from animal passages or from the cryopreserved stocks.

Acknowledgments

This work was supported in part by the National Institutes of Health grants AI136934 (MDP), AI116420 (MDP), AI125770 (MDP), and by a Merit Review Grant I01BX002924 (MDP) from the Veterans Affairs (VA) Program. Maurizio Del Poeta is a recipient of the Research Career Scientist (RCS) Award from the Veterans Affairs Program (IK6 BX005386) and a Burroughs Welcome Investigator in Infectious Diseases.

Conflict of Interest Dr. Maurizio Del Poeta, M.D., is a Cofounder and Chief Scientific Officer (CSO) of MicroRid Technologies Inc., whose goal is to develop new antifungal agents of therapeutic use. All other authors declare no competing interests.

References

1. Bekkering S, Dominguez-Andres J, Joosten LAB et al (2021) Trained immunity: reprogramming innate immunity in health and disease. Annu Rev Immunol 39:667–693. https://doi.org/10.1146/annurev-immunol-102119-073855
2. MacLeod MK, Kappler JW, Marrack P (2010) Memory CD4 T cells: generation, reactivation and re-assignment. Immunol 130:10–15. https://doi.org/10.1111/j.1365-2567.2010.03260.x
3. Akhtar S, Aggarwal N, Demkowicz R et al (2020) *Cryptococcus* and HIV. QJM 113:347–348. https://doi.org/10.1093/qjmed/hcz299
4. Elsegeiny W, Marr KA, Williamson PR (2018) Immunology of Cryptococcal infections: developing a rational approach to patient therapy. Front Immunol 9:651. https://doi.org/10.3389/fimmu.2018.00651
5. Henao-Martinez AF, Beckham JD (2015) Cryptococcosis in solid organ transplant recipients. Curr Opin Infect Dis 28:300–307. https://doi.org/10.1097/QCO.0000000000000171
6. Nanjappa SG, Klein BS (2014) Vaccine immunity against fungal infections. Curr Opin Immunol 28:27–33. https://doi.org/10.1016/j.coi.2014.01.014
7. Tetz G, Collins M, Vikina D et al (2019) *In vitro* activity of a novel antifungal compound, MYC-053, against clinically significant antifungal-resistant strains of *Candida glabrata*, *Candida auris*, *Cryptococcus neoformans*, and *pneumocystis spp*. Antimicrob Agents Chemother 63:4. https://doi.org/10.1128/AAC.01975-18
8. Denning DW, Bromley MJ (2015) Infectious disease. How to bolster the antifungal pipeline. Science 347:1414–1416. https://doi.org/10.1126/science.aaa6097
9. Biswas PS (2021) Vaccine-induced immunological memory in invasive fungal infections – a dream so close yet so far. Front Immunol 12:671068. https://doi.org/10.3389/fimmu.2021.671068
10. Hurtado JC, Castillo P, Fernandes F et al (2010) Mortality due to *Cryptococcus neoformans* and *Cryptococcus gattii* in low-income settings: an autopsy study. Sci Rep 9:7493. https://doi.org/10.1038/s41598-019-43941-w
11. Del Poeta M, Wormley FL Jr, Lin X (2023) Host populations, challenges, and commercialization of cryptococcal vaccines. PLoS Pathog 19:e1011115. https://doi.org/10.1371/journal.ppat.1011115
12. Brena S, Omaetxebarria MJ, Elguezabal N et al (2007) Fungicidal monoclonal antibody C7 binds to *Candida albicans* Als3. Infect Immun 75:3680–3682. https://doi.org/10.1128/IAI.01840-06
13. Casadevall A, Pirofski LA (2018) A therapeutic vaccine for recurrent vulvovaginal candidiasis. Clin Infect Dis 66:1937–1939. https://doi.org/10.1093/cid/ciy188
14. Rivera A, Lodge J, Xue C (2022) Harnessing the immune response to fungal pathogens for vaccine development. Annu Rev Microbiol 76:703–726. https://doi.org/10.1146/annurev-micro-041020-111511
15. Ibrahim AS, Luo G, Gebremariam T et al (2013) NDV-3 protects mice from vulvovaginal candidiasis through T- and B-cell immune response. Vaccine 31:5549–5556. https://doi.org/10.1016/j.vaccine.2013.09.016
16. Devi SJ, Schneerson R, Egan W et al (1991) *Cryptococcus neoformans* serotype a glucuronoxylomannan-protein conjugate vaccines: synthesis, characterization, and immunogenicity. Infect Immun 59:3700–3707. https://doi.org/10.1128/iai.59.10.3700-3707.1991

17. Casadevall A, Pirofski L (2005) Insights into mechanisms of antibody-mediated immunity from studies with *Cryptococcus neoformans*. Curr Mol Med 5:421–433. https://doi.org/10.2174/1566524054022567
18. Datta K, Pirofski LA (2006) Towards a vaccine for *Cryptococcus neoformans*: principles and caveats. FEMS Yeast Res 6:525–536. https://doi.org/10.1111/j.1567-1364.2006.00073.x
19. Wozniak KL, Young ML, Wormley FL Jr (2011) Protective immunity against experimental pulmonary cryptococcosis in T cell-depleted mice. Clin Vaccine Immunol 18:717–723. https://doi.org/10.1128/CVI.00036-11
20. Rella A, Mor V, Farnoud AM et al (2015) Role of sterylglucosidase 1 (Sgl1) on the pathogenicity of *Cryptococcus neoformans*: potential applications for vaccine development. Front Microbiol 6:836. https://doi.org/10.3389/fmicb.2015.00836
21. Maziarz EK, Perfect JR (2016) Cryptococcosis. Infect Dis Clin N Am 30:179–206. https://doi.org/10.1016/j.idc.2015.10.006
22. Normile TG, Chu TH, Sheridan BS et al (2022) Vaccine protection by *Cryptococcus neoformans* Deltasgl1 is mediated by gammadelta T cells via TLR2 signaling. Mucosal Immunol 15:1416–1430. https://doi.org/10.1038/s41385-022-00570-3
23. Normile TG, Del Poeta M (2022) Three models of vaccination strategies against cryptococcosis in immunocompromised hosts using heat-killed *Cryptococcus neoformans* Deltasgl1. Front Immunol 13:868523. https://doi.org/10.3389/fimmu.2022.868523
24. Wang Y, Wang K, Rivera A et al (2023) Development of a heat-killed fbp1 mutant strain as a therapeutic agent to treat invasive *Cryptococcus* infection. Microbiol Spectr 11:e0495522. https://doi.org/10.1128/spectrum.04955-22
25. Liu TB, Xue C (2014) Fbp1-mediated ubiquitin-proteasome pathway controls *Cryptococcus neoformans* virulence by regulating fungal intracellular growth in macrophages. Infect Immun 82:557–568. https://doi.org/10.1128/IAI.00994-13
26. Zhai B, Wozniak KL, Masso-Silva J et al (2015) Development of protective inflammation and cell-mediated immunity against *Cryptococcus neoformans* after exposure to hyphal mutants. mBio 6:e01433. https://doi.org/10.1128/mBio.01433-15
27. Lin J, Pham T, Hipsher K et al (2022) Immunoprotection against *Cryptococcosis* offered by Znf2 depends on capsule and the hyphal morphology. MBio 13:e0278521. https://doi.org/10.1128/mbio.02785-21
28. Normile TG, Rella A, Del Poeta M (2021) *Cryptococcus neoformans* Δsgl1 vaccination requires either CD4(+) or CD8(+) T cells for complete host protection. Front Cell Infect Microbiol 11:739027. https://doi.org/10.3389/fcimb.2021.739027
29. Janbon G, Ormerod KL, Paulet D et al (2014) Analysis of the genome and transcriptome of *Cryptococcus neoformans* var. *grubii* reveals complex RNA expression and microevolution leading to virulence attenuation. PLoS Genet 10:e1004261. https://doi.org/10.1371/journal.pgen.1004261
30. Arras SDM, Ormerod KL, Erpf PE et al (2017) Convergent microevolution of *Cryptococcus neoformans* hypervirulence in the laboratory and the clinic. Sci Rep 7:17918. https://doi.org/10.1038/s41598-017-18106-2

Index

Erin E. McClelland (ed.), *Cryptococcus neoformans: Methods and Protocols*, Methods in Molecular Biology, vol. 2775, https://doi.org/10.1007/978-1-0716-3722-7,

S

T

U

V

W

Y

MIX
Papier aus verantwortungsvollen Quellen
Paper from responsible sources
FSC® C105338

If you have any concerns about our products,
you can contact us on
ProductSafety@springernature.com

In case Publisher is established outside the EU,
the EU authorized representative is:
Springer Nature Customer Service Center GmbH
Europaplatz 3, 69115 Heidelberg, Germany

Printed by Libri Plureos GmbH
in Hamburg, Germany